Accessing the E-book edition

Using the VitalSource® ebook

Access to the VitalBook™ ebook accompanying this book is via VitalSource® Bookshelf – an ebook reader which allows you to make and share notes and highlights on your ebooks and search across all of the ebooks that you hold on your VitalSource Bookshelf. You can access the ebook online or offline on your smartphone, tablet or PC/Mac and your notes and highlights will automatically stay in sync no matter where you make them.

1. **Create a VitalSource Bookshelf account at** ***https://online.vitalsource.com/user/new*** or log into your existing account if you already have one.
2. **Redeem the code provided in the panel below to get online access to the ebook.** Log in to Bookshelf and click the **Account** menu at the top right of the screen. Select **Redeem** and enter the redemption code shown on the scratch-off panel below in the **Code To Redeem** box. Press **Redeem**. Once the code has been redeemed your ebook will download and appear in your library.

Energy Modeling and Computations in the Building Envelope

No returns if this code has been revealed.

DOWNLOAD AND READ OFFLINE

To use your ebook offline, download BookShelf to your PC, Mac, iOS device, Android device or Kindle Fire, and log in to your Bookshelf account to access your ebook:

On your PC/Mac

Go to ***http://bookshelf.vitalsource.com/*** and follow the instructions to download the free **VitalSource Bookshelf** app to your PC or Mac and log into your Bookshelf account.

On your iPhone/iPod Touch/iPad

Download the free **VitalSource Bookshelf** App available via the iTunes App Store and log into your Bookshelf account. You can find more information at ***https://support.vitalsource.com/hc/en-us/categories/200134217-Bookshelf-for-iOS***

On your Android™ smartphone or tablet

Download the free **VitalSource Bookshelf** App available via Google Play and log into your Bookshelf account. You can find more information at ***https://support.vitalsource.com/hc/en-us/categories/200139976-Bookshelf-for-Android-and-Kindle-Fire***

On your Kindle Fire

Download the free **VitalSource Bookshelf** App available from Amazon and log into your Bookshelf account. You can find more information at ***https://support.vitalsource.com/hc/en-us/categories/200139976-Bookshelf-for-Android-and-Kindle-Fire***

N.B. The code in the scratch-off panel can only be used once. When you have created a Bookshelf account and redeemed the code you will be able to access the ebook online or offline on your smartphone, tablet or PC/Mac.

SUPPORT

If you have any questions about downloading Bookshelf, creating your account, or accessing and using your ebook edition, please visit ***http://support.vitalsource.com/***

ENERGY MODELING AND COMPUTATIONS IN THE BUILDING ENVELOPE

ENERGY MODELING AND COMPUTATIONS IN THE BUILDING ENVELOPE

Alexander V. Dimitrov

CRC Press is an imprint of the
Taylor & Francis Group, an **informa** business

CRC Press
Taylor & Francis Group
6000 Broken Sound Parkway NW, Suite 300
Boca Raton, FL 33487-2742

Printed on acid-free paper
Version Date: 20150706

International Standard Book Number-13: 978-1-4987-2320-6 (Pack - Book and Ebook)

Visit the Taylor & Francis Web site at
http://www.taylorandfrancis.com

and the CRC Press Web site at
http://www.crcpress.com

Contents

Preface.....xi
Author.....xiii

1. Introduction: The Buildings' Envelope—A Component of the Building Energy System.....1
1.1 Systematic Approach Applied to Buildings.....1
1.2 Envelope System (Envelope) and Energy Functions Design.....3
1.3 Summary Analysis of the *Building–Surrounding* Energy Interactions.....11

2. Physics of Energy Conversions in the Building Envelope at Microscopic Level.....13
2.1 Idealized Physical Model of the Building Envelope as an Energy-Exchanging Medium (Review of the Literature from Microscopic Point of View).....16
2.2 Conclusions and Generalizations Based on the Survey of Literature Published in the Field.....30
2.3 Design of a Hypothetical Physical Model of Phonon Generation in Solids: Scatter of Solar Radiation within the Solid.....32
2.3.1 Internal Ionization and Polarization Running in Solids (Formation of Temporary Electrodynamic Dipoles).....32
2.3.2 Hypothetical Mechanism of Energy Transfer in the Building Envelope Components.....36
2.3.2.1 Physical Pattern of Energy Transfer within the Envelope Components.....36
2.3.3 Hypothetical Model of Energy Transfer through Solid Building Components: A Model of Lagging Temperature Gradient.....41
2.3.3.1 Model of Lagging Temperature Gradient.....48
2.4 Micro–Macroscopic Assessment of the State of the Building Envelope.....51
2.4.1 Microscopic Canonical Ensemble: Collective Macroscopic State.....51

2.4.2 Introduced Macroscopic State Parameters of the Building Envelope Considered as a Physical Medium of the Electrothermodynamic System 53
2.4.2.1 Temperature Field and Gradient of the Lagrange Multiplier 53
2.4.2.2 Pressure Field 58
2.4.2.3 Field of the Electric Potential: Potential Function and Gradient of the Electric Potential 61
2.4.2.4 Entropy: A Characteristic of Degeneration of the Heat Charges (Phonons) within the Envelope Control Volume 66
2.4.3 Conclusions on the General Methodological Approaches to the Study of an Electrothermomechanical System 73

3. Design of a Model of Energy Exchange Running between the Building Envelope and the Surroundings: Free Energy Potential 75
3.1 Energy-Exchange Models of the Building Envelope 75
3.2 Work Done in the Building Envelope and Energy-Exchange Models 81
3.2.1 Law of Conservation of the Energy Interactions between the Envelope Components and the Building Surroundings 82
3.2.2 Special Cases of Energy Interactions 86
3.2.2.1 Energy Model of Transfer of Entropy and Electric Charges 86
3.2.2.2 Energy Model of Entropy Transfer with or without Mass Transfer 88
3.3 Specification of the Structure of the Free Energy in the Components of the Building Envelope (Electrothermodynamic Potential of the System) 89
3.3.1 Finding the Structure of the Free Energy Function 92
3.3.1.1 Links between Entropy and the System Basic Parameters 95
3.4 Distribution of the Free Energy within the Building Envelope 97
3.4.1 State Parameters Subject to Determination via the Free Energy Function 99

4. Definition of the Macroscopic Characteristics of Transfer 101
4.1 General Law of Transfer 106
4.2 Physical Picture of the Transmission Phenomena 108
4.3 Conclusions 111

5. Numerical Study of Transfer in Building Envelope Components 113
5.1 Method of the Differential Relations 113
5.2 Method of the Integral Forms 119
5.3 Weighted Residuals Methodology Employed to Assess the ETS Free Energy Function 122
5.3.1 Basic Stages of the Application of WRM in Evaluating Transport within the Envelope 125
5.3.1.1 One-Dimensional Simple Finite Element 140
5.3.1.2 Two-Dimensional Simple Finite Element in Cartesian Coordinates 140
5.3.1.3 Two-Dimensional Simple Finite Element in Cylindrical Coordinates 141
5.3.1.4 Three-Dimensional Simple Finite Element 141
5.3.2 Modeling of Transfer in a Finite Element Using a Matrix Equation (Galerkin Method) 142
5.3.3 Steady Transfer in One-Dimensional Finite Element 146
5.3.3.1 Integral Form of the Balance of Energy Transfer through One-Dimensional Finite Element 147
5.3.3.2 Modified Matrix Equation of 1D Transfer 150
5.3.3.3 Transfer through 1D Simple Finite Element Presented in Cylindrical Coordinates 155
5.3.4 Steady Transfer in a 2D Finite Element 160
5.3.4.1 Equation of a 2D Simple Finite Element in Cartesian Coordinates 161
5.3.4.2 Design of Transfer Equation in Cylindrical Coordinates regarding a Three-Noded 2D Finite Element 166
5.3.5 Transfer through a 3D Simple Finite Element 170
5.3.5.1 Design of the Matrix Equation of Transfer in Cartesian Coordinates 170

6. Initial and Boundary Conditions of a Solid Wall Element 175
6.1 Effects of the Environmental Air on the Building Envelope 175
6.1.1 Mass Transfer from the Building Envelope (*Wall Dehumidification, Drying*) 176
6.1.1.1 Processes Running at a Cold Wall ($T_A \geq T_{w_i}$) 177
6.1.1.2 Processes Running at a Cold Wall ($T_w < T_A$) 178
6.2 Various Initial and Boundary Conditions of Solid Structural Elements 179
6.3 Design of Boundary Conditions of Solid Structural Elements 182
6.3.1 Boundary Conditions of Convective Transfer Directed to the Wall Internal Surface 183
6.3.2 Boundary Conditions at the Wall External Surface 185

7. Engineering Methods of Estimating the Effect of the Surroundings on the Building Envelope: Control of the Heat Transfer through the Building Envelope (Arrangement of the Thermal Resistances within a Structure Consisting of Solid Wall Elements) 191
7.1 Calculation of the Thermal Resistance of Solid Structural Elements 194
7.2 Solar Shading Devices (Shield) Calculation 203
7.3 Modeling of Heat Exchange between a Solar Shading Device, a Window, and the Surroundings 208
7.3.1 Mathematical Model 212
7.4 Design of Minimal-Admissible Light-Transmitting Envelope Apertures Using the Coefficient of Daylight (CDL) 213
7.4.1 Energy and Visual Comfort 213
7.4.2 Calculation of the Coefficient of Daylight (CDL) 218
7.5 Method of Reducing the Tribute of the Construction and the Thermal Bridges to the Energy Inefficiency 223
7.5.1 Characteristics of Heat Transfer through Solid Inhomogeneous Multilayer Walls 224
7.5.2 Method Described Step by Step 227
7.5.3 Description of the Energy Standard of the Construction (EE_{Const}) 227
7.5.4 Employment of the Energy Standard to Assess How the Building Structure Affects the Energy Efficiency 229
7.6 Assessment of Leaks in the Building Envelope and the Air-Conditioning Systems 233
7.6.1 Measuring Equipment of the Method "Delta-*Q*" 234
7.6.2 Modified Balance Equation of Leaks in Air Ducts, Air-Conditioning Station, and Envelope 236
7.6.3 Delta-Q Procedure: Data Collection and Manipulation 238
7.6.4 Normalization of the Collected Data 241
7.7 Mathematical Model of the Environmental Sustainability of Buildings 244
7.7.1 General Structure of the Model 244
7.7.2 Selection of an Ecological Standard: Table of Correspondence 248
7.7.3 Comparison of Systems Rating the Ecological Sustainability in Conformity with the General Criteria 255
7.8 Conclusion 258
Acknowledgments 262

8. Applications (Solved Tasks and Tables) ... 263
8.1 Matrix of Conductivity $[K^{(1)}]$ ... 263
8.2 Matrix of Surface Properties $[F^{(1)}]$ ... 264
8.3 Generalized Matrix of the Element Conductivity $[G^{(1)}] = [K^{(1)}] + [F^{(1)}]$ ... 265
8.4 Vector of a Load due to Recuperation Sources $\{f_C^{(1)}\}$ ... 265
8.5 Vector of a Load due to Convection to the Surrounding Matter $\{f_C^{(1)}\}$ ... 266
8.6 Vector of a Load due to a Direct Flux $\{f_{Dr}^e\}$ ... 266
8.6.1 Design and Solution of the Matrix Equation ... 267

References ... 293

Index ... 305

Preface

The following are two basic reasons that justify the researchers' efforts to delve in the nature of energy-related processes and interactions:

1. Needs of our civilization for new and cheap energy carriers
2. The urgent need of econo̒my for primary energy carriers justified by resource or ecology issues

These two contradicting tendencies could be combined via the optimization of energy demand and supply based on plausible physical preconditions and adequate mathematical models and their engineering applications.

The book *Energy Modeling and Computations in the Building Envelope* proposes an innovative model of energy transformations taking place in the building envelope at microwave level. It describes the birth and directed movement of energy particles—phonons and electrons.

A generalized mathematical functional is designed using the physical model, which describes energy transfer in discrete areas of the building envelope. Numerical examples of transfer estimation are also given. The presented models can be used as a basis of the design of algorithms to be implemented in subsequent computer codes.

In addition, Chapter 7 gives a description of the author's engineering practice and methods of the calculation of building envelope components, which would increase building energy efficiency, decrease the consumption of primary energy carriers, and raise the ecological sustainability of construction products.

Alexander V. Dimitrov

Author

Alexander V. Dimitrov is a professional lecturer with 35 years of experience in four different universities. In addition to universities in Bulgaria, Dimitrov has lectured and studied at leading scientific laboratories and institutes, including the Luikov Heat and Mass Transfer Institute and the Belarusian Academy of Sciences in Minsk; the Lawrence Berkeley National Laboratory, Environmental Energy Technology Division; the University of Nevada–Las Vegas (UNLV) College of Engineering; and Stanford University in California.

Professor Dimitrov has conducted systematic research in energy efficiency, computer simulations of energy consumption in buildings, the distribution of air flow in an occupied space, and modeling of heat transfer in building envelope and leaks in the ducts of heating, ventilating, and air conditioning (HVAC) systems. He has defended two scientific degrees: Doctor of Philosophy (PhD) in 1980 and Doctor of Science (DSc) in 2012.

Professor Dimitrov has significant audit experience in the energy systems of buildings and their subsystems. He has developed an original method for evaluating the performance of the building envelope and energy labeling of buildings. He also has experience in the assessment of energy transfer through the building envelope and ducts of HVAC systems. His methodology has been applied in several projects with great success. He has developed a mathematical model for the assessment of the environmental sustainability of buildings, named BG_LEED. He earned his professor degree in "Engineering Installations in Buildings" with the dissertation "The building energy systems in the conditions of environmental sustainability" at the European Polytechnical University in 2012. He has authored more than 100 scientific articles and 8 books (including 3 in English).

1

Introduction: The Buildings' Envelope—A Component of the Building Energy System

1.1 Systematic Approach Applied to Buildings

In today's energy science, specifically in its *building power energy* section, it is ascertained that in the rating of primary energy end users, modern construction and operational building technologies have steadily occupied the impressive *second position,* following industry (transport occupies the third position). These have a different share in the national energy balances, varying within 26%–39%. The existing data on the energy consumption in rapidly developing countries, like India and China, have outlined an upward tendency of energy share in the construction business. In developed countries like those in the EU, the United States, Japan, South Korea, Canada, and Australia, systematic and comprehensive activity has gotten underway with the reduction of the consumption of primary energy generated by *fossil fuel* resources and in a wider implementation of renewable energy sources. Economy of energy so far has been pursued in different ways—from governmental state regulations and subsidies to informal social and even personal efforts of the citizens in their roles as private investors in various energy-saving projects.

Architectural science reflects the public urge to reduce the consumption of primary energy from *fossil fuel* resources. Removal of the effect of classical technologies of energy supply on the global climate in urbanized areas has also been pursued. Change in the architectural concepts of building design dates back to the 1980s. It affected not only the design philosophy and aesthetics but also the *methodological* approaches to the elaboration of related architectural projects.

In conformity with those changes, a building was considered not only as a residential space and storehouse of primary human needs but also as a complex system. Except for regular physiology (need of air, water, food, sexual relations), it was expected to guarantee the occupants' aesthetics, safety, security, intimacy, etc. (Szokolay 1981; Trubiano 2012; Akasmija 2013; Zemella and Faraguna 2014).

The objects of the integral conceptual design are the main building systems:

- *Technological system*: Executing the main building functions
- *Envelope system*: Protecting the building from environmental impacts, with aesthetical function
- *Structural system*: Guaranteeing building strength and reliability
- *Energy system*: Transforming and supplying energy to the end-users (building energy system [BES])
- *Logistics system*: Transporting people, energy sources, materials, water, and waste
- *Entertaining system*: Providing relaxation and entertainment (audio, video, home theater, etc.)
- *Sanitary system*: Guaranteeing hygiene
- *Safeguard system*: Preventing break-in, fire, gas and bio attacks, etc.
- *Building system for automatic control and monitoring* (*BSAC and M*): Performing control, regulation, and management of processes and databases

The internal physical environment (light, heat, air, and sound) could also be considered as a component of a modern building system, focusing on the occupants. In this, consider a building equipped in conformity with modern technology. However, all systems are important for a building's smooth functioning, but the energy system plays the major role, as it *energizes* all building functions of vital importance. It transforms the supplied energy carriers into useful energy, thus guaranteeing comfort of living and operation of all electric appliances.

Within that systematic approach, it is assumed that the *BES* consists of the following major components, conforming to the course of energy flow: *energy center, energy-consuming subsystems,* and *BSAC and M.* As a subcomponent of the energy system, the *building envelope* should also be taken into account, since its components (facades and roof) are also *energy committed*—they function as barriers, filters, or *intelligent membranes* against energy impacts of the surroundings (Димитров 2008b).

The architectural experience of energy systems design, gained in the later part of the 1940s, could be classified following two trends: (1) improvement of the energy efficiency of the building technologies and (2) promotion of the use of natural energy in the overall building energy balances (partially through all the methods of the so-called solar architecture). Yet full coverage of building energy needs has been hard to achieve. Those strategies ask the following questions:

- What will the envelope building system look like and what should its general architectural design be in order to adequately apply the respective strategy?
- What type of envelope should the designers use, based on the evolution analyses of facade energy functions and the progress of architectural state of the art and strategies?

1.2 Envelope System (Envelope) and Energy Functions Design

The building envelope is a special and very important target of architectural design (Dimitrov 2009; Димитров 2011a; Trubiano 2012; Akasmija 2013; Zemella and Faraguna 2014). Firstly, the envelope (including facades) should protect the residential areas against impacts of the surroundings. Secondly, it should also meet important aesthetical requirements, conforming with the surroundings (adjacent buildings, grass plots, etc.). Thirdly, the building envelope should execute important energy-related (energy) functions, since the building systems consume a considerable amount of useful (secondary) energy in order to compensate leakage through the envelope and energy outflows. In its character, the envelope is an inseparable part of the BES. Its consideration (including walls and roof) as a unit of the BES is verified by the energy functions of its components (barriers, filters, or *smart* membranes opposing environmental energy impacts). Moreover, the envelope serves as a medium where spontaneous or controlled energy transformations take place—note that they are accounted for in the total energy balance. It also determines the effectiveness of the entire BES (Hopness 2009). Hence, the functions of the building envelope, including facade and roof, should be integrated (see Figure 1.1).

Historically, architectural design focused on different qualities of the envelope. In the early Middle Ages, it was entrusted with protective functions, and later, in the Renaissance and till the 1970s, it was *responsible* for the artistic appearance of the building (Lstlburek 2009). As already mentioned, only during the recent several decades, due to the oil crises in the Near East and the adequate response of the construction branch within the new world business environment, that *new envelope materials were invented* (including foams; wadding; reflective, selective, and electrochromatic coatings; aerogel plates; and phase-exchanging materials). Together with other newly designed materials, these have been subjected to permanent testing (Ball 1999; Klassen 2003, 2004, 2006; Addington and Schodek 2005; Tonchev

FIGURE 1.1
PV panels integrated in the building envelope: (1) workshop hall; (2) covered railway station; (3) sporting court; (4) block of flats in Paderborn/Gr 1.8 kW; (5) public administrative building; (6) commercial building (mall); (7) outside the townhouse; (8) private residential building.

and Dimitrov 2009) and are successfully incorporated in the envelope structure via nanotechnologies. Hence, a technical basis of the development of envelope extra energy functions has emerged.

As time passed, the architectural ideas on the envelope energy functions evolved passing through the following two stages:

1. Reasonable minimization of energy consumption, including energy transfer through the envelope, while maintaining building indoor comfort (Димитров 2008a; Dimitrov 2014a).
2. Encouraging the incorporation of environmental energy in the building energy mix—for now, this cannot fully cover the building needs of energy (Szokolay 1981).

The envelope type is a special concern of architectural design. In addition, various vitreous elements (windows, shop windows, skylight, etc.) are solid wall* components intended to let the sunlight through. Recently, specialized envelope components have been designed (e.g., photovoltaic panels [see Figure 1.1], hybrid light receivers [see Figure 1.2], or wind turbines [see Figure 1.3]) to satisfy investors' new functional requirements (Hopness 2009).

* According to archaeologists, *palisade–palisade wall* emerged around 900 AC and later evolved into *wooden frame–timber frame*, the prototype of modern envelope. The first buildings with wood envelope dated around 1200 AC (Norway, Stave Church in Borgund, Laerdal). The timber frames were hollow (Wales House in England and Waltrop House in Germany, fourteenth century). In the first decade of the twentieth century, an additional outside paper layer was stuck to the wooden frame (rosin paper). It is considered to be the first insulation against infiltration. Around 1950, batting (fiberglass batt) and later foaming polyurethane filled in the air cavity.

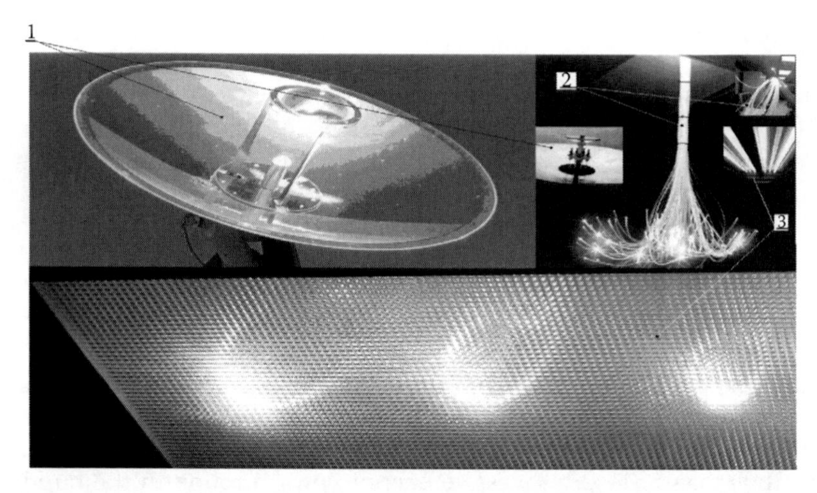

FIGURE 1.2
Building envelope integrated with components of a hybrid lighting system: (1) solar energy receiver; (2) optical cables; (3) lighting fixture operating in daylight.

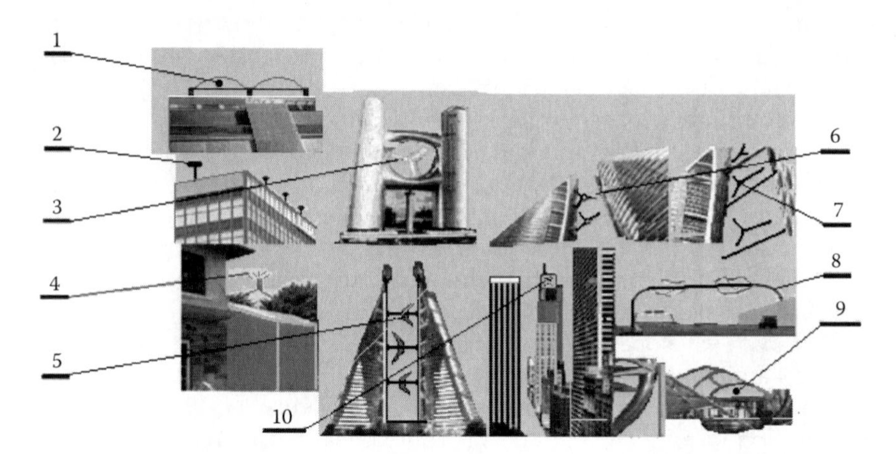

FIGURE 1.3
Building envelope combined with wind turbines: (1) rooftop horizontal propellers; (2) vertical roof wind turbines; (3) integrated wind generation station; (4) stand-alone wind turbine; (5) a series of wind turbines installed between two buildings; (6–7) a series of wind turbines; (8) highway horizontal propellers; (9) Mylars' roof horizontal turbine; (10) powerful rooftop wind turbine.

It is a priori assumed that the building envelope (including building facade) should protect the occupied areas from impacts of the surroundings (Hopness 2009). Also, the envelope design should involve aesthetics guaranteeing a unique appearance, yet in conformity with the adjacent structures. In addition, the building envelope should execute important energy-related functions, since building installations consume a significant amount of energy to compensate penetrating energy flows, which come from the surroundings.

Parallel to the pursuit of general solutions, architects should focus on envelope design during preliminary structural survey. These architects should also consider the envelope as a system of suitably arranged glazed and solid components, while subsequent solutions should involve aesthetical and functional issues of envelope–surroundings interaction. Hence, even the preliminary building design faces specific problems in evaluating the energy interactions and predicting the building's *hunger* for energy. Classical architectural concepts separate the functions of glazed envelope components from those of solid components. The latter play a role similar to "electrical resistors compared to electrical current" (Paschkis 1942; Szokolay 1981), modulating and dephasing temperature amplitudes (in the daytime, solid components absorb excessive energy flows running in the building, redirecting them to the inhabited areas at night). Solid building components (e.g., *Trombe–Michel* wall) can protect the building from overloads in the daytime and save energy at night. Glazed components, not modulating and dephasing external temperature but letting through visible light, possess pure aesthetical functions. They are intended to satisfy indoor thermal and visual comfort and save energy. These current scientific and technological achievements of leading scientific groups and companies indicate that such a *separatist* approach to the envelope elements contradicts the idea of envelope operation as an integral unit, but not, for example, the envelope components (including facades), which are part of a building illumination system (windows, skylights [see Figure 1.1], elements absorbing hybrid light [see Figure 1.3], etc.). Yet these can be combined with a photovoltaic system installed on the southeast, south, and southwest walls or roofs. It has the effect of a polished stone facade, thus attributing to building aesthetics. Moreover, envelope characteristics (flexibility, transparency, weight) are improved and envelope costs drop.

As seen in Table 1.1, the European average price per 1 m^2 of surrounding walls and photovoltaic wall panels is comparable to that of polished stone, which puts in a very strong competitive position (Димитров 2008a).

The energy conversion in the building envelope takes place at microscopic (nano) levels. This imposes the involvement of suitable theoretical, methodological, and technical tools to tackle the arising problems. In order to answer the question "To what extent are the envelope functions in relation

TABLE 1.1

Price per 1 m^2 of Surrounding Walls

Glass	Stone	Polished Stone	Photovoltaic Facade Cells
600 Eu/m^2	700 Eu/m^2	1200 Eu/m^2	200–1000 Eu/m^2

with the operation of the energy system?" it seems necessary to recall that energy enters the building in two ways:

1. Spontaneously from the surroundings—as solar radiation, mechanical (wind) energy, and entropy transfer
2. In an organized manner—owing to energy-logistics initiatives of the management body supply of energy carriers (natural gas, naphtha, or coal) or electricity

Spontaneous energy inflow occurs in three ways: via heat conductivity, radiation, and convection (infiltration/exfiltration). Using the tools of physics of constructions, architectural practice has developed four techniques to oppose energy flows running through the envelope. These employ the following resistors *operating* in the building envelopes:

- *Volumetric* (*Fourier's ones*): Opposing diffusion transfer (see pos. 1 in Figure 1.4a)
- *Capacitive ones*: Opposing the unsteady transfer (see pos. 2 in Figure 1.4b)
- *Reflective ones*: Opposing radiative transfer (see pos. 3 in Figure 1.4a)
- *Aerodynamic ones*: Opposing convective leakage (under dynamic pressure gradient or building stack effect)

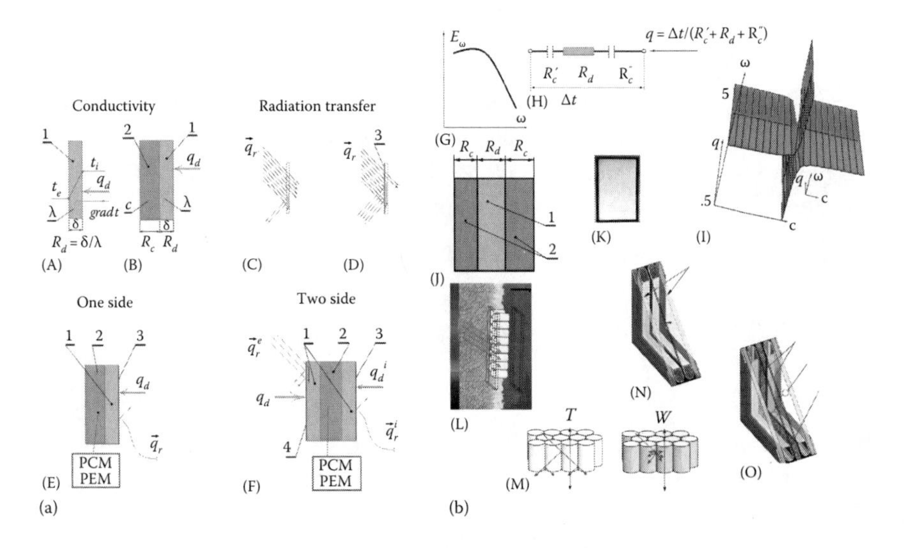

FIGURE 1.4
Two techniques of controlling the energy flux though the components of the envelope: (a) the envelope as an *energy barrier*; (b) the envelope as an *energy filter*, (1) volumetric (Fourier) resistance, (2) capacitive resistance (with or without phase exchange), (3) reflective resistance long-wave spectrum, and (4) wide-spectrum reflective resistance.

Techniques of the design of solid and glazed components of the envelope system used in architectural practices are illustrated in Figure 1.4. These help keep down the energy-accumulating capability and conductivity of the envelope through an increase of the heat resistance of the thick components and reduction of the surface of windows and glazed components to the required ergonomics minimum (Dimitrov 2014); see more details in Chapter 7.

An idea of passive regulation of the flows running to and from the building envelope by modeling its function of an *energy filter* was the subsequent architectural step of understanding the building energy functions (in the 1980s). It was found that the environmental energy flows could be shifted in time and weakened, if needed—see the works of O.E. Vlasov (Власов 1927) and Paschkis (1942) on heat–electrical analogy and the early works of Mackey (Mackey and Wright 1944; Mackey 1946). There, the energy transfer through the envelope component is interpreted as the sum of two components:

$$q = \bar{q} + \tilde{q},$$

where

$\bar{q} = 1.65A_0 \dfrac{0.608(T_m - T_i)}{0.865 + \delta_W/\bar{\lambda}}$ is the component of the specific heat flux, which is constant in time and magnitude depending on T_m, T_i, and the value of the thermal resistance of the building envelope—$\delta_W/\bar{\lambda}$ (other variables are A_0, which is the envelope area of the cross section, thickness δ_W, and thermal conductivity $\bar{\lambda}$)

$\tilde{q} = 1.65A_0 \sum_{n=1}^{\infty} \Delta_n T_n \cos(15n\theta - a_n - \varphi_n)$ is the heat flux variable in time $\theta(12h \Rightarrow \theta = 0)$, which is presented as a superposition of n $(1 < n < \infty)$ harmonics

Δ_n and T_n are the damping ratio and the phase delay coefficient of the nth harmonic $\left(\text{as } \Delta_n = \left(1.41\left(F^2 + G^2\right)^{-1}\right),\ \phi_n = \text{arctg}\left(\dfrac{F - G}{F + G}\right) \right.$ and resp.

$$\left. F = \left(r\frac{q_0}{q_c} + \pi_1 c_C \frac{q_0}{q_c} \right), G = \left(S_0 \frac{q_0}{q_c} + \pi_1 S_C \frac{q_0}{q_c} \right) \right)$$

These coefficients depend on volumetric density, thermal conductivity, and diffusivity of the used building materials ($q_{0,C}$ and $S_{0,C} \rightarrow f(\rho, \bar{\lambda}\ u\ a)$) in an implicit form. This trend of study was continued in more recent times in the works of Mitalas (Mitalas and Stephonson 1967; Mitalas 1968, 1978) and Bogoslavskiy (Богословский 1982).

The *energy-filtering* function of the building envelope shifts in time to the external energy flux to a more favorable moment and weakens its penetration at the designer's discretion. Note that when a particular component has been dimensioned, manufactured, and mounted on a building facade, it was expected to *operate* without a significant change in its functional characteristics throughout the life cycle of the surrounding system, following a steady program (i.e., it is passive). This can be achieved by selection and use of components and materials, with characteristics that are *spectrally determined*. The *spectral determination* of facade structures is performed by a selection of the thermal capacity of the solid structural element and by the volumetric resistance to the conductive flow (see Figure 1.4b). This can also be achieved by the arrangement of special filters on the track of the radiative energy flow (see Figure 1.4e through h), even at the *design* stage. Classical materials are used to vary the thermal capacity of thick envelope components. These are reinforced concrete, ceramic plates, rock material, or their combination having high *thermal mass* and displaying solid–liquid phase exchange at temperatures, typical for the building envelope (about 290–300 K).

The second step of architecture in understanding the energy functions of the envelope is its treatment as an intelligent energy membrane. This approach was promoted by energy simulations performed in the mid-1980s (Clear et al. 2006), launching the idea that buildings have a potential to considerably reduce the annual consumption of illumination and cooling energy (by 20%–30%). This was achieved in residential and public buildings erected in areas with moderate and hot climate, by mounting the so-called intelligent or dynamic windows, which would only let through the solar energy needed for interior illumination. The first prototype of a building with an intelligent envelope was built in 2003 (Lee et al. 2004). The facade consists of windows coated with a thin ceramic electrochromatic film, with characteristics that may be controlled by BSAC and M in a range suitable for the actual climatic area. For the visible spectrum, these characteristics are the transparency coefficient and the coefficient of solar heat accumulation. The illumination system works under the BSAC and M, with sensors that measure the room illumination and the natural solar flux falling on the windows. The inner illumination comfort has been guaranteed independently of the external illumination by controlling the degree of window darkening and automatic switch of the electric fluorescent lamps. Thus, the architectural understanding of the envelope energy function evolved over time. Figure 1.5 shows a scheme of evolution of the facade functions regarding two stages:

1. *First stage*: The envelope executes a passive function opposing external impacts (it is a barrier or filter).
2. *Second stage*: The envelope behavior (its physical characteristics, respectively) can be *fitted* to the impacts due to the surroundings, indicated by BSAC and M. Thus, one can meet the building needs by keeping the inner environment parameters within comfortable limits; the envelope should behave as an *intelligent membrane* (see Figure 1.6) (Димитров 2008b; Tonchev and Dimitrov 2009).

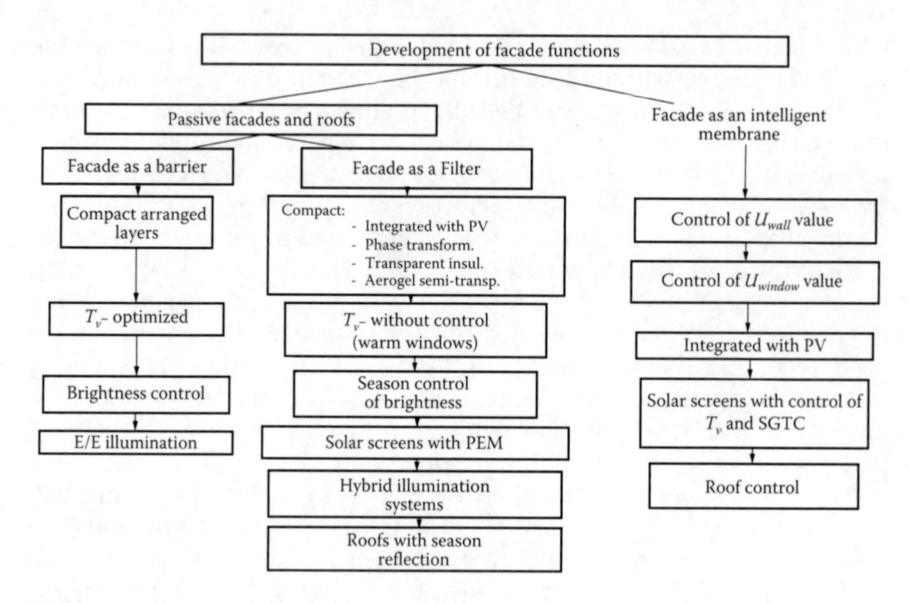

FIGURE 1.5
Scheme of the evolution of the facade functions.

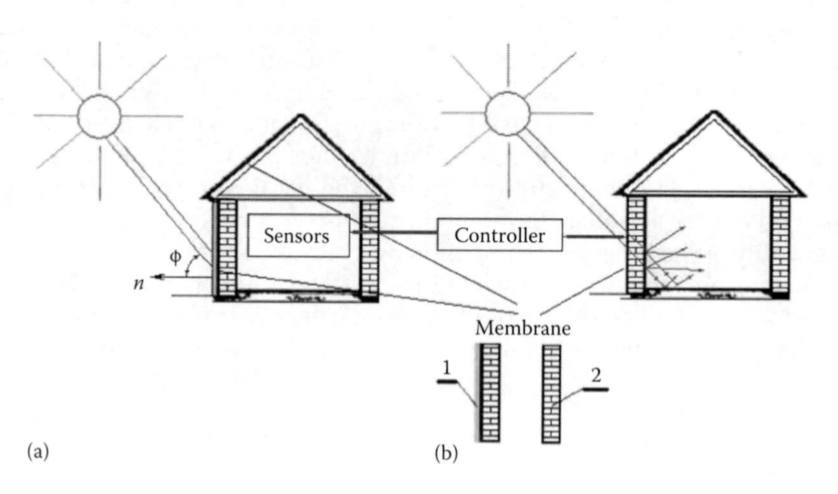

FIGURE 1.6
Scheme of a building envelope, type *intelligent membrane.* (1) *Spectral-sensitive layer;* (2) *transparent* structure carrying the envelope: (a) the facade membrane is *switched on;* (b) the facade membrane is *switched off;* (c) the roof membrane is *switched off.*

1.3 Summary Analysis of the *Building–Surrounding* Energy Interactions

The analytical overview of the applied architectural energy strategies in building envelope design, set forth in Section 1.2, and the outline of envelope energy functions are based solely on macroscopic thermodynamic concepts and laws. Note that other types of energy interactions between the surroundings and the building have been a priori excluded. These are, for example, mechanical, electrical, or electromagnetic effects. Yet the analysis of the latter should concern envelope structural properties or electric conversion capabilities, adopting deformation-strength or purely electrical approaches.

Modern thermophysics does not have at its disposal a common (microscopic and macroscopic) systematic approach and philosophy to assess the energy interactions between a building envelope and the surroundings. As previously mentioned, modern science has found examples where the building envelope was treated as an integrated system, which operates as a coherent mechanism, satisfying building energy demands. Those arguments have allowed the formulation of the following basic objectives of this book:

- To develop a more general idea on the *building–surrounding* energy interactions based on the analysis of microscopic transfer running within the envelope
- To propose a generalized physical model describing these interactions at microscopic levels using macroscopic thermodynamic characteristics
- To design mathematical models predicting the effects of energy interactions with the help of classical analytical tools
- To specify the author's own engineering methodologies of design and inspection of the building envelope, accounting for the respective energy impacts
- To formulate conclusions and recommendations for further research in the field

The results found are presented as follows: Chapter 2 (Section 2.3.3), Chapter 3, Chapter 5 (Section 5.2), and Chapter 7.

2

Physics of Energy Conversions in the Building Envelope at Microscopic Level

An adequate idea on the mechanisms of energy interaction between the surroundings and the building envelope components at microscopic level is of crucial importance for the development of construction science. This is so, since the plausible assessment of the running energy-exchange processes stimulates the development of new design strategies, the introduction of new and intelligent structural materials, and the employment of advanced electronics and automatics in civil engineering. Hence, one can erect various buildings (residential, public, and commercial), while such constructional activity was unconceivable decades ago.

As is known, solid structural materials used in envelopes are classified with respect to their chemical composition as follows: pure metals (metals and nonmetals), alloys (compositions), and insulating systems (Incropera 1974; Incropera and DeWitt 1985, 1996, 2002; Назърски 2004).

Specialized references (Kersten 1949; Carslaw and Jaeger 1959; ASM, 1961; Incropera 1974; Desai et al. 1976; Farouki 1981; Harrison 1989; Rohsenow et al. 1998; Incropera and DeWitt 1996; Ball 1999, 2002; Bird 2007) prove that enough empirical data on the energetic characteristics and properties of various substances have been collected in the databases during the last 100–130 years, thus creating a mine of information on envelope materials. Their aggregate state can be solid, liquid, or gaseous depending on the current state parameters. Besides pure mechanical properties, material capabilities to absorb (reflect), transfer (capture), or accumulate energy are of crucial importance for envelope designers. Yet, the physical properties (and energy-exchange properties in particular) of materials strongly depend on their operational aggregate state.*

Building materials currently involved in architectural projects (Klassen 2003, 2004, 2006) have various energy characteristics. Figure 2.1 illustrates their ability to transfer heat and shows the range of the heat conductivity coefficient $\bar{\lambda}$, W/mK. The results prove the capacity of various materials to receive, accumulate, and transfer heat. It is seen that a material's aggregate

* However, in various constructional operations, we use materials in all the three physical states. Yet, solid materials are mostly involved in the fabrication of building envelope components.

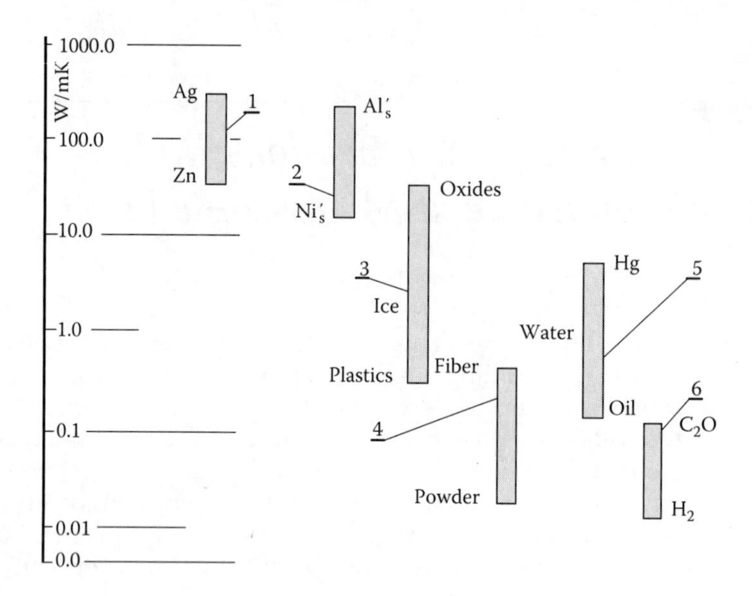

FIGURE 2.1
Range of variation of the conductivity coefficient of materials used in thermotechnics, regarding their aggregate state and chemical composition. (1) pure metals; (2) alloys; (3) non-metals; (4) insulation system; (5) liquids; (6) gases.

state is in strong relation with the heat conduction, and the coefficient of thermal conductivity λ, W/mK* varies within the following ranges:

- For solids—$0.05 \leq \lambda \leq 500$ W/mK.
- For fluids—$0.02 \leq \lambda \leq 5$ W/mK.

Therefore, the ability of solids to transfer heat under normal conditions is 4–100 times larger than that of fluids due to solid, smaller intramolecular distance (Incropera and DeWitt 1985). Similar proportions are valid for some other characteristics of solids (see Table 8.1).

Modern science explains two levels (macroscopic and microscopic) of energy-related capabilities of solids (the latter being insulators, conductors, smart filters, and filters of different energy impacts). Thus, engineering models become complete and accurate, and the predicted material behavior is in a better agreement with the experimental evidence. The wide variety of technical means, including phenomenological models at a microscopic level, subsequent theoretical approaches, and classical and modern computational methods, is of strategic importance in the engineering design of buildings and building components. It is an object of this book.

* The coefficients of thermo- and electroconductivity λ and Ω are directly proportional to each other conforming to the Wiedeman–Franz law— $\lambda = L_0 \cdot T \cdot \Omega$.

The representative energy-related characteristics of a building material at *macroscopic level*, except for the already discussed coefficient of thermal conductivity $(\bar{\lambda}, \text{W/mK})$, are the coefficient of electrical conductivity $(\bar{\Omega}, \text{Wm/A}^2)$, specific heat capacity c_V, J/m³ K, and diffusivity $\alpha = \bar{\lambda}/\rho c_V$.

It is known that a proportionality relation between the first two quantities exists (according to the law of Wiedeman*/Franz, 1853 year (Bejan 1953)):

$$\bar{\lambda} = \bar{\Omega} \text{ Lo } T. \tag{2.1}$$

Here $\text{Lo} = (\pi^2/3)(k_B/e^0)^2 = 2.45 \times 10^{-8}$, V/K² is the universal Lorentz constant for all metals and T is the operational temperature. The use of relations $\bar{\lambda}/\bar{\Omega} = (2.45 \times 10^{-8}\ T)$ or $\bar{\Omega} = \bar{\lambda}/(T\ \text{Lo})$ resulting from the earlier formula significantly facilitates the computations.

The first important factor affecting material capacity to transfer and accumulate heat is the operational temperature (T), and the subsequent empirical relation reads as follows:

$$\lambda = a_P T^2 + a_I \cdot T^{-1}.$$

Here a_p and a_I are coefficients experimentally found for various materials. It is established that the thermal conductivity coefficient attains maximal value (λ_{max}) when the operational temperature is $T = (a_i/2a_p)^{0.333}$ (Bejan 1953). The latter varies for different materials. It increases for materials with strongly expressed impurity-scattered conduction and larger† coefficient a_I (because the width of the band gap in different materials is affected by the operational temperature, and thereby, it affects the intensity of the emitting electrons in the free zone). The energy properties of the objects may depend on other macroscopic properties, such as the effective electric potential, pressure, electromagnetic tension, enthalpy, entropy, and the energy level of Fermi, Gibbs, Helmholtz, Landau, Duhem, and others.

The capability of solid building materials to transfer, absorb, filter, or control external energy (i.e., control the envelope volume) is assessed via physical models of transfer (at a macroscopic level, such models were designed by Fourier (1801), Fick (1855), and Ohm (1826)‡ (Benjamin 1998; Atkins 2007). Those models reflected subsequent laws of heat, mass, and electrical transfer discovered and popularized in the nineteenth century. They are the basis of significant part of modern engineering computational tools in the field—see (Bosworth 1952; Nottage and Parmelce 1954; Buchberg 1955; Brisken and Reque 1956; Carslaw and Jaeger 1959; Bird et al. 1960; Rohsenow and

* G.H. Wiedeman (1826–1899).

† This phenomenon is explained by the fact that the operational temperature T affects the width of the banned area and the emission of free electrons. The banned area according to modern physics is located between the valence band and the free area.

‡ J.L.J. Fourier (1768–1830), G. Ohm (1789–1854), and A. Fick (1828–1901).

Choi 1961; Kutateladze 1963; Boelter et al. 1965; Luikov 1968; Klemens 1969; Stephenson and Mitalas 1971; Eckert and Darke 1972; Welty 1974; Buftington 1975; Kreith and Black 1980; Chapman 1981; Lienhard 1981; Todd and Ellis 1982; Wolf 1983; Butler 1984; Kakas 1985; Seem et al. 1989; Brorson et al. 1990; Thomas 1992; Bejan 1993; Gebhart 1993; Kakas and Yener 1993; Stocker 1993; Taine and Petit 1993; Morely and Hughes 1994; Poulikakos 1994; Holman 1997; Cengel 1998; Pitts and Sisson 1998; Wilkinson 2000; Kreith and Bohn 2001; Nilsson and Riedel 2007).

2.1 Idealized Physical Model of the Building Envelope as an Energy-Exchanging Medium (Review of the Literature from Microscopic Point of View)

The development of quantum mechanics and low-temperature physics in the late nineteenth century made possible the elucidation and study of the mechanisms of energy transfer at a microscopic level. Hence, it is now known that energy transfer results from two physical phenomena*:

1. Generation of lattice vibration waves caused by quasiparticles (known as phonons†) and their scatter.
2. Migration of free electrons and their interactions with phonons—both particles act in combination or separately.

Consider a solid building envelope with dominating insulation and semiconduction properties. Then, the energy transfer through it is mainly due to waves of lattice vibration. In pure metals, however, predominant is the second effect outlined earlier, while in alloys, the contribution of the two phenomena is commensurable (Incropera 1974; Tzou 1992, 1995, 1997).

As shown in the literature (Özisik and Schutrum 1960; Joseph and Preziosi 1989; Joseph and Preziosi 1990; Tzou 1992; Duncan and Peterson 1994; Tien and Chen 1994), three physical models are employed to describe energy transfer at a microscopic level: *a two-step model* created by M. Kaganov in 1957 (Kaganov et al. 1957) and later refined by C. Anisimov in 1974 (Anisimov et al. 1974) and Qiu and Tien in 1993 (Qiu and Tien 1993); a *model of scattering phonons,* created by Guyer and Krumhansi in 1966 (Guyer and Krumhansi 1966) and improved by Joseph and Preziosi in 1989 (Joseph and Preziosi 1989) and Tzou in 1995 (Tzou 1995) (it was promoted in 1995 as the model of *lagging*

* The thermodynamic laws concerning an idealized crystal body are valid for all real crystal and amorphous structures (amorphous solids have a randomly ordered lattice, and thermodynamics states that one should account for their residual entropy) (Tien and Lienhard 1985).

† Phonons—quasiparticles belonging to the class of ***bosons***.

behavior); and *a model of phonon radiative transfer*, designed by Majumdar in 1993 (Majumdar 1993) and thematically oriented to the mathematical description of conductive heat transfer in thin dielectric films.

Using generalizations of the physical models discussed herein, an outline has been created for the specific physics of energy transfer in building envelope components and describe transfer by means of a *hypothetical model of lagging gradient*. Prior to that, however, it is necessary to present in short the physical prequel that served as a methodological basis of the development of energy transfer models. According to modern science, solid building materials are composed of (Tien and Lienhard 1985)

- Periodically arranged atomic structures called lattice or network
- Free electrons (electronic gas)

The theoretical bases of energy conversion in solids involve an idealized physical model called *crystal lattice structure*.* It is widely considered to be the simplest crystal system, for example, a cubic crystal lattice, characterized by isometric geometrical parameters ($a = b = c \approx 10^{-10}$ m and $\alpha = \beta = \gamma = 90°$) (Tien and Lienhard 1985).

By the end of nineteenth century and in the early twentieth century, the theory of heat transfer and thermodynamics dominated over the so-called kinetic thermal theory (Lee 2002), in the core of which lies the concept of material continuum. Then, however, numerous physical experiments at low temperature ($T < 100$ K) proved that solids ceased to behave in conformity with the postulates of the kinetic theory (Lee et al. 1963; Lee 2002). The most drastic departure from the predictions of that theory was the behavior of bodies having accumulated heat (displaying the so-called accumulation capacity), assessed by the *specific heat capacity*—C_V, J/m³ *K*. It has been found (see Figure 2.2) that C_V depends on T in the temperature range $0 < T < 100$ K, and the dependence is nonlinear.

For temperature close to the absolute zero ($T \to 0$ *K*), the specific heat capacity is zero. Then, C_V increases with the climbing in temperature, and at $T/\theta_D \geq 2, 1$ ($T > 100$ *K*),† it asymptotically approaches a constant value (C_V = 6.4 cal/g-mol K). That value was experimentally found by Dulong and Petit in 1819—C_V = const [$C_V = 3N_A k_B/m$ = const($C_V = 3R \approx 6.4$ cal/g-mol K)].

The establishment of the nonlinear dependence of C_V on temperature generated the first crisis in the understanding of energy accumulation in solids, since the kinetic gas theory was not able to explain it. (Einstein 1905, 1907, 1934; Lee et al. 1963; Kelly 1973; Van Carey 1999) and then Debye (1910, 1928)‡

* The crystal lattice is a matrix formed by basic crystal cells. Various configurations between atoms are possible in the crystal lattice-230 in number. They are grouped in 32 categories and reduced to 6 basic schemes—isometric, tetragonal, orthorhombic, hexagonal, monoclinic, and triclinic (Tien and Lienhard 1985).

† θ_D—Debye temperature.

‡ Peter Debye (1884–1966)—Nobel Prize winner in chemistry, 1936.

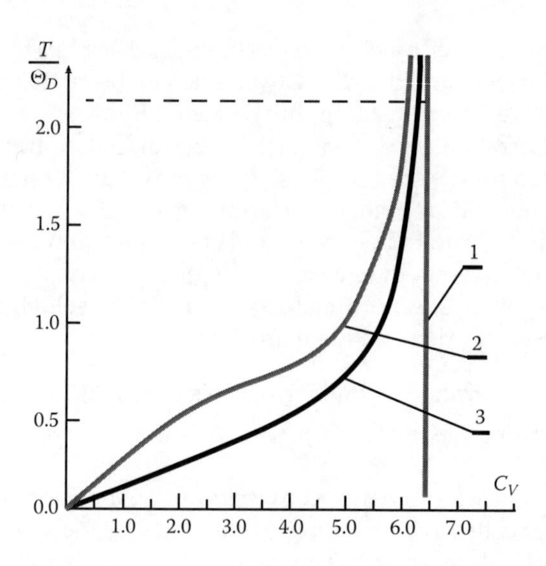

FIGURE 2.2
Variation of the specific heat capacity at low temperature: (1) approximation by means of Dulong/Petit law; (2) approximation by means of the Einstein model (quadratic approximation); (3) approximation using Debye model (the approximation is in agreement with the experimental evidence for Pb, Ag, KCl, Zn, NaCl, Cu, Al, CaF_2, and C).

were the first that managed to overcome those difficulties. They assumed that discrete (quantum) processes took place in a discrete medium (consisting of atoms and molecules). Thus, they laid a new foundation of the theory of heat transfer,* that is, the study of heat transfer based on the postulates of the quantum mechanics and statistical thermodynamics.

As already stated, the physical model of Einstein/Debye was adopted as a starting methodological basis of describing the nature of the energy interactions between the building envelope components and the surroundings. The following is a short summary of the main points of that approach:

- Rigid bodies are built of recurring atomic structures with different topology—about 70% of metals and alloys have face-centered cubic (fcc), body-centered cubic (bcc), or hexagonal crystal (cph) lattices (Tien and Lienhard 1985).
- Due to the existence of substantial interatomic forces, the atoms can perform only three-dimensional (3D) oscillations around a relative

* The entire solid composed of N atoms is modeled as a $3N$ multidegree harmonic oscillator, whose atoms (oscillators) vibrate autonomously with identical frequency ν_E, that is, longitudinal harmonics with one and the same frequency propagate along the axes of a Cartesian coordinate system.

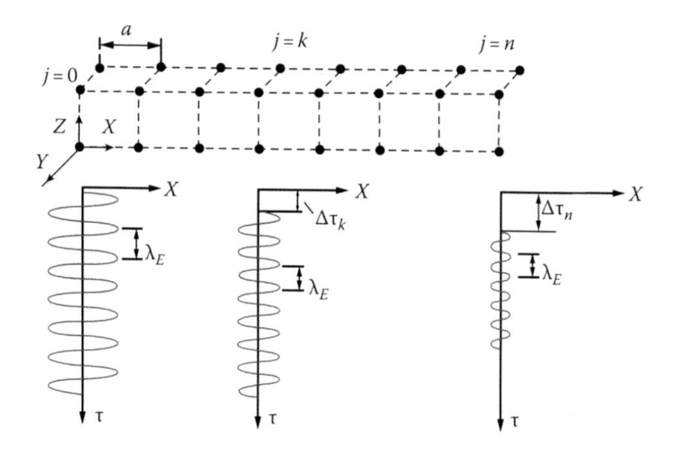

FIGURE 2.3
Longitudinal vibrations of atomic structures (Einstein waves).

equilibrium position (atoms in solids are *deprived* of two of the three types of motion, in contrast to atoms in gases and liquids).

- The entire rigid body composed of N atoms is modeled as a multilevel 3N harmonic oscillator in which each atom (oscillator) autonomously vibrates longitudinally along the three axes of a Cartesian coordinate system with identical frequency (so-called characteristic frequency—ν_E).
- Oscillation amplitudes of the individual atoms depend on their position in space and obey the distribution of Maxwell/Boltzmann (Van Carey 1999).

Without delving into details, note that energy transfer in solids involves two types of high-frequency wave motion (Van Carey 1999):

- *First type*: Energy transfer via longitudinal oscillations of atomic structures or so-called waves of Einstein (see Figure 2.3).

  Oscillations have a wavelength λ_E smaller than the basic size of the grid "***a***" (see Figure 2.3). In this basic case, all lattice atoms vibrate longitudinally along the directory of transfer, and with identical characteristic frequency ν_E (it is identical for all atoms with a natural frequency of the harmonic potential—we assume that it is constant for each atom), but with different amplitudes. Moreover, their vibration is unsynchronized (disorderly). The same mechanism of energy transfer predominates at medium and high temperature ($T > 100$ K) or ($T/\theta_D > 2.2$), in all nonmetallic solids and alloys. The characteristic frequency ν_E depends on material structure and chemical composition.

The energy is transferred in discrete portions (quanta), multiples to the quantum numbers $n(n = 1,2,3,\ldots)$.* A. Einstein proposed the following expression of its assessment:

$$e_{Ph} = (n+0.5)h\nu_E,$$

where

h is the Planck constant ($h = 6.63 * 10^{-34}$ J·s)

ν_E is the frequency of atom oscillation ($\nu_E = c/\lambda_E$, λ_E—wavelength, c—speed of light)

The internal energy of the whole body is found as a sum of the energy of all components of the vibrating system:

$$\left(U - U_0 = h\sum_{1}^{3N} \nu_j \left[e^{h\nu/kT} - 1 \right]^{-1} = 3Nk_B T\bar{n}(\nu) \right).$$

Vibration of atoms with a fixed frequency of multiple integers (quantum number) is interpreted as owing to phonons (these are quasiparticles charged with vibration energy). A simple 1D model of a harmonic oscillator illustrates the earlier considerations. Its motion is described by Newton's equation†:

$$m_j \frac{d^2 x_j}{dt^2} = C_f \left(x_{j+1} + x_{j-1} - 2x_j \right).$$

The solution is sought in the form $x_j = x_0 \cdot e^{i(jKa+\omega\tau)}$ (Van Carey 1999), and the integral found produces the relation between oscillation angular frequency $\omega(\omega = 2\pi \cdot \vartheta_E)$ and the magnitude of the wave vector—wave number $K(K = 2\pi/\lambda_E)$ (see Figure 2.4 illustrating the solution of the differential equation). The behavior of the atomic lattice depends on the type of *impacts* penetrating the boundary of the control surface (see Figure 2.4). If the amplitudes of the external disturbances are *smaller* than the interatomic distance "**a**," lattice atoms oscillate *autonomously and independently*, that is, propagation of Einstein longitudinal waves takes place.

Physically relevant integrals are only those that predict the vibration of atoms, whose wave numbers are multiples of the grid basic size "**a**." Oscillation damping Δ is a function of the wave number and

* $(n + 0{,}5) = \bar{n}(\nu)$ later called function of phonon distribution within the frequency area.

† The linear restoration force acting within the crystal lattice has an electromagnetic character. Depending on bonds, crystal lattices are classified into four groups: lattices with covalent bonds, lattices with ionic bonds, lattices with metal bonds, and lattices with molecular bonds (Tien and Lienhard 1985).

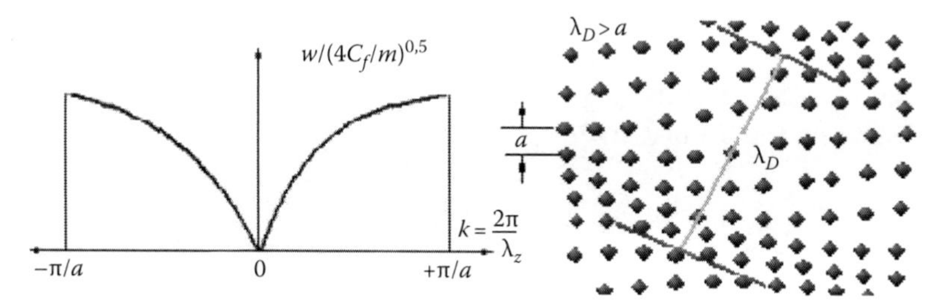

FIGURE 2.4
Relation between the angular frequency and the wave number.

grid basic size—$\Delta = x_{j+1}/x_j = e^{-ika}$. At higher wave numbers ($K > 2\pi/a$), the structure starts moving as an elastic membrane, and the atom oscillations *correlate* with each other. The entire lattice behaves as a flexible membrane (i.e., propagation of the Debye waves takes place). This effect was proved by Debye.

- *Second type*: energy transfer via transversal waves (waves of Debye), where the lattice oscillates as a flexible membrane.

Those grid transversal waves arise when solid boundaries undergo arbitrary energy fluctuations, forcing the atoms to move with identical frequency (and wavelength larger than the basic size of the grid—$\lambda_D > a$), but performing a correlated motion. Then, motion of the entire array resembles motion of a flexible membrane—see Figure 2.4. Energy transfer via transversal waves dominates at low temperatures ($0 < T/\theta_D \leq 1.2$, θ_E—characteristic temperature of Debye—see Table 8.11). At a higher temperature ($T/\theta_D > 1.2$), transfer energy becomes commensurable to that transferred via longitudinal waves (according to Debye, the ratio is 1:2, which is confirmed by the excellent agreement between his model and the experimental evidence). Heat transfer in a solid body as suggested by Debye that takes place with a maximum frequency (cutoff frequency) ν_D, which depends on the solid volumetric density N/V_0 (see data in Table 8.11 as an illustration), is as follows:

- $\nu_D = \bar{c}\left(\frac{3}{4\pi}\frac{N}{V_0}\right)^{0.333} = 5.32 * 10^8 \cdot \bar{c}(\rho/M)^{0.3333}$—for a 3D solid with density ρ kg/m³ and molecular mass M, kmol (since $N/V_0 = \rho(M/N_A)$, while the Avogadro number N_A is constant per 1 kmol c).
- $\nu_D = c^2 \cdot \frac{N}{\pi \cdot A_0} = 0.32c^2\left(\frac{N}{A_0}\right)$—for a 2D solid with circumferential area A_0 (Tien and Lienhard 1985).

According to Van Carey (1999), ν_D should be interpreted as a limit frequency below which the solid behaves as an elastic continuous membrane. Yet, a detailed analysis of the mathematical formulas shows that it is the essential maximum frequency of energy transfer. A combined energy transfer takes place within a solid, via transversal and longitudinal waves, whose frequency distribution function is defined by Debye, having the following form (Van Carey 1999):

$$g(\nu) = 4\pi \cdot V_0 \left(\frac{2}{c_t^3} + \frac{1}{c_l^3} \right) \cdot \nu^2 = \frac{12\pi \cdot V_0}{\bar{c}^3} \cdot \nu^2$$

Here c_t and c_l are velocities of propagation of the longitudinal (Einstein) and transverse (Debye) waves, while $\vec{c}(c)$ is the mean wave (sound) velocity (for 2D and 3D bodies), which depends on solid physical properties (density, coefficient of the volumetric expansion/contraction, Poisson coefficient, etc.).

The comparative evaluation of Debye and Einstein transfer frequencies shows that they differ significantly from each other. For example, for silver ($\rho = 10{,}500$ kg/m^3), the characteristic Debye frequency is $\nu_D = 4.37 * 10^{12}$ Hz, while according to Einstein, it is $\nu_E = 2.85 * 10^{12}$ Hz. (Blackman found reliable characteristic frequencies of transfer, which differed from those of Debye and Einstein. Yet, Einstein's results still remain actual; Saad 1966). The frequency spectra of the vibration of a solid 3D lattice are shown in Figure 2.5, according to the approximations adopted by Einstein and Debye. They are compared to Fine's calculations on tungsten. In 1925, Bohr, Heisenberg, and Jordan Pascal interpreted the method of Debye in terms of quantum mechanics (DeBroglie 1953). They assumed that the energy $E = \bar{n}h\nu$ of the oscillator with frequency "ν" is the energy comprising "n" discrete energy forms.

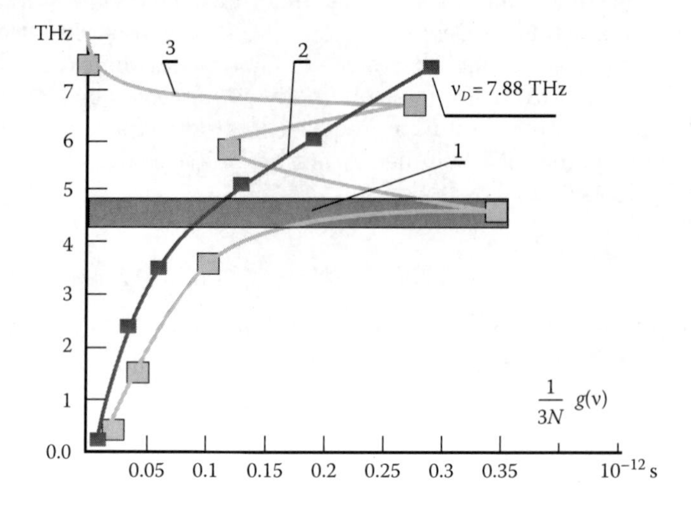

FIGURE 2.5
Comparison of the frequency spectra: (1) approximation of Einstein; (2) approximation of Debye; (3) prediction of Fine.

Modern studies assume that the capacity of solids to transfer energy depends on the frequency ν of vibration transfer. The renowned Russian physicist I. E. Tam* interpreted those vibrations as resulting from pseudo-particle vibration—vibration quanta of solid atoms (see Figure 2.4) called *phonons*,† which were the main energy carriers (Stowe 1984; Tzou 1997; Baierlein 1999; Van Carey 1999). Tam's theory was universally recognized.

Similar to photons, phonons are pseudo-particles belonging to the group of bosons, and they are described via Bose–Einstein statistics (Stowe 1984). For a particular frequency ν, they have the same eigen energy as photons: $\langle E_{Pn} \rangle = \bar{n}(\nu)h\nu_E$, where $\bar{n}(\nu)$ is the mean number of phonons for frequency $\nu = \nu_E$ and at temperature T (for frequency ν_E, the typical phonon in the model of Einstein frequency would carry energy $e_{Ph}^{E} = h\nu_E$ only).

An important feature of the distribution function is that the number of phonons varies proportional to the absolute operational temperature T (the number of phonons in the interpretation of Einstein increases with an increase in temperature).

Phonons interact with corpuscles (electrons, atoms, and molecules) building the body. They are divided into two groups: acoustic and optical (Harrison 1989; Challis 2003; Strosicio and Dutta 2005). It is assumed that thermal phonons are intermediate particles being *responsible* for energy transfer in solids, and moreover, they can be *generated or destroyed* by arbitrary energy fluctuations at the solid surface or within the solid (see Section 2.2).

The calculated values of ν_D, $\theta_D = \nu_D h/k_B$, and $\bar{c}$ given in Table 8.1 can be interpreted as characteristics of phonon motion taking place within basic metals conforming to the Mendeleev table. Calculations are performed using Debye temperature values published in literature (Tien and Lienhard 1985) and the coefficients of thermal conductivity $\bar{\lambda}$ and diffusivity $\alpha = \bar{\lambda}/\rho c_p$ (Incropera and DeWitt 1985). The results are given in an ascending order, conforming to the increase in the mean velocity of phonon propagation in solids c. It reads as follows:

$$c_m = \theta_D \frac{k_B(4\pi/3)^{0.3333}}{h(\rho N_A/M)^{0.3333}}.$$

Since c_m is an integral characteristic of phonon transfer in the solid lattice at microscopic level, it is worth finding its correlation with the coefficients of thermal conductivity $\bar{\lambda}$ and diffusivity α which are macroscopic characteristics. This can be done by analyzing the effect of the *second basic carrier* of energy in solid materials—the electrons—estimating their capacity to change the amount of energy accumulated in solids $u \rightarrow du \left(du = c_p \frac{du}{dT} \right)$.

* I. E. Tam was the 1938 Nobel Prize winner in physics.

† Phonons belong to the group of *Bosons*, which are elementary particles with integer spin (other members of the boson group are photons, pions, gluons, and W- and Z-bosons).

Moreover, transfer intensity depends on the mechanical motion of electrons as free material objects within the body.

As already noted, a mechanism of energy transfer, alternative to the vibration of the crystal structure (phonon scatter), is assumed to exist in solids and especially in metals (Scurlock 1966; Klemens 1969; Bejan 1993). It is known as an isotropic scatter of electrons (*impurity scatter*) or electronic thermal and electrical conductivity. To get an idea on its importance for the operational conditions of the building envelope, we present a short survey of the foundations of the *zone theory* of atomic structure (Kittel and Kroemer 1980; Inoue and Ottaka 2004; Martin 2004). It was developed by the Russian physicist G.D. Borissov on the basis of postulates formulated by the Danish physicist Nils Bohr (1913, 1958). The theory is a *basis* of the *hypothesis of interaction between a solar photon flux and the envelope material.*

Bohr' model (Bohr 1913, 1958; Einstein 1934; Tipler and Llewellyn 2002; Tipler 2004) sets forth the conditions of stable existence of elementary particles within the structure of an atom, consisting of an atomic nucleus and an electron envelope. Pursuant to that model, if an *atomic structure is isolated,* the internal forces (Coulomb force $\vec{R}_C$ and the centrifugal force $\vec{R}_Z$) acting on the electrons counterbalance each other (their resultant is zero $\vec{R}_0 = \vec{R}_C + \vec{R}_Z = 0$), while the electrons run around the nucleus along particular discrete orbits and with constant kinematic characteristics. The values of the electron orbit radii $r_n (n = 1,2,3\ldots)$ are discrete depending on the nucleus mass determining the "Z" number, that is, the sum of protons and neutrons compounding the nucleus. The electron orbit radius reads as follows:

$$r_n = \frac{4\pi n^2 \hbar^2 \varepsilon_0}{Z e^{0^2} m_e}, \tag{2.2}$$

where

- n is an orbit index ($n = 1,2,3\ldots$); $\hbar = h/2\pi$ ($h = 6.6260755 * 10^{-34}$, Js—Plank constant)
- ε_0 is the dielectric constant ($\varepsilon_0 = 8.85418781762 * 10^{-12}$, As/Vm)
- e^0 is the electron electric charge ($e^0 = 1.60217733810^{-19}$, As)
- m_e is the electron mass ($m_e = 9.1093897 * 10^{-31}$, kg)

According to the zone theory of Borissov, the electrons of an atom occupy different orbits, while a limited number of electrons may run along a certain orbit depending on the distance to the nucleus. For instance, 2 electrons at the most can occupy the first orbit (n = 1), 8 electrons—the second orbit, 18 electrons—the third orbit, 32 electrons—the fourth orbit, etc. Electrons do not entirely populate the last and most distant energy orbit known as *valence band.* According to the number of electrons available there (n_{VZ}), the atomic structures are ordered in groups according to Mendeleev's table* (1869)—see

* D. Mendeleev (1834–1907).

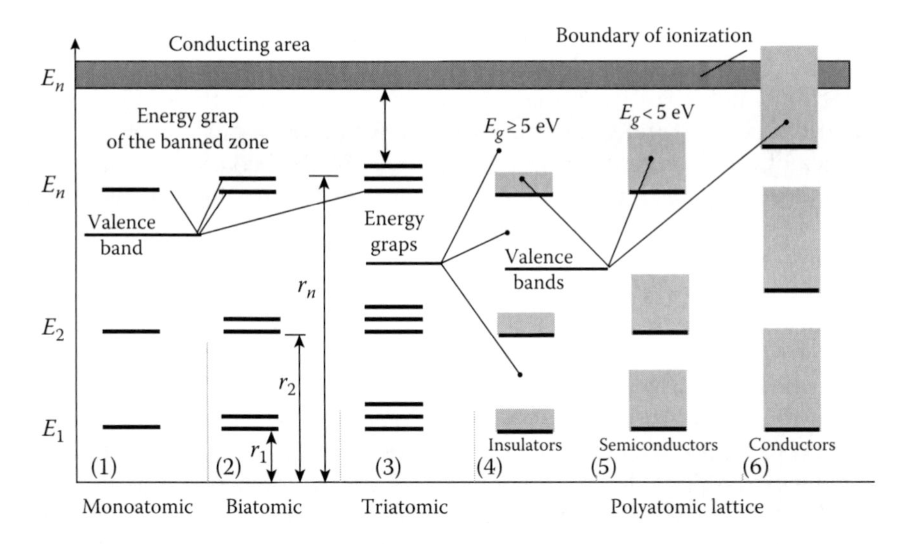

FIGURE 2.6
Location of the electron orbits and energy bands in the atomic structures of solids: (1) Monoatomic; (2) biatomic; (3) triatomic; (4) insulators; (5) semiconductors; (6) conductors.

Table 8.1. For instance, the fourth group comprises atoms with four free electrons occupying the valence band (silicon—Si, germanium—Ge, tin—Sn, etc.), the fifth group—atoms with five free electrons, etc. According to the location of the valence band with respect to the conducting area, solids are divided into insulators, semiconductors, and conductors (see Figure 2.6). A large energy *gap* between the valence band and the free area exists in insulators. It is called the *banned area*, and electrons need energy larger than 5 eV to overcome it. In conductors, however, those two energy orbitals merge. Hence, electrons can easily leave the valence band and start running within the free area under low-energy external impacts. Similar electron behavior is observed in semiconductors, too, but electrons need significantly larger amount of energy ($E_g \leq 5$ eV).

Electron angular speed and energy depend on the orbit index (n) and the nucleus mass (Z). Those quantities are calculated as follows:

$$\omega_n = \frac{z^2 e^{0^4} m_e}{n^3 \hbar^3} (4\pi\varepsilon_0)^{-2}. \tag{2.3}$$

$$E_n = \frac{z^2 e^{0^4} m_e}{n^3 \hbar^3} \left(4\sqrt{2}\pi\varepsilon_0\right)^{-2}. \tag{2.4}$$

(Electron angular speed and energy build with the increase in the nucleus mass ($z\uparrow$) or decrease in the electron orbit radius ($n\downarrow$)). Conforming to the zone theory, substances with complex molecular structure (two and more different atoms) have also a more complex electron envelope, since the electron orbits are packed within group orbits called *energy band orbits or energy bands*—see Figures 2.5 and 2.6d through f).

Electrons alone can leave their orbit, emancipate from the stable atomic structure and exist independently. They move autonomously over the spatial area of the valence band and within the so-called conducting area (see Figure 2.6). To pass into that area, they must overcome the energy barrier of the *banned area*, which is the last energy gap between the valence band and the conducting area. The process of electron march from the valence band to the free area is known as *lattice internal ionization*.

When the electrons within a body enter the conducting area and form a so-called electron gas, they start Brownian (non-organized) or directed (electric current) motion (see Figure 2.6). Nuclei, deprived of electrons, convert into positive ions (anions), and a *gap or vacancy* replaces a missing electron. It behaves as a positron or a positive electric charge. Free electrons interact with the grid anions during motion within metals. Hence, if a metal with mass 1 kmol belonging to the first group (see Mendeleev's table) contains N_A atoms, the accumulated energy will not be $(3/2)N_A k_B$, J but less (Kittel and Kroemer 1980; Hook and Hall 1991; Kittel 1996).

We *propose at macroscopic level* a generalized relation between the parameters of phonon motion (using the mean speed of wave propagation c_m and those of energy transfer, studying the most common engineering solids (including coefficients of thermal conductivity and diffusivity, specific heat capacity, and volume density). The results are plotted in Figure 2.7.

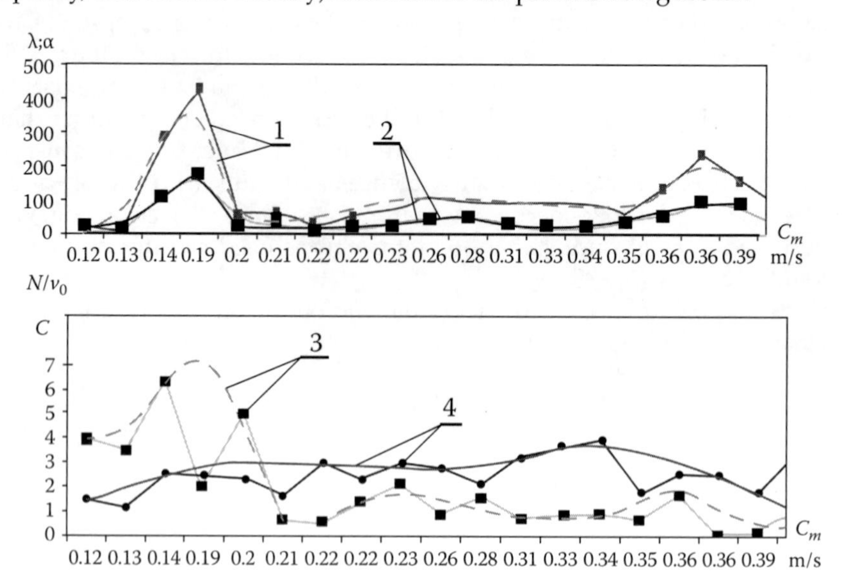

FIGURE 2.7
Variation of the thermal characteristics of solids depending on the velocity c_m: (1) coefficient of thermal conductivity; (2) coefficient of diffusivity; (3) volume density N/v_0; (4) thermal capacity $C = \rho c_p$.

The numerical data in Table 8.11 and their graphical interpretation (see Figure 2.7) prove that (1) the mean speed of phonon motion c_m varies in wide ranges (from $c_m = 0.122$ m/s for lead to $c_m = 0.78$ m/s for beryllium) and (2) a complex relation exists between speed c_m and the coefficients of thermal conductivity and diffusivity $\bar{\lambda}$ and α (small values of $\bar{\lambda}$ and α correspond to small values of c_m). A local extremum is observed within the range of c_m $0.14 < c_m < 0.2$ m/s. It corresponds to metals with excellent transfer properties such as gold, silver, and platinum, displaying relatively low speed of propagation of Debye's waves—$0.14 < c_m < 0.2$ m/s and at the same time—large coefficients of thermal conductivity and diffusivity—$317 \leq \bar{\lambda} \leq 429$ W/mK and $116 * 10^6 \leq \alpha \leq 134 * 10^6$ m²/s. A similar second local extremum occurs at higher speeds $0.34 \leq c_m \leq 0.42$ m/s (see Table 8.11). Such a relation is also valid for the second assessed quantity the diffusivity coefficient.

Our analysis proves that the transfer speed depends on the variation of the solid volume density N/v_0 (see graph 3 in Figure 2.7)—the local extrema of $\bar{\lambda}$ and α correspond to solids with volume density $N/v_0 = 7 * 10^{21}$ and v. A common tendency exists where the speed of propagation of Debye's waves increases with the decrease in the volume density N/v_0. Results prove that the effect of wave propagation speed on the specific amount of accumulated energy is weak within the speed range ($C \approx 2.8 \times 10^6$ J/m³ K). The most significant deviation is observed at $c_m = 0.33$ m/s (corresponding to nickel $C = 3.95 \times 10^6$ J/m³ K) (see graph 4 in Figure 2.7).

The discovery that solids with suitable qualities specified by the volume density ($N/v_0 = \rho M/N_{A)}$ have the highest values of the transfer characteristics of solids ($\bar{\lambda}$ and α) is a curious factology. The knowledge gained can be useful in the design of a hypothetical model of interaction between constructional materials and a solar photon flux (see Section 2.3).

At present, references (Touloukian and Ho 1972, 1976; Desai et al. 1976; Incropera and DeWitt 1996, 2002) state that the energy transfer capacity of solids (mainly nonmetallic solids) depend on transfer frequency. The latter is interpreted as an original frequency of interaction between lattice atoms. The way of lattice structuring is of crucial importance—well-structured materials have good transfer properties (compare, for instance, crystal to amorphous silicon and amorphous carbon to diamond). Crystal nonmetals (diamonds or beryllium dioxide, for instance) are as good conductors as the best metal conductors (aluminum, for instance) (Incropera 1974).

At microscopic level, three different models based on hypothetical mechanisms were designed in the recent 40 years. These are as follows:

1. *Model of Kaganov* involving a two-step mechanism of interaction between phonons and electrons:
   a. *First step*—heat transfer is initiated by directed motion of electrons arriving from the free area. Their motion (thermal) within the crystal lattice is characterized by a specific internal energy

$u_{EI} = c_e T_e$ and, respectively, by an internal kinetic potential (temperature of the electron gas T_e) and local Lagrange multiplier β_L.

b. *Second step*—the kinetic energy of the moving electrons passes to the lattice atoms, which start vibrating (i.e., generating phonons) with frequency ν depending on the equilibrium temperature T_e. Energy transfer from the running electrons to the lattice is known as a *relaxation transfer*, and the amount of energy is estimated by the following expression:

$$\Delta E = G(T_e - T_I), \tag{2.5}$$

where

$G = \frac{\pi^2}{6}\frac{m_e n_e \bar{c}^2}{\tau_R^e T_e} = \frac{\pi^4}{18\lambda}(n_e \bar{c} k_B)^2$ is the coefficient of electron fusion with the atoms of the crystal lattice

τ_R is the time of relaxation of the electron flux (electron energy passes to the lattice atoms via special *T*-waves)

m_e is the mass of the free electrons participating in the heat flux

$n_e = f(n_A, \theta)$ is the electrons per unit volume (electron density)

$\bar{c}$ is the sound velocity $\left(\bar{c} = \frac{k_B}{2\pi h}(6\pi^2 n_A)^{-1/3}\theta_D\right)$

θ_D is the temperature of Debye

λ is the coefficient of conductivity

k_B is the coefficient of Boltzmann

The coefficient G has been measured in the year 1990 by Brorson et al. (1990), and it is shown in Table 8.1.

Condition $T_e \gg T_I$ should hold even though the energy capacity of the electron *gas* is by 1 ÷ 2 times lower than that of the lattice. Here, electrons play the role of energy carriers and stimulators of the phonon motion within the lattice, and they do not serve as energy accumulators.

2. *Model of Geier and Kumhasi*—designed 10 years after Kaganov's model.

The model states that when the oscillating crystal lattice attains *equilibrium* (temperature T_I and internal energy u_I), it starts emitting and scattering phonons, creating heat flux with two components:

- Migrating phonons with sound speed $\bar{c}^2$, forming the so-called elastic coalition. That flux is proportional to the crystal lattice gradient of temperature T_I.
- Phonons forming plastic coalitions (similar to friction losses).

The phonon system accumulates energy.

As mentioned earlier, the model of Guyer and Kumhansi was modified in 1989 by Joseph and Preziosi (1989) and by Tzou (1995). An attempt to unite the latest version of the scattering phonon model with Kaganov's model is made by introducing two types of phonon system relaxation $\left(\tau_R^{P_n}\right)$:

- Relaxation of normal fusion of phonons, keeping the phonon system momentum (elastic coalition) with relaxation time $\tau_N = \frac{5}{9} \times \frac{C_I}{C_e}$.
- Relaxation where the coalition loses momentum (umklapp process—plastic or coalition), and the coalition time is $\tau_R = \frac{3\bar{\lambda}}{\bar{c}^2(C_I - C_E)}$.

Here, $\bar{c}$, C_l, and C_e are of a sound velocity, lattice heat capacity, and capacity of free electrons, respectively, taking part in energy transfer to the lattice.

3. *The mechanism* (proposed by Majumdar in 1993) of energy transfer via phonon emission is the *most interesting one* from a phenomenological point of view, regarding the envelope components. It is characteristic for typical constructional materials where the conduction area is short of free electrons (insulating materials, semiconductors, or dielectrics). Solving the Boltzmann 1D linear equation of transfer of the type

$$\frac{\partial f_\nu}{\partial \tau} + \bar{c}_x \frac{\partial f_\nu}{\partial x} = \left(\frac{\partial f_\nu}{\partial \tau}\right)_{Scattering} \approx \frac{f_\nu^0 - f_\nu}{\tau_R},$$

Majumdar proved that the intensity of heat transfer due to phonon emission in an *acoustic medium composed of acoustically thin walls* depends *solely* on temperature distribution in a priority direction, along axis Ox, for instance. Here, f_ν is the distribution function of phonons with vibration frequency ν equal to f_ν^0 at equilibrium, $\bar{c}_x$ is a projection of the sound speed on the directrix of the heat flux Ox, and τ and τ_R are current and relaxation times of the phonon system during emission. We integrate over the frequency range of oscillations of the phonon system—$0 < \nu \leq \nu_D$—where ν_D is the *cutoff* frequency of Debye.

The results found are interpreted in (Tzou 1992, 1995, 1997) as mechanisms of phonon emission, since the heat flux q for thin films with $L_0(L_0 < 10 \times 10^{-6}$ m) reads as follows:

$$q = \frac{4\sigma}{3(L_0/I)+1}(T_I^4 - T_0^4),$$

and it is similar to the equation of Stefan–Boltzmann concerning radiation of a perfect black body (here L_0 is width of the electrothermomechanical system (ETS) control volume, while l is the phonon free path between two successive coalitions, and σ is the Stefan–Boltzmann constant).

2.2 Conclusions and Generalizations Based on the Survey of Literature Published in the Field

Specific models are used nowadays to assess energy transfer at a *microscopic level*. They are designed on the basis of two types of virtual physical mechanisms described in a number of studies (Kaganov et al. 1957; Guyer and Krumhansi 1966; Stowe 1984; Incropera and DeWitt 1985; Tien and Lienhard 1985; Joseph and Preziosi 1989; Tzou 1992, 1995, 1997; Majumdar 1993; Qiu and Tien 1993; Tien and Chen 1994; Baierlein 1999; Van Carey 1999; Bailyn 2002, Strosicio and Dutta 2005; Srednici 2007). All models postulate the presence of a priory existing temperature gradient (for instance, between electrons with temperature T_e and phonons with temperature T_l or within the crystal lattice itself—T_I = grad[$T_I(X)$]) without formulating hypotheses of its origin.

Classical thermodynamics and heat transfer assume that the energy flux vector is due to the availability of a temperature gradient. Both vectors are synchronized in time. If de-phasing occurs, the vector of energy transfer will always lag with respect to the temperature gradient vector. This however holds true in *closed* thermodynamic systems only. Majumdar (1993) theoretically proved that there exists a third mechanism of energy transfer in solid constructional materials. It operates at the expense of radiated phonons, and it is effective for acoustically thin dielectric films only. Majumdar's mechanism has not been studied in detail (Tzou 1997), although it is of interest in models of heat exchange between the building envelope and the interior. The micro-elucidation of the mechanism of energy transfer by means of *autonomously* vibrating atomic structures is unsatisfactory. It is assumed that either oscillations run a priori or they are due to imported energy fluctuations, without explaining the nature of those phenomena. All considered studies (excluding Majumdar 1993) propose *physical description of transfer in a metallic structure* with a significant number of electrons located in the free area. Hence, the respective mathematical models cannot reflect the specific features of the building envelope material. Note that metals are scarcely used in envelopes, but if so, their function is not to transfer energy (Назърски 2004; Димитров 2006). Those considerations suggest that one needs a reconstruction of the authentic operational conditions to describe energy transfer in the envelope, accounting for the physical properties of the envelope materials and designing *an adequate physical model* of micro-transfer. The mathematical modeling of energy transfer traditionally involves two equations—the first one concerns the specific features of the virtual mechanisms of transfer while the second one is the energy balance. The equations comprise the first- and second-order partial derivatives of temperature with respect to the spatial coordinates and time (Димитров 2011). The two groups of models, *applied to steady transfer*, provide results identical with those of the Fourier classical model, since the time derivatives of temperature are zero $\left(\frac{\partial T}{\partial \tau} = \frac{\partial^2 T}{\partial \tau^2} = 0\right)$.

All known mathematical models of energy transfer in solids at microscopic level concern closed systems only, which have specific characteristics (admissible forms of quantized energy, degrees of freedom of motion, inertial characteristics). Note that *quasiparticles* propagate in those systems, being phonons with characteristic vibration frequency and propagating with equalized mean speed. Most cases consider the internal states of the body only, including energy accumulation.

A number of properties of solids are accounted for at microscopic level—the physical volume, volume density (N/v_0), isometric characteristics of the crystal lattice (for instance, the basic dimension a), and the number of electrons within the valence band. Yet, the effect of the width of the banned area on material capabilities to convert and transfer energy is not considered. There are no studies of the *energy fluctuations* occurring at the solid surface and how they affect the system parameters *at microscopic level*. Information about the mechanisms that produce (generating) phonons in the solid components of the envelope (they are considered immanent granted) should be applied only to the process of interaction with the electrons of the free zone (in Kaganov's model) and their dispersion in their coalition (in Guyer and Krumhansi's model). Moreover, mechanisms at microlevel, related to electron interactions with atoms and ions of the crystal lattice, are poorly studied. There are not enough data on the link between their genesis and the crystal lattice interaction with other boson particles, such as solar photons penetrating the atmosphere and falling on the building envelope. Hence, some materials' *macro* characteristics are casually related to specific only to *micro*processes. This, for instance, is the capability of solids to change their internal energy via the variation of the phonon frequency characteristics—Einstein and Debye waves or the effect of temperature on transfer, including material coefficients of thermal conductivity and diffusivity or the specific heat capacity. Microscopic *generation and scatter* of phonons in solids and the change of some macroscopic characteristics, such as *entropy, enthalpy, and functions of Helmholtz and Gibbs*, are poorly elucidated.

The main task of structural thermodynamics and thermal conduction is to compound an *adequate physical picture* of transfer running in the building envelope at *microlevel*. Moreover, it should involve the properties of the used constructional materials, and the researchers should design adequate physical and mathematical models of energy transfer and relate *micro quantities to macro ones*. In this respect, modern computational methods of the engineering assessment of energy transformations taking place in a complex integrated system, such as the building envelope, should be involved. A second step is the actualization of the engineering methods of envelope design in conformity with the applied strategy of energy consumption (Димитров 2008). The building envelope should be treated as an *open* thermodynamic system, and the in- and outflowing energy fluxes should be analyzed and assessed at microlevel.

2.3 Design of a Hypothetical Physical Model of Phonon Generation in Solids: Scatter of Solar Radiation within the Solid

2.3.1 Internal Ionization and Polarization Running in Solids (Formation of Temporary Electrodynamic Dipoles)

As noted, the thermodynamic analysis at *microlevel* provides *a generalized* pattern of atomic structure vibration. It is assumed that significant interactions between ionized atoms run within the crystal lattice. However, each atom can oscillate around its equilibrium position, but to behave *normally*, its oscillation amplitude should be significantly limited (Einstein 1907, 1934; Bohr 1913, 1958; Van Carey 1999). To our knowledge, the mechanism of oscillation *trigger* and the conditions of its operation are not entirely elucidated. *Our task is to analyze the character of interaction* between the surroundings and the building envelope. In what follows, we will include them in the energy interaction balance, pursuant to the first law of thermodynamics. We *propose here a hypothetic* physical model of phonon generation within crystal lattice structures.

The analysis of *lattice internal ionization* proves that electrons can leave the atomic structure under external energy impacts that would decrease the Coulomb force and increase the centrifugal forces, thus violating the force equilibrium between the atom nucleus and the electron envelope. The probability that electrons of the valence band would pass to the conduction area depends on the width E_g of the *banned area* and the amount of external energy *injected* by high-energy particles (photons—*Ph*), thermal phonons (charges—*Pn*), or electric charges (electrons—*EI*).

Ionization occurs in solids depending on their atomic structure. As found in *insulators* (see Figure 2.6d), the width E_g of the *energy band* is large and practically insurmountable. An electron jump from one to another level is possible only if significant *energy portions* are *injected* to electrons (about 13.59 eV) in order to overcome the energy barriers between the electron orbits. In nature, however, electrons run along identical orbits. This is due to the lack of highly charged photons, which would hit the atom (only a small number of high-energy photons with wavelength $\lambda < 100$ nm that reach the Earth may *elevate* electrons from one to another orbit band).

The electrons in the *valence band* behave similarly, as well. Hence, no electron *jump* from the valence band to the conduction area is observed, and the two-way overcome of the banned area is incidental and owing to photons with high-energy charge ($E_g > 5$ eV and wavelength $\lambda < 248$ nm). As a result, the conduction area remains *empty*. Practically, there are no free electrons, which could be excited to perform direct or Brownian motion. The situation may change applying extreme external power impacts, but only under *breach* of the respective insulator.

Opposite to insulators, substances belonging to the second group and called *conductors* (see Figure 2.6f) have an *energy band*. Practically, it is easily surmountable. Since energy gaps are *narrow*, the electron migration between them needs too small portions of external energy—even thermal phonons can do the job ($\lambda > 700$ nm)—see Figure 2.8. Moreover, merge or overlap between the valence band and the free (conduction) area is sometimes observed in those substances. Hence, *there is enough number of free electrons* (forming an electron cloud—see Figure 2.6d) to perform electric or thermal motion, not needing additional power impacts to be applied on the atomic structure.

There is a third group of substances called *semiconductors* (for instance, Si, Ge, Sn, as well as alloys Ga As, In Sb, Hg Se, In P, Ga P, Cd Te, Zn, Se, Zn, Te, etc.). They have *narrow* banned areas, and free electrons can *populate* the usually empty conduction area under suitable external conditions. Their energy threshold of overcoming the banned area is bound by control. Hence, it could be easily overcome by applying natural energy impacts, including material exposure to thermal photons of the solar spectrum (see Figure 2.8).

The flux of solar radiation consists of photons carrying different energy charges, but photons belonging to the visible part of the spectrum are prevailing in number. The solar flux falls on the building, and photons interact with the envelope components. Three are the forms of their interaction at macrolevel—absorption, reflection, and conduction.

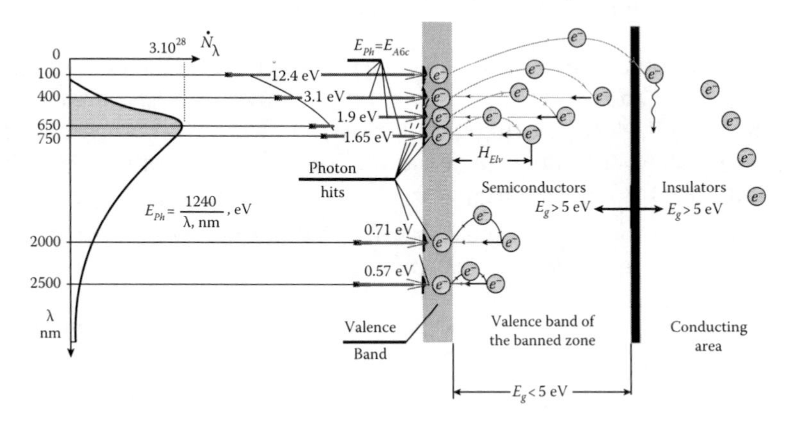

FIGURE 2.8
Examples of banned zone overcome.

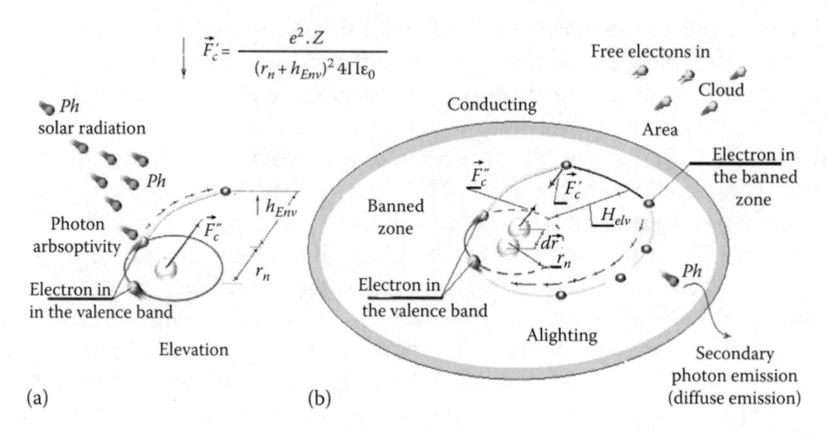

FIGURE 2.9
A mechanism of absorption and emission of photons. (a) Elevation; (b) Alighting.

The interaction of a certain photon with the atomic structure of an envelope component at *microlevel** will depend on the structure and isometric characteristics of the crystal lattice, volume density (N/V), number of electrons in the valence band (n_{VB}), width of the banned area, angle of *attack*—ϕ, and phonon characteristics (flux velocity $J_{\lambda,\pi}$, wavelength, and energy charge, respectively).

If a solar flux meets the building envelope, photons with different energy charges reach the atoms and their electron orbits, including atoms of the valence band. The probability that photons would attack and fall on the electrons of the valence band is a function of the velocity $J_{\lambda,\pi}$ of the flux meeting the normal collecting surface π. As seen in Figure 2.8, photons of the visible part of the spectrum would most likely meet electrons of the valence band due to their large number. The electrons colliding with photons absorb energy $\varepsilon_{A6c}(\varepsilon_{A6c} = E_{Ph} = hc\lambda^{-1})$, jump to a higher orbit, and their angular speed decreases (Equations 2.4 and 2.5). In particular, the *hit* electrons of the valence band have two choices depending on the absorbed photon energy ε_{A6c} (see Figure 2.9):

1. If condition $\varepsilon_{A6c} \geq E_g$ holds, the electrons overcome the energy barrier of the banned area E_g, and reaching the conduction area, they stay as free electrons. Together with other electrons (in the electron cloud), they are involved in thermal or electrical motion (current) thanks to the energy excess.
2. Electrons stay in the banned area due to a lack of energy sufficient to overcome the energy barrier, since $\varepsilon_{A6c} < E_g$.

* A. Compton (Nobel Prize winner in physics 1927) discovered experimentally in 1923 the effect of interaction between the atomic structures and a photon flux (gamma flux). Scatter of the photon flux occurs as a result, and photon energy intensity changes together with photon wavelength (energy orbital). Reynolds (1965) registered atom transformation in 1965, proving that atoms in an excited state, even at low temperature, always pass to a more stable state by emitting one or more photons with frequency ν and energy $E_{ph} = h \cdot \nu$ (so-called quantum jump down). The released energy propagates in the form of quasiparticles (photons).

Regarding semiconductors in natural conditions, the overcome of the energy barrier (the first case discussed earlier) is most unlikely. As for metals, that tendency is highly probable. Note that in the recent 1960s, researchers focused their efforts on finding mechanisms and techniques to control the energy barrier of the banned area ($E_g \Rightarrow$ Var), employing passive or active technologies of increasing the speed of the flux of electrons capable to overcome it (in semiconductors).

The most common events reflect the second case—electrons would elevate within the banned area (without leaving it) to a certain height ($h_{EI\nu}$) depending on the absorbed photon energy and then would return back to the valence band under the action of a Coulomb force. The study of that type of interaction between photons and valence band electrons is relatively incomplete, since researchers focus on the elaboration of technologies of successful electron *elevation* to the conduction area. The electrons remaining in the banned area are considered as *waste*, and their energy is treated as loss or poor heat (Bird 2007). Those phenomena resemble *spring shrinkage-extension*. We assume that they constitute the basis of the *generation of mechanical vibrations of the envelope atomic structure*, without envelope ionization.

Note that violation of the electric charge equilibrium between the atom nucleus and the electron envelope takes place for a while, resulting in the occurrence of a compensating force to restore the equilibrium. The occurring spatial displacement of the nucleus is negligible as compared to that of the charged electrons capable of reaching the free area. Yet, it is multiplied by the larger nucleus mass in the form of a released potential energy of the crystal structure. The energy is a function of the squared deviation of the oscillating atoms from their equilibrium position ($U = U_0 + 0,5\,kr^2$ (Stowe 1984)).

A special electric eccentricity occurs during *shrinkage and extension* of an atomic structure, if the latter is in equilibrium and electrically neutral. Thus, it can be interpreted as a *rotating dipole* of electrically equivalent forces—the positive charge is concentrated at the nucleus, and the center of the negative electric charge moves with respect to its equilibrium position. The dipole generates an electromagnetic moment pulsing in time with a specific frequency, magnitude, and directory. Its magnitude is (Stowe 1984)

$$\Delta\mu = [L(\tau) - L_0]\frac{e^0}{2m},$$

where

$[L(\tau)-L_0]$ is the variation of the electron kinetic moment during charging
e^0/m is the ratio between the electric charge and the electron mass

Consider once again the processes running in the building envelope. Then, the earlier considerations prove that the *polarization of the atomic structure* (without its ionization) under the effect of solar photons is the *most probable mechanism of emergence* of independent atomic oscillators. Oscillation frequency and amplitude would depend on the quality of the incoming photons and on the *hospitality* of the atomic structures (expressed by the number of electrons in the valence band and width of the banned area). As shown in 2.4.2.4, the oscillation amplitudes will also depend on the location of the atomic structure within the crystal lattice topology.

2.3.2 Hypothetical Mechanism of Energy Transfer in the Building Envelope Components

2.3.2.1 Physical Pattern of Energy Transfer within the Envelope Components

The *electron* behavior discussed later is of special scientific interest. Its consideration is needed to elucidate the mechanism of energy propagation deep in the envelope. The processes running in the envelope determine the conditions of its successful operation as an *intelligent membrane* responding to the energy impacts of the surroundings. Qualitative transformation of energy transfer takes place in the envelope—the prevailing 1D energy flux (along the direction of the direct solar radiation) is transformed into a *3D flux of isotropically scattered photons* running in the envelope. It is called photonic gas (Incropera 1974). The processes under consideration are similar to those running in the atmosphere, and they are thoroughly discussed in literature (Duffve and Beckman 1980; Szokolay 1981). In contrast to atmospheric scatter occurring deep into the atmosphere, those processes develop *within a very thin layer* of a solid constructional material. Thickness layer is estimated to a few millimeters or even micrometers of the envelope facial surface.

The atomic model of Bohr (see Figure 2.9a) considers stable motion of electrons in a central power field under the action of Coulomb force $\vec{F}_C$:

$$\vec{F}_C = k\frac{e^0}{r^2}\cdot\frac{\vec{r}_0}{r},$$

where $k = \dfrac{e_{nucleus}}{4\pi\varepsilon_0}$; $(e_{nucleus}Ze^0 = 2.044.Z, \text{MeV})$ is a proportionality coefficient (model constant).

The dynamics of electron motion in a central power field is described by the following equations (Бъчваров 2006):

- $\dfrac{d^2}{dt^2}\left(\dfrac{1}{r}\right)+\dfrac{1}{r} = F_C^r$; $\left(F_C^r = k\dfrac{e^0}{r^2}-\text{projection of the Coulomb force}\right)$—equation of Binet.
- $0.5\left(\dot{r}^2 + r\dot{\varphi}^2\right) - \dfrac{fm_{nucleus}}{r} = E_0$—energy integral.
- $\left(E_0 = T_0 + \Pi_0 = \dfrac{\vec{c}_{t+0}^{\,2}}{2} - \dfrac{fm_e m_{nucleus}}{r_0}\right)$—field potential that can be found at $\tau = 0$.

- $m_e r^2 \cdot \dot{\varphi} \cdot \vec{b}_0 = m_e \dot{S} \cdot \vec{b}_0 = \vec{C}^*$—electron kinetic moment, a vector collinear with the binormal $(\vec{b}_0)$ and perpendicular to the plane formed by vectors $\vec{F}_C$ and $\vec{c}$, while $\dot{S}$ is the electron areal velocity—see Figure 2.9a).

As already noted, the atomic structure of the envelope facial surface undergoes continuous impacts of solar photons, consisting of photon absorption by the electrons. Those *energized* electrons cease their stable motion along steady orbits and *jump* to a higher orbit owing to their energy charge. If the amount of absorbed energy is large enough, they leave the atomic structure. The earlier equation also predicts such an effect if constant E_0 increases. Otherwise, electrons return to their steady orbit *releasing* the absorbed energy via secondary photon emission. The latter marks the initiation of an isotropic scatter of the directed photon flux within the solid body.

Figure 2.10 illustrates the *hypothetic model* proposed herein and tackling the transformation of the directed flux of solar photons into photons moving in all directions.

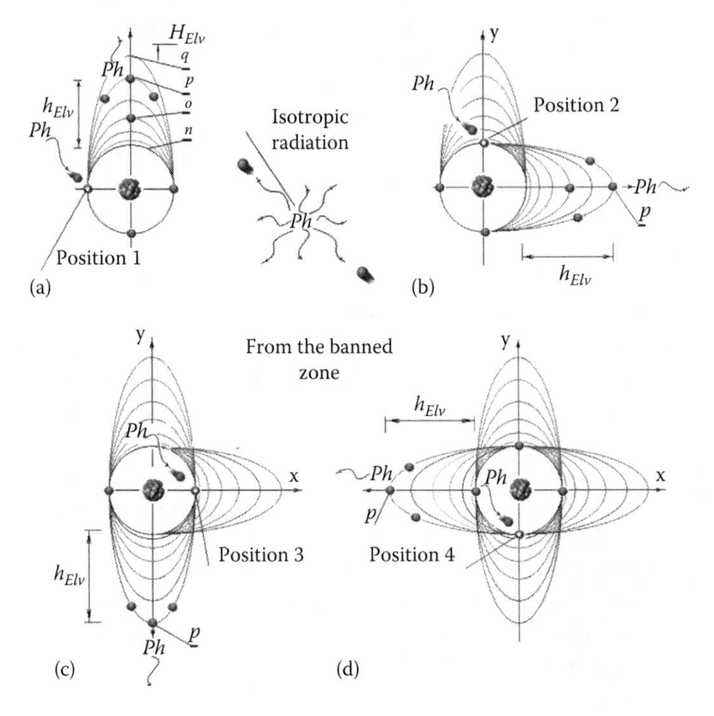

FIGURE 2.10
A model of scattering photon emission from the banned zone. The direction of the secondary radiation is determined by the location of the electron when it reaches the perihelion of the trajectory in the banned zone. Indicated exemplary initial positions of the electrons (from 1 ÷ 4) in which they absorb solar photons. The trajectories in the forbidden zone are simulated with APS "Derive 5": (a) position 1, (b) position 2, (c) position 3, and (d) position 4.

The mechanism operates as follows: solar photons having reached the envelope *bombard* the electrons of the valence band of an atom belonging to the facial surface of a facade element. It is assumed that the photon flux parameters (spectral energy intensity I_λ, J/sm^2 and flux speed J_λ, бp/sm^2) are statistically averaged with respect to the building exposure. The probability of a photon hitting an electron of the valence band is specified by the photon flux speed J_λ, бp/sm^2 in the vicinity of the recipient (the atomic structure), the radius and geometry of the valence band, the spatial orientation of the electron orbits with respect to the photon flux vector, and the number of electron within the valence band. Due to the continued radiation, the electrons of the valence band successfully absorb photons and energize. However, the probability that a particular (marked) electron would experience such an impact depends on its current orbital location, geometry, and orientation, and on the number of photons crossing the projection of the orbit area on a surface normal to the transfer directory.

An electron, having absorbed a photon, *ascends to a banned area* (see Figure 2.9a) along a trajectory specified by the amount of photon energy. Yet, the energy charge is less than that needed to overcome the banned area. The absorbed photon energy $\varepsilon_{\text{Aбc}}$ aids the electron to elevate to height $h_{El\nu}$. Yet, this is not sufficient to overcome the Coulomb force of attraction, and the electron is forced to return to the valence band.

The electron trajectory within the banned area is specified by the magnitude of the received photon impulse (the probability that the electron would be charged by a photon with different energy charge is determined by the speed of the photon flux J_λ). Hence, a certain electron will acquire a trajectory depending on the absorbed *photon energy* (the wavelength and the Lagrange factor) (see Figure 2.9a). The resulting parameters of electron motion within the banned area are varying time of elevation and stay and varying height v of elevation to the respective orbits—*o*, *p*, *q*, etc.—see Figure 2.10a).

The second stage of motion is when the electron descends to the valence band (see Figure 2.9b—descent) under the action of Coulomb force. Then, *the photon leaves the electron* and starts propagating isotropically, along a random directory—see Figure 2.9f), that is, in all directions.

Figure 2.10b through d show the probable orbits of an electron belonging to the valence band of an atom and charged by a photon having met the electron at three random locations specified by angle φ_{0_i} ($\pi/2$, π and $3\pi/2$, respectively—see Figure 2.10d). As seen in Figure 2.10b through d, the perihelion of the electron orbit in the banned area is also de-phased with respect to that shown in Figure 2.10a. It corresponds to the random moment of electron charge by a solar photon. Then, the position at which the photon leaves the electron is also de-phased with respect to the reference case (see Figure 2.10a). Thus, a unified mechanism of isotropic (in all directions) distribution of photons in solids comes into being. However, the atomic structures of solids do not provide identical conditions of operation of that mechanism. The *most favorable position* is occupied by atomic structures exposed to photons, whose flux has the largest speed (see Figure 2.8). Those photons would excite electron motion within the banned area, while the atomic structures form a physical interface between solids and the surroundings.

From a probability point of view, the electrons of the atomic valence bands most often and successfully *jump* to the conduction or banned areas, since solar photons of the visible and the infrared spectra most often bombard and successfully hit them. Physically, this results in *a high-frequency vibration of the electrons* of the valence band within the banned area as well as vibrations of the *entire atomic structure*. The physical parameter recording the high kinetic activity of the valence band electrons is the current value of *temperature* T_A in the vicinity of *point A* of the atomic structure under consideration. It increases with the increase in the kinetic energy of electron oscillation within the banned area.

Photons with one and the same energy (E_{Ph} = const) and wavelength (λ = const) generate oscillations of the valence electrons in the banned area, assessed by temperature $T_A \rightarrow T_A^{\lambda}$ of the following scalar form:

$$T_A^{\lambda} = \psi_t^{\lambda}\left\lfloor J_{\lambda}, E_{Ph}, K_B, \phi, r_{n=VZ}, \bar{\rho}_E(\bar{\rho}_E)_{K_B>1}\right\rfloor.$$

Similarly, function V_A^{λ} describing the dependence of the electric potential on the local conditions in the vicinity of *point A* has the same structure but for $K_{FZ} \rightarrow K_{FZ} < 1$,

$$V_A^{\lambda} = \psi_V^{\lambda}\left\lfloor J_{\lambda}, E_{Ph}, K_{FZ}, \phi, r_{n=VZ}, \bar{\rho}_E(\bar{\rho}_E)_{K_{FZ}<1}\right\rfloor,$$

where

J_{λ}, бp/sm^2 is the speed of photon flux with wavelength λ, nm having fallen on a certain component of the facade (theoretically, $J_{\lambda} \rightarrow J_{\lambda} = 3.02 * 10^5 \lambda^{-4}(e^{2.64.10^{-6}/\lambda} - 1)^{-1}$).

$K_{FZ} = E_g/E_{Ph}$ is the *barrier coefficient of the banned area*, the ratio between the width of the banned area (E_g, eV) and the energy ($E_{Ph} = hc\lambda = 1240/\lambda$, nm), v, of a photon having charged an electron of the valence band. The barrier coefficient K_{FZ} gets a value $K_{FZ} > 1$ for photons whose energy does not suffice the valence electrons to overcome the banned area.

ϕ, *rad* is the angle of *attack*. This is the angle concluded between the directory of a solar photon flux and the normal to the surface of the respective envelope component. The angle of attack ϕ varies in time, and it depends on building latitude and longitude and orientation, current time, and Earth's position with respect to the Sun (see Figure 1.6).

$\rho_E(\bar{\rho}_E)$ is the mean surface (linear) density of the valence band electrons

$$\left(\rho_E = n_{VZ} / \pi \cdot r_{n=VZ}^2, \rho_E = n_{VZ} \Big/ \int_{2\pi} r_{n=VZ} \cdot d\phi\right).$$

n_{VZ} is the number of valence band electrons (equal to the number of the group of Mendeleev's table, which the chemical element belongs to $n_{VZ} = N_{AG}$).

$r_{n=VZ}$ is the radius of the trajectory of the valence electrons.

Consider structurally defined relations $T_A^\lambda(\lambda)$, $V_A^\lambda(\lambda)$ of a certain material, distribution $E_{Ph} = f(\lambda,\tau)$ with respect to the normal of the envelope facial surfaces, cos(ϕ), and the width of banned area E_g. Then, the integral effect of the entire photon flux ($\lambda_1 \leq \lambda \leq \lambda_2$) on the temperature and electric potential at *point A* of the facial surface of each envelope component will read as follows:

$$V_A = \int_{\lambda_1}^{\lambda_2} V_A^\lambda(\lambda)\cdot d\lambda \quad \text{and} \quad T_A = \int_{\lambda_1}^{\lambda_2} T_A^\lambda(\lambda)\cdot d\lambda.$$

The electric potential and temperature of points of the facial surface specify values of the generalized potential (Equation 3.42) participating in the boundary conditions of transfer will be discussed in Chapter 6.

Temperature is an integral macroscopic physical quantity, subject to measurement, and *classical thermodynamics* treats it as a basic state parameter or a *generalized force* yielding transfer of thermal charges (phonons) in massive bodies. Cascade isotropic transfer of photons takes place deep in those bodies. Figure 2.11 illustrates the mechanism of photon scatter and charging within the structure of the atomic lattice of the constructional material. The most numerous atomic structures in a massive body, operating as *centers of scatter,* are located right at the photon-absorbing surfaces (so-called recipient surfaces). Therefore, depending on the frequency with which valence band

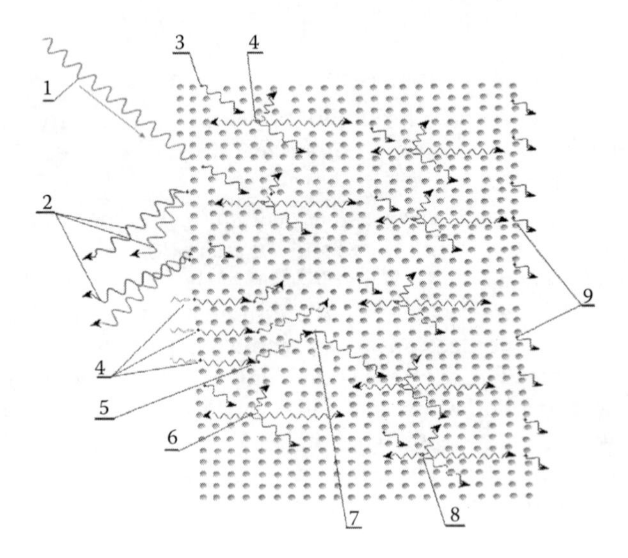

FIGURE 2.11
Models of the behavior of phonons during photon irradiation of constructional materials: (1) a direct incoming solar flux; (2) photons emitted out of the body by the electrons in the banned area (reflected flow); (3, 4) photons carried inside the body by the electrons in the banned area; (5–8) isotropically de-energizing phonons emitted inside the body; (9) photons emitted to the interior (traversing).

electrons absorb photons (depending on the electron number and speed of the photon flux), the electrons will launch to the banned area. Since the numerous atomic structures scatter secondary emitted photons, the probability of charging the atomic valence bands decreases with the increase in the penetration depth. The better the location of an atomic structure from the *recipient* surfaces, *the less likely the hits by secondary photons. Respectively, the electrons will hardly ascend to the banned area, and electron oscillation will be less intensive and with less energy while the local temperature will be lower.*

Summarizing the earlier arguments, the metaphoric description of the *proposed* mechanism of scatter of unidirectional photon fluxes states that the solar photons use electrons as *lifts or carrier rockets* to move within the banned area. Having reached the perihelion of their trajectories, they leave them and rush in an arbitrary direction. Three are the results of that phenomenon:

1. Transformation of photon unidirectional motion (see pos. 1 in Figure 2.11) into a spontaneous scattered radiation of photons takes place (see pos. 6 and 8 in Figure 2.11). It takes the form of a secondary (isotropic) emission in all directions, including emission directed to the surroundings. Part of the solar energy absorbed by the body returns as a *reflected* flux to the surroundings (see pos. 2 in Figure 2.11). These are photons emitted by electrons, which belong to atoms of a surface structure or to neighboring atoms. Yet, the electrons have moved within the banned area, but the spontaneous emission have been directed to the surroundings.
2. The spontaneous photon emission is transformed into a cascade of photon absorption. Photons have been isotropically emitted by electrons belonging to atomic structures located deeper in massive envelope components. Electrons of internal atomic structures isotropically emit photons, part of which are directed to the building interior (see pos. 9 in Figure 2.11—transparent photons).
3. Oscillations start in the atomic structures characterized by high frequency and *short* amplitude. This is in fact generation of phonon motion in solids.

2.3.3 Hypothetical Model of Energy Transfer through Solid Building Components: A Model of Lagging Temperature Gradient

The survey of references shows that researchers consider solids as open thermodynamic systems exchanging quantized (electromagnetic) energy with the surroundings, thus giving the following idea of energy transfer in solids. Solar photons, having fallen on the building envelope, affect the atomic structures belonging to the first rows of the crystal lattice and cause ionization or polarization. Moreover, two related effects are observed—atomic structures ionize (generating an electron gas) or *start vibrating* and polarize (phonons/thermal charges emerge). The original solar photons undergo qualitative modification (they scatter under the discussed secondary emission, and their frequency and energy decrease). Yet, $\bar{\lambda}$ increases respectively, but part of the photons, although modified, successfully move through the solid body and form a photon gas* (Debye 1928; Incropera 1974) within the internal environment—the building interior).

During habitation of the control volume, photons interact with the atomic structures of the crystal lattice performing *two types of work*. The first one is *electrical* where photons *knock* electrons out of the valence band, while the second one is *thermal* where they generate phonon/thermal charges depending on the physical characteristics of the medium—number of electrons in the valence band, width of the banned area, molecule mass and density, and current values of the energy parameters.

Consider a monolayer component of the building envelope with a thickness L_0 and a cross section A_0 ($A_0 = 1$ m²), whose normal is directed opposite to the falling solar flux. Then, one can assess the *total amount of energy* within the control volume using the following equation of balance:

$$\langle E_{Pn}^{0}\rangle + \langle E_{Pn}^{In}\rangle + \langle E_{EI}^{0}\rangle = \langle E_{Pn}^{Out}\rangle + \langle E_{Pn}^{Gen}\rangle + \langle E_{Pn}^{Gen}\rangle + \langle E_{EI}^{Gen}\rangle,$$

where

$\langle E_{Pn}^{0}\rangle$ and $\langle E_{Pn}^{Gen}\rangle$ are the energies of *phonons* (residual and generated).

$\langle E_{Ph}^{In}\rangle, \langle E_{Ph}^{Gen}\rangle$, and $\langle E_{Ph}^{Out}\rangle$ are the energies of *photons* (incoming, generated within, and leaving the control volume).

$\langle E_{EI}^{0}\rangle$ and $\langle E_{EI}^{Gen}\rangle$ are the energies of *electrons* (residual and emitted under internal ionization) inhabiting the control volume.

To describe in detail the respective components of the earlier equation, it is useful to use the so-called function of the total microstate energy Z†

---

* At scatter, frequency of the secondary emission is $\nu_{Sec} = \nu\left[1 + (1-\cos\alpha)\frac{h\nu}{m_e c^2}\right]^{-1}$ or photon wavelength increases by $\lambda' = \frac{\lambda_{Sev} - \lambda}{(h/m_e c^2)}$, where α is the angle of photon flux scatter. A. Compton, Davisson, and Germer registered in 1927 a secondary photon emission from a solid (a nickel plate) hit by an electron flux (Van Carey1999). Radiation with wavelength $\lambda = 0{,}0486$ nm was registered, which corresponded to the quasiparticle predicted by Broglie in 1924 and moved within a massive body with speed 0.05 m/s—part of the light speed in vacuum $\left(c_{Ph} = 0.05c = 0.15 * 10^8 \text{ m/s}\right)$

† The function Z_N of the total microstate energy of an ideal gas consisting of N molecules takes the form $Z_N = \frac{1}{N!}\frac{1}{2^{3N}}\left(\frac{2mk_B T v_0^{2/3}}{\pi.\hbar^2}\right)^{3N/2}$, while that of the Einstein solid $Z_N = \left[\sum \exp(-E_j)\right]^N = \left[\frac{\exp(-\beta\hbar\omega/2)}{1-\exp(-\beta\hbar\omega)}\right]$, $\omega = \sqrt{k/m}$.

(partition function/Zustandssumme—state sum after Planck) (Baierlein 1999):

$$Z = \prod_{E_j} \left\{ Z^{(j)} \right\} = Z^{(1)} * Z^{(2)} * \ldots * Z^{(j)} * \ldots * N^{(N)}$$

$$= \sum e^{-\beta E_1 (1)} * \sum e^{-\beta E_2 (1)} * \ldots * \sum e^{-\beta E_j (j)} * \ldots * \sum e^{-\beta E_N (N)},$$

where

$Z^{(j)} = \sum e^{-\beta \varepsilon_i (j)}$ is a function of the total energy of the jth particle of the control volume occupying the ith respective energy orbital.

$E_j = \sum_{n_{iX}} \sum_{n_{iY}} \sum_{n_{iZ}} \alpha^2 \left(n_{iX}^2 + n_{iY}^2 + n_{iZ}^2 \right)$ is energy of the jth particle of the control volume in the respective ith energy orbital ($\alpha^2 = (\hbar^2/2\ m)\ (\pi/V_0^{1/3})^2$, and a n_{iX}, n_{iY}, and n_{iZ} are whole positive numbers, including zero).

$\beta = 1/(k_B T)$ is the Lagrange multiplier*.

Where technically possible, the function of the total microstate energy Z is used to calculate the function of the canonical (standard) distribution of the probability $P(\psi_i)$ of whether the energy E_i of a certain electron would belong to the range $\varepsilon_i < E_i < \varepsilon_i + d\varepsilon_i$. Moreover, function Z is also used to calculate the distribution function of the state density $D(E)$, temperature T, Fermi potential ε_F, functions of Gibbs, Landau, Helmholtz, or the free energy potential ψ used in Section 2.3.

The probability that a certain particle would land in the ith respective orbital is specified by the so-called distribution function (Lee 2002):

- For bosons, $f_{BE}(E_i, \mu, t) = \dfrac{1}{\exp\left[\beta\left(E_i - \varepsilon_F\right)\right] - 1}$ (function of Bose–Einstein).†
- For fermions, $f_{FD}(E_i, \mu, t) = \dfrac{1}{\exp\left[\beta\left(E_i - \varepsilon_F\right)\right] - 1}$ (function of Fermi–Dirac).

They differ from each other by the sign before the unit in the denominator only. Disregarding that difference, the *distribution function* reads as follows:

$$f(E_i, \mu, t) = \exp(-\beta \cdot E_i) * \exp(\beta \cdot \varepsilon_F) = \frac{e^{\beta \varepsilon_F}}{e^{\beta E_i}}.$$

* The term "Lagrange multiplier β" is rooted in the method of Lagrange multipliers employed to solve the equation of thermodynamic probability and derive the distribution law of Maxwell–Boltzmann, $N_i = \alpha e^{-\beta \varepsilon_i}$, which specifies the most probable number of particles possessing an energy level ε_i (Saad 1966).

† Gives the exact number of quasiparticles in an orbital.

Here, $\varepsilon_{Eff} = (E_i - \varepsilon_F)$ is the effective energy (ε_F Fermi potential).

The components of the energy balance equation E_{Ph}^{In} and E_{Ph}^{Out} take a traditional form when we assess the integral amount of photon energy considering frequency in the range $0 \le \nu < \infty$ and introducing the macroscopic distribution of the state density $D(\nu) - \langle E_{Ph} \rangle = \int \langle e_{Ph} \rangle_{\nu} D(\nu)\, d\nu$.

The photon flux energy can be alternatively estimated by using the *total energy function* of the microstate of the entire flux Z_{Ph}, which can be found as a sum of the energies of all microstates $E_{Ph}^{Env}(E_{Ph}^{Env} = \bar{n}(\nu)\hbar \cdot \nu)$ (Lee 2002):

$$Z_{Ph} = \sum_{s=0}^{\infty} \exp(-\beta \cdot \bar{n}(\nu)\hbar\nu) = \frac{1}{1 - \exp(-\beta\hbar\,\nu)} \tag{2.6}$$

(The Lagrange multiplier $\beta = 1/k_B T_{Enter}$ in that cases assumes different values depending on the control surface temperature T_{Enter} ($T_1 = 6000$ K or $T_2 = 297$ K).)

Assuming that there exists a normal particle distribution around a certain value, the particle number $\bar{n}(\nu)$ for frequency ν will be (Lee 2002)

$$\bar{n}(\nu) = -\left(\frac{\partial}{\partial \alpha} \ln Z_i\right)_{\beta,\nu_0} = \frac{1}{\beta}\left(\frac{1}{\partial \mu} \ln Z_i\right)_{\beta,\nu_0},$$

while the fluctuation (the mean quadratic deviation) can be assessed via the expression $\mathrm{Var}(n(\nu)) = \left(\frac{\partial^2}{\partial \alpha^2} \ln Z_i\right)_{\beta,\nu_0}$. That formula, used to assess the boson number, reads $\bar{n}(\nu) = (e^{\beta(E_i - \varepsilon_F)} - 1)^{-1}$, while its form for fermions is $\bar{n}(\nu) = (e^{\beta(E_i - \varepsilon_F)} + 1)^{-1}$.

In its turn, the energy of the form with frequency ν will be calculated via the product $\langle E_{Ph} \rangle_{\vartheta} = \bar{n}(\nu)h\nu$. The energy density (the number of EM forms in the frequency range $\nu \div \nu + d\nu$) takes the form $\left\langle \begin{matrix}\text{Number of}\\ \text{EM forms}\end{matrix} \right\rangle = D_{EM}(\nu)d\nu = \left(\frac{8\pi}{c^3}\nu_0\right)\nu d\nu$ (Baierlein 1999). Then, members related to the photons participating in the equation of total *energy* balance will be written as

$$\langle E_{Ph} \rangle = \int_0^{\infty} (\bar{n}(\nu)h\nu) D_{EM}(\nu) \cdot d\nu = \int_0^{\infty} \left[1/\exp 4.8 * 10^{-9}(\nu/T_{1,2}) - 1\right] h\nu \left(\frac{8\pi}{c^3}\right) \nu_0 * \nu^2 d\nu =$$

$$= \int_0^{\infty} \left(\frac{8\pi h}{c^3} \frac{\nu^3}{e^{4.810^{-9}\frac{\nu}{T_{1,2}}}}\right) v_0 d\nu.$$

The integral of the earlier expression, used to assess the incoming flux of photons $\langle E_{Ph}^{In} \rangle$, absorbed by the facial surface of the control volume with temperature $T_{Boun}^{1} = T_1$, reads as follows:

$$\langle E_{Ph}^{In} \rangle = \left(\frac{8\pi^5 k_B^4}{15c^3 h^3} \nu_0 \right) * T_1^4.$$

Due to phonon radiation, photons are emitted in the control volume v_0, and the flux is $\langle E_{Ph}^{Gen} \rangle$. It scatters in all directions as schematically shown in Figure 2.12. The intensity of photon emission with fixed frequency (ν)—$\vec{I}_\nu$—is projected on the axes of the Cartesian coordinate system $\left(\vec{I}_\nu = I_\nu^X \vec{i} + I_\nu^Y \vec{j} + I_\nu^Z \vec{k} \right)$, and axis $O\vec{x}$ is directed along the normal to the envelope. The component I_ν^X (along $O\vec{x}$, for instance) can be written as

$$I_\nu^X = \overline{c}_x \cdot e_{Ph} = \overline{c} \cdot \cos \alpha_I \cdot h\nu,$$

where

$\overline{c}$ is the photon transfer speed in a specific medium
$\overline{c}_x$ is the transfer speed along the respective direction
e_{Ph} is the eigenvalue of photons with a fixed frequency (ν)

The intensity of phonon emission along the chosen axis $O\vec{x}$, but for the entire form of emitted photons with frequency (ν) and probability function of distribution within the form $\overline{n}(\nu) - \langle e \rangle_{Ph} = \overline{n}(\nu) h\nu$, is found as

$$I_{\langle e \rangle Ph}^X = \overline{c} \cdot \cos\alpha_1 \cdot \overline{n}(\nu) h\nu,$$

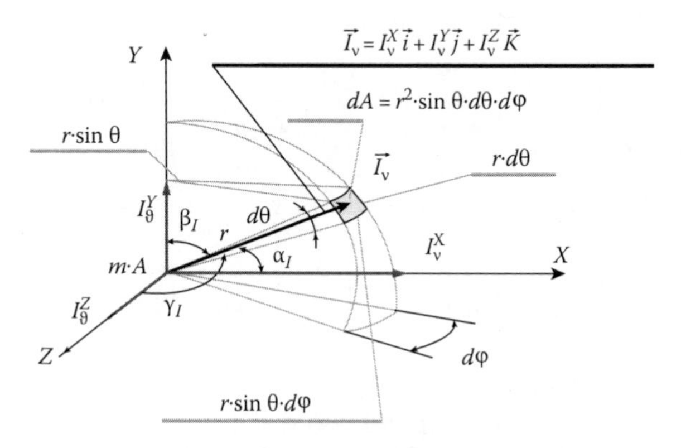

FIGURE 2.12
Diagram of the propagation of photons emitted by phonons.

while for all energy orbitals within the frequency range $0 \le \nu \le \nu_D$ with density of states $D(\nu) \cdot f(\nu)$, it reads as follows:

$$I^{X}_{\langle e\rangle Ph} = \sum_{0}^{\nu_D} \bar{n}(\nu) \cdot h\nu \cdot D(\nu) \cdot f(\nu) \cdot \bar{c} \cdot \cos\alpha_{I} \cdot \Delta\nu$$

Energy transfer within the stereo-metric angle Ω (see Figure 2.12) can be assessed via

$$Q = \sum_{\Omega=4\pi} I^{X}_{\langle e\rangle Ph} \Delta\Omega.$$

Consider phonons generated by the facial control surface (with local temperature T_1). Then, the energy that they emit has the following integral form (Majumdar 1993):

$$\left\langle E^{Gen}_{Ph} \right\rangle = \left(\frac{8\pi^5 k_B^4}{15c^3 h^3} V_0' \right) T_1^4.$$

For the photon flux flowing out of the control volume $\left\langle E^{Out}_{Ph} \right\rangle$, we use a similar form putting the temperature of the control surface $T^1_{Boun} = T_2$ in the earlier equation:

$$\left\langle E^{Out}_{Ph} \right\rangle = \left(\frac{8\pi^5 k_B^4}{15c^3 h^3} V_0' \right) T_2^4.$$

Introduce a special form of the distribution function Z to find the energy of the photon flux $\left\langle E^{0}_{Ph} \right\rangle$ flowing through the control volume. It is similar to Planck's function but accounts for scatter due to internal polarization and ionization. That transformation is possible, employing *scatter* models in the time–space or frequency continuum, for instance:

$$Z = \sum_{j} \exp(-e_j/k_B T) = \sum_{0}^{\infty} \left(e^{-nh\upsilon/kT}\right) = \sum_{0}^{\infty} \left(e^{-h\upsilon/k_B T(x)}\right)^n = \frac{1}{1 - e^{-h\upsilon/k_B T(x)}}$$

for $0 \le x \le \delta, \left(Ox = x \cdot \vec{i} \quad \vec{i} - \text{normal of the isotherms}\right);$

$$Z = \sum_{j} \exp(-e_j/k_B T) = \frac{1}{1 - \mathrm{e}^{h\nu(E_g, n_{XZ}, K_{FZ})/k_B T}}.$$

Here, the numerator of the exponent $\nu = \nu(E_g, n_{Vz}, F_{FZ})$ is a function of frequency damping. The latter depends on the macroscopic characteristics of the solid body, including the coefficient K_{FZ} that characterizes the passability of the banned area and specifies the number of electrons in the valence band.

Regarding the time–space continuum, the following integral function expresses the energy of the photon flux passing through the control volume:

$$\langle E^0_{Ph} \rangle = \int_0^{\infty} (\bar{n}(\nu)h\nu) D_{EM}(\nu) \cdot d\nu$$

$$= A_0 \int_0^{\delta} \left(\int_0^{\infty} \left[1/\exp 4.8 * 10^{-9} (\nu / T(x)) - 1 \right] h\nu \left(\frac{8\pi}{\bar{C}^3_{Sol}} \right) \nu^2 d\nu \right) dx$$

$$= A_0 \int_0^{\delta} \left(\int_0^{\infty} \left(\frac{8\pi h}{c^3} \frac{\nu^3}{e^{4.810^{-9}\frac{\nu}{T(X)}} - 1} \right) d\nu \right) dx.$$

We can perform integration if the parameters of the thermodynamic system are specified at macroscopic level, for instance, via the functional $T = T(x)$ (considering a given temperature T_1 of the first bounding surface of the control volume and knowing the temperature gradient in the solid). Then, the functional will have the following form:

$$T = T_1 - \lim \left(\frac{T_1 - T_2}{\delta} \right)_{\delta \to 0} \cdot dx = T_1 - \text{grad}(T) dx.$$

The next *component* of the energy balance equation concerns the electrons of the control volume. Conditionally, the electron energy consists of two parts: energy of inherited electrons $\langle E^0_{EI} \rangle$ and energy of *released* electrons $\langle E^{Gen}_{EI} \rangle$ supplied during the internal ionization that runs in the envelope (we rewrite here the expression $\langle E^{CV}_{EI} \rangle = \int_0^{\infty} \langle \varepsilon_{EI} \rangle \cdot D_{EI}(\varepsilon_{EI}) \cdot f(\varepsilon_{EI}) d\varepsilon$).

The total number of free electrons moving as an electron gas within the envelope control volume with mass 1 kmol will be equal to the value of the integral of the ordered state density (the product $D(\varepsilon_{EI})$ x $f(\varepsilon_{EI})$) (Van Carey 1999):

$$N^{Gas}_{EI} = \int_0^{\infty} D(\varepsilon_{EI}) x f(\varepsilon_{EI}) \cdot d\varepsilon.$$

Electrons leaving the valence band under internal ionization and electrons returning after having interacted with the ionized atoms during recombination will also participate in that sum:

- $N^{leave}_{EI} = K_{FZ} * n_{Vz} N = K_{FZ} n_{VZ} N_{A\rho} \dfrac{A_0}{M} dx$—for electrons leaving the valence band due to internal ionization
- $N^{return}_{EI} = N_{EI} * f(N_{EI})$—for recombining electrons

Here, $K_{FZ}(K_{FZ} = E_g/\bar{n}(\nu)h\nu)$ is the *barrier coefficient of the banned area*, n_{VZ} is the number of electrons in the valence band, ρ is the specific density and molar mass M of the body, N_A is the Avogadro number, and A_0 is the cross-sectional area of the control area (area length dx varies within the interval $0 \le x \le \delta_W, m$). Thus, the current number of free electrons within the control area reads as follows:

$$N_{EI} = N_{EI}^{Gas} + N_{EI}^{leave} - N_{EI}^{return}$$

$$= (1 - g(N))\int_0^{\infty} D(\varepsilon_{EI})xf(\varepsilon_{EI}) \cdot d\varepsilon + K_{FZ}xn_{Vz}N_{A\rho}\frac{A_0}{M}dx.$$

It will depend on the barrier coefficient, specific density, and molecular mass of the body, as well as on the number of valence electrons (note that not all electrons of the valence band leave the atomic structure; Tzou 1997). The free electron distribution in the considered open system is not uniform within the whole control volume. In areas where the barrier coefficient is less than **1**, a higher-volume concentration of electrons will be observed. These are control volume areas where condition $\bar{n}(\nu)h\nu > E_g$ holds. They are close to the entrance (facial) control surface.

2.3.3.1 Model of Lagging Temperature Gradient

The scheme in Figure 2.13 illustrates the physics of interaction between the energy flux of the surroundings and the envelope components. The hypothetical mechanism adopted in the present study operates in two regimes depending on the value of the barrier coefficient $K_{FZ}\left(K_{FZ} = E_g/\bar{n}(\nu)h\nu\right)$:

1. If $K_{FZ} < 1$, atom ionization takes place. The valence band releases electrons (thermal electron flux emerges in the free area). The thermal electrons attain a specific temperature (T_e) conforming to their internal energy (u_e), and the temperature is found via $T_e = u_e/C_e$. (Here C_e is the thermal capacity of the electron gas.) Next, interaction between electrons of the electron flux and atoms of the material crystal lattice starts, increasing lattice temperature (T_l). The thermal flux in the control volume, directed from the electron gas to the lattice, can be estimated via the expression $q_e = G(T-T_l)A_0 \cdot L_0$, where G is a factor of interaction between the electrons and the lattice. Temperature T_e attained by the free electrons in this case depends on the intensity of the interaction between the flux of *attacking* photons and the valence band electrons, whose number depends in its turn on the atomic structure (respectively, on its position in Mendeleev's table). One can find the inflowing flux of photons and the valence band electrons via $T_e = T_{PH} - q_{Ph-e}/(F_B L_0 A_0)$—here F_B is a factor of interaction between

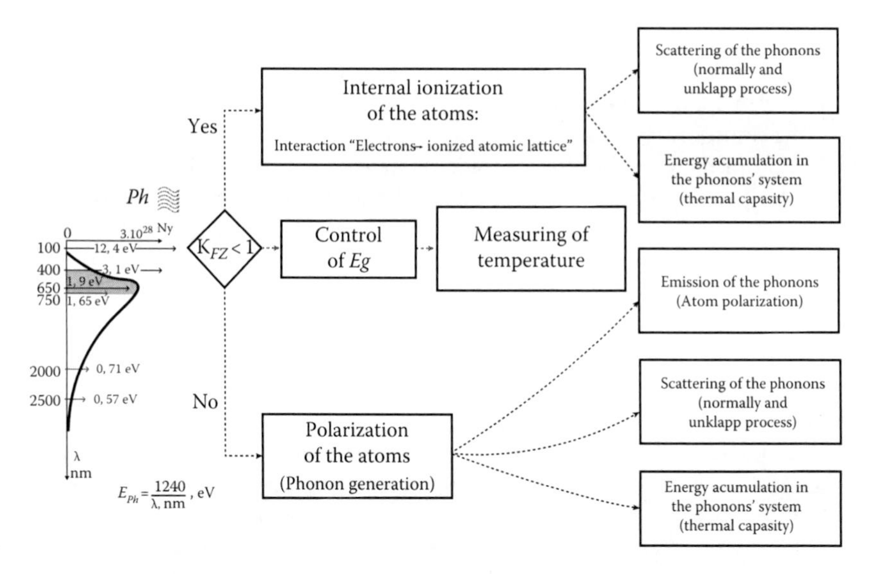

FIGURE 2.13
A scheme illustrating the interaction energy between the environment and the building envelope.

the inflowing photon flux and the valence band electrons, similar to factor *G*. The photon flux loses efficiency, since free electrons absorb the majority of photons during ionization. Phonons emerge during ionization, and subsequent interaction between electrons and ions takes place. They scatter within the lattice of the building envelope according to three modes: Phonon–photon interaction with and without impulse loss and phonon accumulation in the lattice (see Figure 2.13).

2. If $K_{FZ} > 1$, only atoms belonging to the lattice facial area *polarize*. If *solar* photons charge valence band electrons, the latter starts moving around the nucleus but along new orbits. Phonons emerge during that process, which in their turn emit photons (see Figure 2.13) absorbed by atomic structures deep in the body. Thus, a cascade mechanism of *phonon generation* and subsequent *photon emission* toward the atomic lattice take place, yet without atom ionization (the number of electrons in the free area is insignificant). The flux of secondary emitted photons scatters within the structure of the envelope resulting in *atom vibration*. The generated lattice vibrations (phonons) scatter in their turn following the discussed mechanism (see Figure 2.13). They lose energy (degenerate*) to a local equilibrium specified by the limit values of the Lagrange multiplier ($\beta_{Bound} = 1/k_B T_{Bound}$).

* System degeneration goes stronger if the number of energy orbitals increases, and the particles have a wide choice of occupying an orbital. Thus, the number of microstates with one and the same energy charge increases (Saad 1966).

We propose a generalization of the earlier considerations. Consider non-isolated systems with boundaries penetrated by energy, which is quantized in the form of bosons (photons or phonons) or fermions. Then, variation of lattice temperature and subsequent emergence of temperature gradient (directed opposite to the flux of quasiparticles) are observed. Note that they are *a result* of the energy exchange, and not its prime cause.* The leading factor in the envelope components is the radiation-quantized heat exchange, while *the temperature field emerges as a result of* energy penetration. The lag of the temperature gradient with respect to the thermal flux in the building envelope depends on a number of factors reflecting the wall structure (materials, overall dimensions, layer arrangement—see Figure 1.4). Last but not least, *the temperature gradient lag* will depend on the controlled emission of valence band electrons to the free area, including change of the banned area width or barrier coefficient value. Hence, the physical pattern of the running processes described herein can be formulated as *lagging of the temperature gradient with respect to the thermal flux* or *model of the lagging temperature gradient.* Energy propagation deeper in the control volume takes place in the form of phonon transfer *from areas* of phonon emergence (identified by the increased temperature) *to areas* of phonon degeneration (areas of thermodynamic equilibrium where the temperature of the control volume is the lowest one). Phonons can be interpreted as peculiar *thermal* charges, which scatter within the control volume, distributing their energy between the atomic lattice and other phonons.

A typical *thermal charge* (phonon) occurs as a result of the interaction between the inflowing photon flux and the atomic structure. Its energy is determined by the physical properties of the atomic lattice and the type of the input energy impact. However, the degeneration of the typical *thermal charge* under real physical conditions depends on the characteristics of the bounding surfaces of the control volume. Entire degeneration may be attained if the value of the Lagrange multiplier $\beta = (k_B T)^{-1}$ approaches infinity ($\Rightarrow\infty$), that is, medium temperature T reaches the absolute zero.

At macroscopic level, heat transfer through constructional solid materials building the envelope *can be compared* to conduction of electricity. It can be treated as a directed motion of thermal charges (see Section 4 and Subsection 2.1—Equation 2.1—the law of Wiedeman–Franz). Yet, the *difference* between carriers of electricity and heat in solids (electrons and phonons) is *essential.* It consists in that the *thermal charges* (phonons belonging to the class of bosons) are quasiparticles that possess prevailing wave properties and lose *their identity while scattering within the occupied material volume.* This is in contrast to *electrons* (belonging to the class of fermions), which have expressed

* According to the classical theory of transfer, the thermal flux and the temperature gradient vary simultaneously (i.e., their phases coincide). Yet, the theory postulates that the temperature gradient is the prime cause of thermal flux emergence. Heat wave theory (Tzou 1997) proves that the thermal flux and the temperature gradient de-phase owing to system relaxation under various energy interactions.

corpuscle properties and keep their mass during the entire process of transfer under normal conditions. Generalizing the proposed *hypothetical model*, we may conclude that energy transfer in the building envelope proceeds in different forms corresponding to the admissible forms of motion of energy carriers.

According to modern science, $\overline{n}_{(v)}$ phonons may occupy a certain energy orbital (with carrying frequency $\overline{n}_{(v)}$). The phonon number in different energy orbitals is specified by the local value of the *Lagrange multiplier* β_L [$\beta_L = (k_B \cdot T_L)^{-1}$], where k_B is the Boltzmann constant, or by the local temperature T_L. At low values of the Lagrange multiplier (under high temperature, respectively), phonons occupy *high-energy orbitals,* while standing waves with small length carry them.

When the Lagrange multiplier increases (the local temperature $T_L \Downarrow$, respectively), phonons leave the high-energy orbitals and pass to low-energy orbitals (long-standing waves). Phonon motion between permitted energy orbitals is characterized pursuant to the laws of quantum mechanics, depending on the energy generated and distributed in the ETS control volume. Phonons (pseudo-particles) are capable to interact with the material particles—electrons (fermions), atoms, and molecules—and charge them with energy. They are divided into two groups: thermal and acoustic ones. It is assumed that they result from arbitrary energy fluctuations occurring at the control volume boundary. The thermal phonons with a frequency of order $(3 \div 8)$ *THz* are the ones responsible for energy transfer in solids.

2.4 Micro–Macroscopic Assessment of the State of the Building Envelope

2.4.1 Microscopic Canonical Ensemble: Collective Macroscopic State

As discussed in Section 2.3, there are particles (atoms, electrons, molecules) and bosons (phonons and photons) enormous in number within the internal structure of the solid elements of the building envelope, and their number multiplied exceeds $m * 10^{26}$. Consider that their individual behavior can be predicted via methods and algorithms based on laws and theorems of the quantum mechanics, and their distribution in the time–space continuum—by integration of the Schrodinger equation. Yet, *their transitory disposition* is a random function determined by the internal energy U, the total number of discernable particles N, volume v_0, and the initial and boundary conditions imposed on the control area. However, knowing those parameters is not sufficient in clarifying the processes running in the components of the building envelope. This means that the reflection of the behavior of the individual energy carrier (photon, phonon, electron, ion, etc.) is not enough, and one should follow the behavior of the whole set of statistically discernible energy carriers (bosons or fermions). Hence, one should address the powerful apparatus of the statistical thermodynamics to do the job. As is known from (Aston and Fritz 1959; Tribus 1961; Fast 1962; Lee et al. 1963; Reif 1965; Saad 1966; Sonntag and Van Wylen 1966; Schrodinger 1967; Münster 1969; Melehy 1970; Kestin and Dorfman 1971; Silver 1971; Jancovici 1973; Kelly 1973; McQuarrie 1973; Ландау and Лифшиц 1976; Smith 1982; Wark 1983; Ziegler 1983; Stowe 1984; Tien and Lienhard 1985; Hill 1986; Lawden 1986; Sears and Salinger 1986; Hang 1987; Stocker 1988; Lay 1990; Lucas 1991; Carrod 1995; Grenault 1995; Greiner et al. 1995; Shavit and Guffinger 1995; Stöcker 1996; Tzou 1997; Zemansky 1998; Van Carey 1999; Carter 2001; Attaard Phil 2002; Bauman 2002; Lee 2002; Tipler and Llewellyn 2002; Cheng 2006; Kondeppud 2008; Carter 2009), the object of study of the statistical thermodynamics is the behavior of systems consisting of a large number of elementary objects. Each such system passes through a number of microscopic states a_j during a current time period (τ). However, states a_j are transitory dispositions of the particles that constitute the system and have specific behavior and distribution. Consider an unbalanced system. Then, conforming to the second law of thermodynamics, particle number $\tilde{\Gamma}$ arbitrarily increases until system arrival at equilibrium. The whole set of $\tilde{\Gamma}$ microscopic states a_j is called *microscopic canonical ensemble.* Transition from one to another microscopic state takes place *automatically* and at equal intervals of time ($\approx 10^{-15}$ s or at constant frequency $\nu = 1/\tilde{\Gamma} \approx 10^{15}$ Hz) under the action of forces (Lay 1990). Suppose that a system enters and exits a microscopic state. Then, the use of measuring equipment* to monitor its entrance–exit *operation* will result in a very long set of data on the microstates of the system $\langle A(j) \rangle$. The mean value of those characteristics, assumed as a characteristic of the equilibrium state, reads as follows:

$$\langle A \rangle_i = \frac{1}{\hat{N}} \sum A(j)_i,$$

where

$\hat{N}$ is the total *entrance* number until attaining equilibrium

$\langle A \rangle_i$ denotes the averaged records of the measuring equipment at the jth microscopic state

It turns out that a set of microscopic states exists, whose particles have identical behavior and state characteristics $\langle A \rangle_i = \text{const}$. That set forms the so-called collective macrostate (or macrostate no. 1). Parallel to that macrostate, other macrostates exist, as well, but they consist of microscopic states whose number is considerably smaller as compared to their number in the

* The measuring instruments perform the averaging of the characteristics of each microstate, and the assessments obtained adopt a macroscopic character.

collective macrostate. Those parallel macrostates are called minoritarian macrostates no. 1, 2, ...

In fact, the system enters each set of macrostates, but its longest *stay* is in the *collective* macrostate. Hence, the record of the characteristics of the collective macrostate includes not only information on particle behavior there, but also information on particle behavior in the minoritarian macrostates. We call those averaged characteristics *macroscopic state parameters* of a thermodynamic system (in particular, these are the parameters of the building envelope, which is taken as a physical medium).

2.4.2 Introduced Macroscopic State Parameters of the Building Envelope Considered as a Physical Medium of the Electrothermodynamic System

The hypothetic mechanism of scatter (degeneration) of phonons in solids described in Section 2.3.3 is invisible for the human sensors. Yet, it can be visualized by *a record (measurement)* of the distribution of the envelope state parameters (temperature, at the first place) and pressure and electric potential if needed (they are assumed as *basic system parameters*). The sets of their values in the physical space occupied by the building envelope define the so-called fields of temperature, pressure, and electric potential.

Besides those parameters bound to direct measurement, so-called calculation state parameters are used to assess the macroscopic energy characteristics of the envelope. They comprise entropy, functions of Gibbs, Landau, and Helmholtz (known also as *free energy* functions), as well as energy. Their theoretical basis lies in the assessment of the thermal work done in the control volume of the building envelope. In what follows, we shall verify the use of estimations performed by means of *macroscopic calculations*. The latter more fully describe the energy-exchange (energy) processes *running in the building envelopes*. They are methodically based on the processes of transfer described in Section 2.3.3 and on estimations of the work done by the generalized forces within the building control volume.

*2.4.2.1 Temperature Field and Gradient of the Lagrange Multiplier**

The temperature distribution in the control volume of a system is formally described by a scalar function of the spatial coordinates of a point and time. Generally, the function of the temperature field has the following form:

$$T = f_1(x, y, z, \tau),$$

$$T = f_2(r, \theta, z, \tau), \quad \text{and}$$

$$T = f_3(r, \theta, \lambda, \tau),$$

* Notation β in the equation of Maxwell/Boltzmann.

in Cartesian, cylindrical, and spherical coordinates, respectively. The following special types of temperature fields exist:

- Steady (not depending on time τ)
- Plane (depending on two spatial coordinates)
- One-dimensional (depending on one spatial coordinate)

Figure 2.14 shows plots of several temperature fields in 2D areas: (1) a field in a vertical cross section of a flat wall (see Figure 2.14a); (2) an angle bordered by air (see Figure 2.14b); (3) temperature fields in a horizontal cross section of an envelope (see Figure 2.14c); and (4) a field in a vertical cross section of a wall and in a flour plate bordered by soil (see Figure 2.14d).

The temperature field characteristics refer to a conditional regime of wall *functioning* where temperature T_i of the building interior is higher than temperature T_e of the surroundings. Functions f_1, f_2, and f_3 defining the temperature field are continuous.

Figure 2.15a illustrates temperature at specific points lying on the surface of a wall with dimensions $L_0 = 4{,}60$ m and $h = 3$ m, whose structure is shown in Figure 1.4a. It is found by performing an energy analysis of a real structure erected in 1986 (Dimitroff et al. 2003).

Visual mapping of the temperature field can be effectively performed by plotting spatial surfaces of identical temperatures. These are the so-called isothermal surfaces. The temperature field can be presented by means of

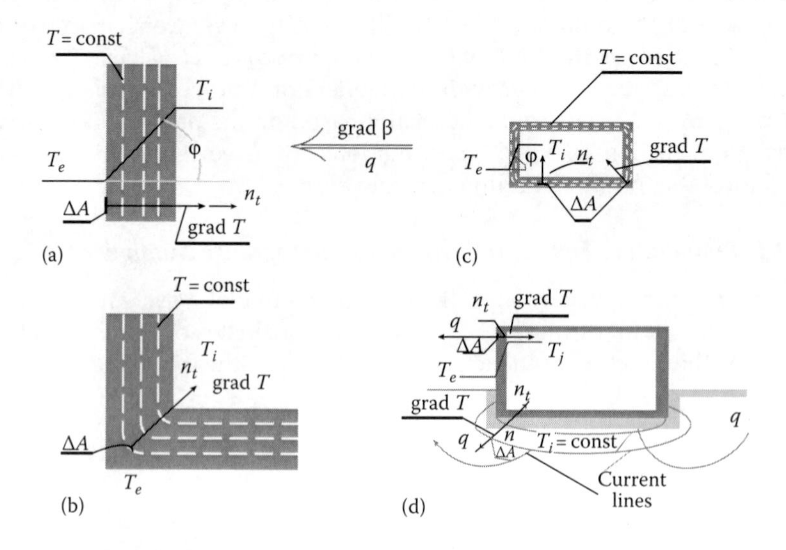

FIGURE 2.14
Distribution of temperature fields in 2D areas: (a) in a flat wall; (b) in the corner; (c) in a horizontal section of the envelope; (d) in a vertical section of the envelope.

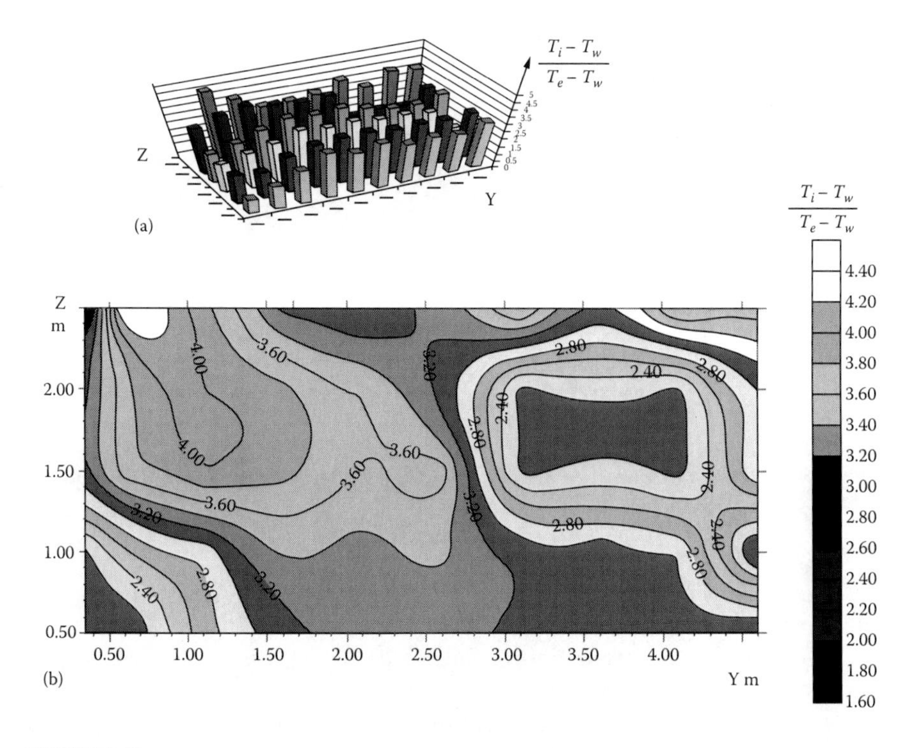

FIGURE 2.15
Distribution of a temperature field on a flat vertical external wall: (a) Discrete values at points on the surface of the wall; (b) isotherms of the field.

families of isothermal surfaces as shown in Figure 2.15b, where isotherms of the temperature field shown in Figure 2.15a are plotted.

Note that only one isothermal surface can pass through a space point occupied by an *ETS*. Hence, considering the general case of an *ETS*, the isotherms are closed spatial surfaces, or they coincide with the boundaries of a certain volume. At each space point A_i (x_i, y_i, z_i), which an isothermal surface T_i passes through, one can define a normal vector $\vec{n}_T$, which is positively directed toward the higher temperatures—see Figure 2.15a and b:

$$\vec{n}_T = n_{T_x}\vec{i} + n_{T_y}\vec{j} + n_{T_z}\vec{k} \tag{2.7}$$

in a Cartesian (rectangular) coordinate system.

$$\vec{n}_T = r\vec{n}_{T_r} + \theta\vec{n}_{T_\theta} + z\vec{n}_{T_z} \tag{2.8}$$

in a cylindrical coordinate system.

$$\vec{n}_T = r\vec{n}_{T_r} + \theta\vec{n}_{T_\theta} + \lambda\vec{n}_{T_\lambda} \tag{2.9}$$

in a spherical coordinate system.

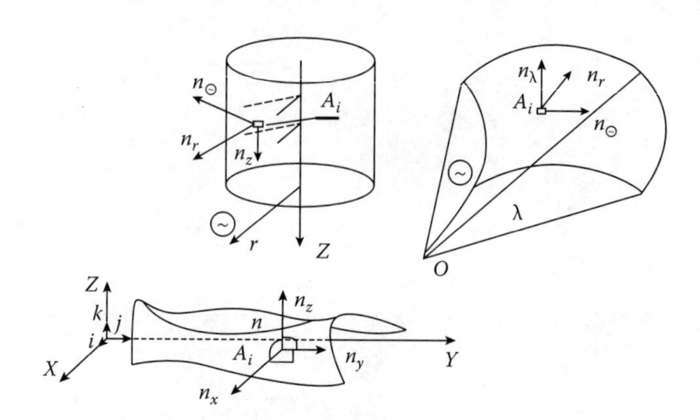

FIGURE 2.16
Isolines of the field characteristics distribution in the Cartesian, cylindrical, and spherical coordinate systems.

Here,

$\vec{i}$, $\vec{j}$, and $\vec{k}$ are components of the normal vector $\vec{n}_T$ on the axes of a rectangular coordinate system $Oxyz$.

$\vec{n}_{T_r}$, $\vec{n}_{T_\theta}$, and $\vec{n}_{T_z}$ are components of the normal vector on the axes of a cylindrical coordinate system $O\,r\,\theta\,z$.

$\vec{n}_{T_r}$, $\vec{n}_{T_\theta}$, and $\vec{n}_{T_\lambda}$ are components of the normal vector $\vec{n}_T$ on the axes of a spherical coordinate system $O\,r\,\theta\,\lambda$ (see Figure 2.16).

Temperature variation along the normal vector $\vec{n}_T$ is one of the most important characteristics of the temperature field. It specifies the temperature gradient, and it is defined by the expression

$$\text{grad}\, T = \lim_{\Delta\vec{n}_T \to 0} \frac{\Delta T}{\Delta \vec{n}_T} = \frac{\partial T}{\partial \vec{n}_T} = \nabla T. \tag{2.10}$$

It acquires the following forms in different coordinate systems (see Equation 2.7, 2.8, or 2.9):

$$\text{grad}\, T = \frac{\partial T}{\partial \vec{n}_T} = \nabla_D T = \vec{i}\,\frac{\partial T}{\partial x} + \vec{j}\,\frac{\partial T}{\partial y} + \vec{k}\,\frac{\partial T}{\partial z}$$

$$\text{grad}\, T = \frac{\partial T}{\partial \vec{n}_T} = \vec{n}_{T_{Cyl}} \nabla_C T = \vec{n}_{T_r}\,\frac{\partial T}{\partial r} + \vec{n}_{T_\theta}\,\frac{1}{r}\frac{\partial T}{\partial \theta} + \vec{n}_{T_z}\,\frac{\partial T}{\partial Z}$$

$$\text{grad}\, T = \frac{\partial T}{\partial \vec{n}_T} = \vec{n}_{T_{Sp}} \nabla_{Sp} T = \vec{n}_{T_r}\,\frac{\partial T}{\partial r} + \vec{n}_{T_\theta}\,\frac{1}{r}\frac{\partial T}{\partial \theta} + \vec{n}_{T_\lambda}\,\frac{1}{r \cdot \sin\theta}\frac{\partial T}{\partial \lambda}$$

where

$$\nabla_D = \vec{i}\frac{\partial}{\partial x} + \vec{j}\frac{\partial}{\partial y} + \vec{k}\frac{\partial}{\partial z};$$

$$\nabla_C = \vec{n}_{T_r}\frac{\partial}{\partial r} + \vec{n}_{T_\theta}\frac{1}{r}\frac{\partial}{\partial \theta} + \vec{n}_{T_z}\frac{\partial}{\partial z};$$

$$\nabla_{Sp} = \vec{n}_{T_r}\frac{\partial}{\partial r} + \vec{n}_{T_\theta}\frac{1}{r}\frac{\partial}{\partial \theta} + \vec{n}_{T_\lambda}\frac{1}{r \cdot \sin\theta}\frac{\partial}{\partial \lambda}$$

are Hamiltonians.

The temperature field gradient grad T is a vector directed along the normal $\vec{n}_T$ (the direction of temperature increase is positive), and it is perpendicular to the isotherms. It is expressed as the derivative of temperature with respect to $\vec{n}_T$.

The gradient magnitude is different at different points A_i belonging to an isothermal surface, and it depends on the distance to a neighboring isotherm. (The gradient largest value is that corresponding to the smallest distance between neighboring isothermal surfaces. If the isothermal surfaces are equidistant, the gradient will have one and the same magnitude at different points.) It is assumed that grad T is evenly distributed in each subarea ΔV_0 of a solid element, and the gradient is defined by the volume derivative

$$\nabla T = \lim_{\Delta V_0 \to 0} \oint_{A_0} T \cdot d\nu_0 / \Delta\nu_0.$$

The use of temperature (and its gradient grad T) as a basic macroscopic state parameter of the building envelope is methodically verified by the link between temperature and the variation of the internal energy within a local area of the control volume. Using the temperature field topology, one can assess (1) whether processes of internal ionization and polarization run in a subarea $\Delta\nu_0$ of the control volume (where free electrons are released and phonons emerge, and energy *accumulation* takes place); or (2) whether this is a transitional subarea where energy transfer takes place accompanied by increase in system entropy (origination of system degeneration).

The extreme importance of *temperature* as a basic state parameter becomes notable when exposing its link to the local value of the Lagrange multiplier, whose behavior is reciprocal to that of temperature $\lfloor \beta_{Loc} = (k_B T)^{-1} \rfloor$—it increases as the temperature decreases. The theoretical values of temperature in the components of the building envelope vary within limits $6000 \geq T \geq 0\,K$ (the first value corresponds to the *internal temperature* of photons arriving at the envelope control surface and to the temperature of the Sun). Respectively, the Lagrange multiplier β, corresponding to the range of temperature variation of phonons that have scattered within the body, will vary within limits $0.12 \times 10^{-20}, J^{-1}(\approx 0) < \beta \leq \infty\ J^{-1}$ (the last limit value is attainable theoretically, only). The distribution of the *local values* of the Lagrange multiplier β_{Loc} within the envelope control volume *carries essential information* on the pattern of interaction between the envelope components and the external energy impacts penetrating its boundary. Equilibrium of the control volume is attained when conditions $\beta_{Bound} = \beta_{Loc}$ are locally satisfied at its boundaries. Besides, the change of the value of the local Lagrange multiplier β_{Loc} reflects the *change* of the *phonon number* $\bar{n}(\nu)$ in each energy orbital. It also configures *the phonon distribution functions* $f_{Ph}(E_j,\mu,T) = \dfrac{1}{\exp\left[\beta_{Loc}(E_j - \varepsilon_F)\right] - 1}$ and the *function of the total energy* of the microstate $Z(j) = \sum e^{-\beta_{Loc}E_j^{(j)}}$.

In its turn, the distribution of the vector function grad β (reciprocal to (grad T)) within the envelope control volume *more rationally* specifies the priority direction of phonon degeneration and microscopic transfer (the transfer vectors are collinear and *unidirectional* with grad β—see Figure 2.14).

2.4.2.2 Pressure Field

Pressure is defined by the equality

$$p = \lim_{\Delta A_0 \to 0} \sum_{A_0} F_n / \Delta A_0,$$

where

$F_n(F_n = F_n^T + F_n^d)$ is the normal component of forces of collisions between the moving particles of the working body (fluid)—see Figure 2.17

F_n^T is a component of F_n due to its thermal motion

F_n^d is a component of F_n due to its mechanical motion

ΔA_0 is an elementary area

A_0 is the area of the wall element

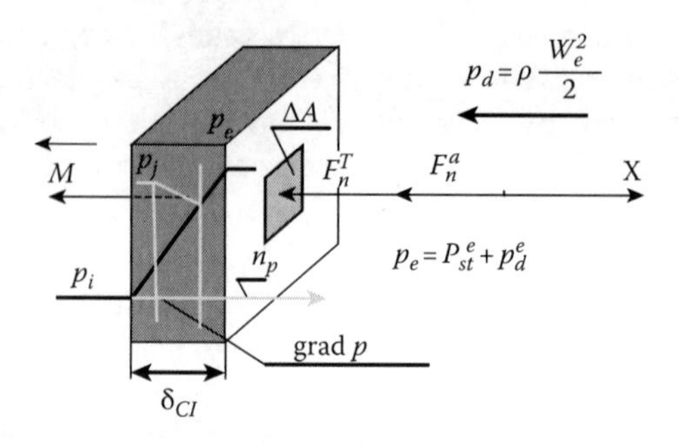

FIGURE 2.17
Types of pressure onto a vertical wall.

If the working body does not execute mechanical motion, pressure is defined by the expression

$$p_{st} = \sum_{A_0} F_n^T / A_0.$$

Pressure is a physical quantity characterizing the intensity of normal forces applied by the working body on the wall surface (in this particular case, a steam–air mix is the working body, and it applies pressure on the sides of the wall element).

Considering atmospheric pressure at sea level under standardized conditions (the so-called atmospheric pressure p_a), pressure is divided into absolute pressure p, overpressure p_{ov}, and vacuum pressure p_v:

$$\left[p_{ov,v} = \pm\left(p - p_a\right)\right].$$

The full pressure at a specific point consists of static pressure p_{st} and dynamic pressure p_d:

$$p = p_{st} + p_d,$$

and the dynamic pressure is set forth by the expression $p_d = 0.5\ \rho \cdot W_e^2$, where W_e is the velocity of the working fluid in a non-disturbed flow (i.e., pressure applied to the surface of the vertical wall element $W_e = 0$; there, the static pressure p_{st}^{bl}, called static pressure at the point of *blockage*, is equal to the full pressure in a non-disturbed flow $[p_{st}^{bl} = p]$).

Pressure field is described by the following functions:

$$p = p_1(x, y, z, \tau)$$
$$p = p_2(r, \theta, z, \tau)$$
$$p = p_3(r, \theta, \lambda, \tau),$$

in Cartesian, cylindrical, and spherical coordinates, respectively.

Similarly to the temperature field, pressure field is presented by isobaric surfaces (also called level surfaces). They are loci in the space occupied by the *ETS*, where pressure value is constant.

The pressure field gradient is defined by a formula similar to formula (2.10):

$$\text{grap}\ p = \lim_{\Delta\vec{n}_p \to 0}\left(\frac{\Delta p}{\Delta\vec{n}_p 2}\right) = \frac{\partial p}{\partial\vec{n}_p}, \tag{2.11}$$

where $\vec{n}_p$ is the normal vector of the isobars.

Using the Hamiltonians ∇_D, ∇_C, and ∇_{Sp} written in terms of the direction cosines of the unit vector $\vec{n}_P$ ($\vec{n}_{P_x}, \vec{n}_{P_y}, \vec{n}_{P_z}$; $\vec{n}_{P_r}, \vec{n}_{P_\theta}, \vec{n}_{P_z}$; and $\vec{n}_{P_r}, \vec{n}_{P_\theta}, \vec{n}_{P_\lambda}$) and expressed in Cartesian, cylindrical, and spherical coordinate systems, we have

$$\nabla_D = \vec{n}_{p_x} \frac{\partial}{\partial x} + \vec{n}_{p_y} \frac{\partial}{\partial y} + \vec{n}_{p_z} \frac{\partial}{\partial z}$$

$$\nabla_C = \vec{n}_{p_r} \frac{\partial}{\partial r} + \frac{1}{r} \vec{n}_{p\theta} \frac{\partial}{\partial \theta} + \vec{n}_{p_z} \frac{\partial}{\partial z}$$

$$\nabla_{Sp} = \vec{n}_{p_r} \frac{\partial}{\partial r} + \frac{1}{r} \vec{n}_{p\theta} \frac{\partial}{\partial \theta} + \frac{1}{r.\sin\theta} \vec{n}_{p\lambda} \frac{\partial}{\partial \lambda}$$

and the definition equality (Equation 2.10) takes the following forms:

$$grad\ p = \frac{\partial p}{\partial \vec{n}_p} = \nabla_D p = \vec{n}_{p_x} \cdot \frac{\partial p}{\partial x} + \vec{n}_{p_y} \cdot \frac{\partial p}{\partial y} + \vec{n}_{p_z} \cdot \frac{\partial p}{\partial z}$$

$$grad\ p = \frac{\partial p}{\partial \vec{n}_p} = \nabla_C p = \vec{n}_{p_r} \cdot \frac{\partial p}{\partial r} + \frac{1}{r} \vec{n}_{p\theta} \cdot \frac{\partial p}{\partial \theta} + \vec{n}_{p_z} \cdot \frac{\partial p}{\partial z}$$

$$grad\ p = \frac{\partial p}{\partial \vec{n}_p} = \nabla_{Sp} p = \vec{n}_{p_r} \cdot \frac{\partial p}{\partial r} + \frac{1}{r} \vec{n}_{p\theta} \cdot \frac{\partial p}{\partial \theta} + \frac{1}{r.\sin\theta} \vec{n}_{p\lambda} \cdot \frac{\partial p}{\partial \lambda}$$

(The direction of pressure increase is taken to be the positive direction of the normal vector $\vec{n}_P$.)

Pressure gradient is a vector directed along the normal and perpendicular to the isobaric surfaces. It indicates pressure variation in the same direction.

*The distribution of the macroscopic state parameters pressure** and *its gradient* grad p holds information on the *total internal energy* within the control area and on the *direction* of macroscopic corpuscular transfer of molecules, ions, and other fermions, including electrons and even photons,† as soon as they display individual behavior.

If $\vec{n}_P$ (grad p, respectively) is directed along Ox (it is assumed that $p_e > p_i$ in Figure 2.17), then the flux of transferred substance $\vec{M}$ is directed from the surroundings to the building interior, that is, *opposite to* grad p. Finally, it is an important characteristic of a multiphase working medium, since it is directly referred to the definition of medium current phase state.

* $\langle p \rangle = \beta^{-1}(\partial \ln Z/\partial v_0)_{\mu,\beta}$. For instance, pressure of an ideal gas reads $p = (\partial\psi/\partial v_0)_{T,v_0} = 2U/3v_0$ (Lee 2002). Here Ψ is Landau's free energy.

† Photon pressure is $p = U/3v_0$ (Lee 2002).

2.4.2.3 Field of the Electric Potential: Potential Function and Gradient of the Electric Potential

We clarified in our survey (Section 2.1) the importance of the energy momentum needed by the valence band electrons to overcome the banned area and convert into electric charges freely roaming within the lattice. In solids, they form the so-called electron gas (Fermi 1936, 1938; Dodge 1944; Anisimov et al. 1974; Naidu and Kamataru 1982; Stowe 1984; Roald 1986; Harrison 1989; Finn 1993; Morely and Hughes 1994; Benjamin 1998; Zemansky 1998; Griffith 1999; Patterson Walter 1999; Van Carey 1999; Challis 2003; Martin 2004; Halliday and Walker 2005; Strosicio and Dutta 2005; Bird 2007; Nilsson and Riedel 2007), since like gas molecules, the individual electrons have three degrees of kinematic freedom, move in all directions, and contribute to the change of the body internal energy. Note, however, that the distance between them is 1.2 $\hat{A}$.

The classical theory states that the energy of the electron gas within the control volume depends on the available number of electrons in the free area N_{El} and on the mean temperature of the entire electron cloud T_{El}, while the following formula is used to estimate energy*:

$$E_{Heat}^{Class} = \frac{3}{2} N_{El} T_{El} k_B \left(= \frac{3}{2} N_{El} x \beta_{El}^{-1} \right)$$

and β_{El} is the mean value of the Lagrange multiplier for the electron gas.

It can be easily proved that the electron gas in solids (including metals) is strongly *non-degenerated* in real conditions (Tien and Lienhard 1985). This means that the number of electrons is larger than that of the free orbitals, which they could occupy (the fermions can occupy only one energy orbital). The condition of attaining a state of *degeneracy* is fulfilled under very high temperature ($T_{El} \gg 2.10^6$ K), that is, at very low values of the Lagrange multiplier ($\beta_{El} \Rightarrow 0$), which the electron gas cannot attain in real conditions. Hence, usually only a small part of the available electrons remain free, while most of them are *captured* in their high-energy orbitals around the nuclei of the lattice atoms.

Note that several generations of researches have been tackling a large variety of related problems. These are, for instance, formulations of a theory of transition of electrons from steady orbitals to free motion, mechanisms of formation of the *cloud* of free electrons, character of the electron potential, outline of other characteristics such as *free path*, energy transfer, and electron interaction with other particles and with the lattice. More details can be found in (Faulkner 1998). Yet, E. Fermi† (1936, 1938), who found the minimal

* The assessment of quantum effects proves that the energy of the electron gas can be calculated via a more appropriate formula (Stowe 1984): $E = 0.6\, N_{El}\varepsilon_F\, \lfloor 1 + 4.112\, (\beta_{El}\varepsilon_F)^{-1} \rfloor$.

† Enrico Fermi (1901–1954г.), Nobel Prize winner in physics, 1938.

energy needed to keep the electrons on a steady orbit (*Fermi level*—ε_F), is one of the most prominent researchers in the field.

In conformity with his theory, the value of the minimal energy ε_F reads (Stowe 1984) as follows:

$$\varepsilon_F = \frac{h^2}{2m_{EI}}\left(\frac{3N_{EI}}{8\pi V_0}\right)^{3/2} = \frac{T_F}{k_B}.$$

Here T_F is the Fermi temperature (see Table 8.1).

The distribution of electrons in the energy orbitals depends on electron effective energy $\varepsilon_{Eff} = (\varepsilon_{EI} - \varepsilon_F)$ needed by electrons to leave the steady orbitals and convert into free-moving electric charges (i.e., this is the energy that electrons need to overcome the banned area). In the distribution function $\overline{n}_{EI} = f(V)$, the electric potential is defined as the ratio between the effective energy $\varepsilon_{Eff} = (\varepsilon_{EI} - \varepsilon_F)$ and the electron electric charge is e^0 = 1.6021773310–19, As), while $V = (\varepsilon_{EI} - \varepsilon_F)/e^0$). The distribution function acquires two values (see Figure 2.18).

- $\overline{n}_{EI}(V) = 0$ is for an electron in a steady orbital ($V \leq 0$).
- $\overline{n}_{EI}(V) = 1$ is for an electron moving within a free area ($V > 0$).

If ($V \leq 0$), all electrons in the valence band occupy the high-energy orbitals around the nuclei of the lattice atoms. Besides, there are no electrons in the free area ($\overline{n}_{EI} = 0$). If the electrons are energized by a quantum impact (for instance, photon absorption) or by another form of variation of the Lagrange multiplier β_{Loc}, they leave their steady orbits. This is accounted for by the plots of the function $\overline{n}_{EI} = f\left(V = \frac{\varepsilon_{EI} - \varepsilon_F}{e^0}\right)$, shown in Figure 2.18.

If V belongs to the interval $0 \leq V \leq E_g/e^0$, only part of the electrons leave the valence band and enter the free area, and the distribution function takes the following form:

$$\overline{n}_{EI} = 1 - \frac{1}{e^{\beta e_{Eff}} + 1} = \frac{1}{e^{-\beta e_{Eff}} + 1}.$$

Its plot is called *Fermi-tail* conforming to the adopted physical terminology, and the tail length depends on the value of the averaged Lagrange multiplier $\beta_{EI} = (k_B T)^{-1}$. Other valence band electrons keep moving around the nucleus but along nonsteady orbits, as discussed in Section 2.3. Then, a mechanism of internal polarization and phonon excitation switches on.

If $V \geq E_g/e^0$, all electrons of the valence area pass to free energy orbitals and form an electron cloud within the free area. Then, the distribution function $\overline{n}_F$ is $\overline{n}_{EI} = 1$. It is reasonable to note that part of the individual electrons, having joined the electron cloud, return to their *high-energy* orbitals around

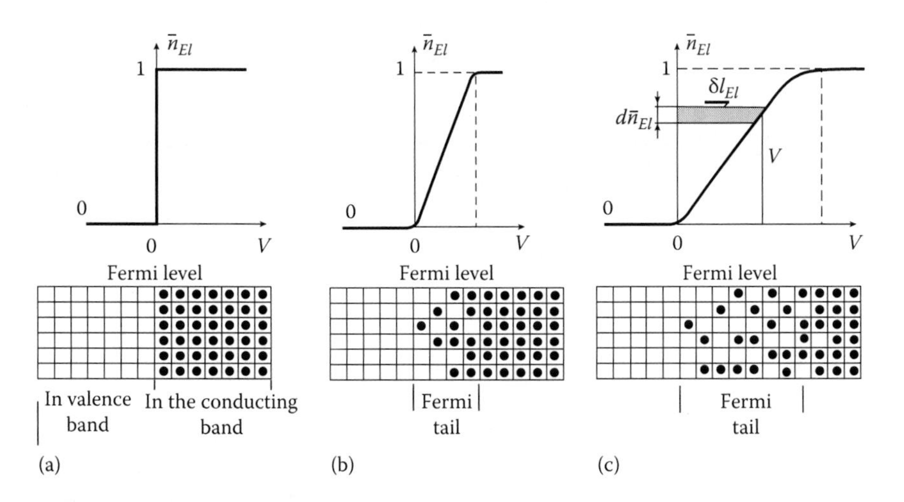

FIGURE 2.18
Distribution of electrons: (a) transition without *Fermi-tail;* (b) transition with a short *Fermi-tail;* and (c) transition with a long *Fermi-tail.*

the nuclei of the lattice atoms, being attracted by the nuclei positive charge. Note that phonons also emerge during that process of interaction between the atom lattice and the returning electrons, and part of the kinetic energy of the electron cloud passes to the lattice.

Figure 2.18 shows three types of transfer:

1. Transfer without a *tail* where electrons pass to the free area with minimal energy, since the number of free energy orbitals is significantly larger (in metals, for instance).
2. Transfer with a short *tail* where the increase in the number of free energy orbitals is inversely proportional to the Lagrange multiplier typical for semiconductors.
3. Transfer with a long *tail* requiring significant amount of energy, as is in insulators.

The number of electrons in the free area is found by integrating the following expression:

$$dN_{El} = \left(\frac{dr^3 \cdot dp^3}{h^3} \right) \cdot \bar{n}_{El},$$

where $\frac{dr^3 \cdot dp^3}{h^3}$ is the distribution of the energy orbitals.

Using the relation $0,5p^2/m_{El} = e^0$ and performing integration over the control volume v_0 and along all directions of the electron momentum vector, we find the following form of dN_{EI}—change of the number of electrons in the electron gas ($m_{EI} = 9.1093897 * 10^{-31},\ kg$):

$$dN_{EI} = \left[\frac{4\pi v_0 (2m_{EI})^{3/2}}{h^3}(e_{EI})^{0.5} de_{EI}\right]\bar{n}_{EI}.$$

The energy of the electron gas at absolute zero ($T = 0$ K) is assessed via the expression

$$E_{Heat}^{EI} = \int e_{EI} dN_{EI} = \frac{3}{5} N_{EI} x \varepsilon_F,$$

where $N_{EI} = \frac{8\pi V_0 (2m_{EI})^{3/2}}{5h^3} \varepsilon_F^{5/2}$ is the number of electrons in the electron gas in the control volume.

Note that the relative amount of energy carried by the electrons at mean temperature (normal for the envelope) and at high temperature is not significant in itself (it is below 1% according to some estimations; Stove 1984). Yet, the role of electrons to excite vibrations within the lattice by interacting with the atoms is significant and essential, especially in metals. This very deduction implies estimation of the amount of energy transferred by electric charges. In what follows, we call that energy *electrical work*.

The electrical and mechanical works done are equivalent, and they agree with identical thermodynamic laws (Fermi 1936).* That assertion is based on the existence of physical laws proving that the mechanical energy is directly converted into electrical energy and vice versa (see, for instance, the electrodynamic induction and the piezoelectric effect). Thus, the electrical work de_0 needed to change the electric charge de_0 is assessed by the following expression (Fermi 1936; Sychev and Shier 1978):

$$\delta I_E = \vec{F}_E \cdot d\vec{e}_{EI},$$

where $\vec{F}_E$ is the electric driving force.

As is known from the theory, that work depends on the trajectory of charge motion (Schelkunoff 1963). If however we use the electric potential $V(x, y, z)$, which is a scalar function of the spatial coordinates (Schelkunoff 1963), the work needed to shift an electric charge is assessed only by the difference between the values of $V(x, y, z)$ at the two end points of the trajectory.

* The link between the electric driving force E, the internal energy U, and temperature T is specified by the Helmholtz equation (Fermi 1938): $E - TdE/dT = U$, while Van't Hoff's equation produces the ratio between the mechanical work $\mathbf{L}$ and the internal energy: $L - TdL/dT = -dU$

As theory assumes, the electric potential at a point is defined as a ratio between the following two scalar quantities:

1. Work $I_{EA,\infty}$, *J* за needed for moving the electric charge from point **A** to an infinitely distant spatial point (→∞)
2. Electrical quantity e_{EI}, C of the charge:

$$V_A = \frac{I_{EA,\infty}}{e_{EI}}, \quad \frac{J}{C} = \text{volt}.$$

(It is assumed that the potential at an infinitely distant *point B* (→∞) is zero, i.e., $V_\infty = 0$). The distribution of *V* in the physical space is similar to that of temperature (see Equations 2.7 through 2.9):

$$V = V_1(x, y, z, \tau),$$

$$V = V_2(r, \theta, z, \tau), \quad \text{and}$$

$$V = V_3(r, \theta, \lambda, \tau).$$

From a geometrical point of view, equipotential surfaces most clearly represent the definition areas of the scalar functions V_1, V_2, and V_3.

Similar to Sections 2.4.1 and 2.4.2, the following equation specifies the gradient of the scalar function of the electric potential:

$$\text{grad } V = \lim_{\Delta\vec{n}_V \to 0}\left(\frac{\Delta V}{\Delta\vec{n}_V}\right) = \frac{\partial V}{\partial\vec{n}_V}, \tag{2.12}$$

where $\vec{n}_V$ is a vector normal to the equipotential surfaces (surfaces with identical electric potential). Using the Hamiltonians ∇_D, ∇_C, and ∇_{Sp}, written in terms of the direction cosines of the unit vector $\vec{n}_V$ ($\vec{n}_{V_x}, \vec{n}_{V_y}, \vec{n}_{V_z}$; $\vec{n}_{V_x}, \vec{n}_{V_\theta}, \vec{n}_{V_z}$; and $\vec{n}_{V_u}, \vec{n}_{V_\theta}, \vec{n}_{V_\lambda}$) and expressed in Cartesian, cylindrical, and spherical coordinate systems

$$\nabla_D = \vec{n}_{Vx}\frac{\partial}{\partial x} + \vec{n}_{V_y}\frac{\partial}{\partial y} + \vec{n}_{V_z}\frac{\partial}{\partial z}$$

$$\nabla_C = \vec{n}_{V_r}\frac{\partial}{\partial r} + \frac{1}{r}\vec{n}_{V\theta}\frac{\partial}{\partial\theta} + \vec{n}_{V_z}\frac{\partial}{\partial z}$$

$$\nabla_{Sp} = \vec{n}_{V_r}\frac{\partial}{\partial r} + \frac{1}{r}\vec{n}_{V\theta}\frac{\partial}{\partial\theta} + \frac{1}{r\cdot\sin\theta}\vec{n}_{V\lambda}\frac{\partial}{\partial\lambda},$$

the definition equality (Equation 2.11) takes the following forms:

$$\text{grad } V = \frac{\partial V}{\partial \vec{n}_V} = \nabla_D V = \vec{n}_{V_x} \cdot \frac{\partial V}{\partial x} + \vec{n}_{V_y} \cdot \frac{\partial V}{\partial y} + \vec{n}_{u_z} \cdot \frac{\partial V}{\partial z}$$

$$\text{grad } V = \frac{\partial V}{\partial \vec{n}_V} = \nabla_C V = \vec{n}_{V_r} \cdot \frac{\partial V}{\partial r} + \frac{1}{r} \vec{n}_{V_\theta} \cdot \frac{\partial V}{\partial \theta} + \vec{n}_{V_z} \cdot \frac{\partial V}{\partial z}$$

$$\text{grad } V = \frac{\partial V}{\partial \vec{n}_V} = \nabla_{Sp} V = \vec{n}_{V_r} \cdot \frac{\partial V}{\partial r} + \frac{1}{r} \vec{n}_{V_\theta} \cdot \frac{\partial V}{\partial \theta} + \frac{1}{r \cdot \sin\theta} \vec{n}_{V_\lambda} \cdot \frac{\partial V}{\partial \lambda}.$$

(The direction of potential increase is taken to be the positive direction of the normal vector $\vec{n}_V$.)

Considering this particular case, the gradient vector is directed along the normal vector, and it is perpendicular to the isopotential surfaces, as proved by the potential variation, which is in the same direction. It is an important characteristic of the potential field in depth of the envelope components, especially close to their facial surfaces, since it directly concerns the definition of the electric driving force acting in the integrated systems.

2.4.2.4 *Entropy: A Characteristic of Degeneration of the* **Heat Charges (Phonons)** *within the Envelope Control Volume*

The physics of interaction between the building envelope and the environment at microlevel (see Section 2.3.3, Figure 2.15) proves that the thermodynamic system passes through $\tilde{\Gamma}_j$ number of microstates. To assess the degree of variation of the envelope energy state, we assume the Boltzmann–Plank form of *entropy* (Lee 2002) as an indicator of *degeneration of phonons* in the course of their scatter, that is,

$$S_i = k_B \ln\left(\tilde{\Gamma}_i\right), \quad \text{J/K}.$$

The entire change of the system energy status can be assessed by means of the following expression:

$$S_{Eq} = \frac{1}{n_M} \sum_{i=1}^{n_M} S_i.$$

Here $k_B = 1.38 \cdot 10^{-23}$, J/K is the Boltzmann constant, and $\{\tilde{\Gamma}_i\}$ is the matrix of the *found number* of microstates per each current macrostate out of (n_M) macrostates.

As is a priori known (Lee 2002), entropy conforms to the following objective laws:

- Entropy never decreases spontaneously (by itself) in nonequilibrium systems, but only increases (the second law of thermodynamics).
- All thermodynamic systems (ETS) obey the basic principle stating that if a system occupies some of its minoritarian macrostates, it spontaneously passes to its collective (dominating) macrostate.
- Entropy attains its maximum when the system attains equilibrium with the surroundings; the rate of entropy increase depends on the *path* passed by the system (so-called thermodynamic transition processes).

If the volume of the control area is $V_0(\nu_0 = A_0\delta_W = 1.0 * \delta_W, m^3) \Rightarrow A_0 = 1, m^2$, *energy scatter within the volume* (so-called position entropy s_{ν_0}; Lee 2002) results in entropy increase. A *second source of entropy increase* $s \Rightarrow s_T$ is the change in the frequency of phonon radiation, that is, the *decrease in phonon eigen energy* depending on the Lagrange local multiplier (β_{Loc}) and on *the temperature local value* S_{ν_0} (Lee 2002). *Finally*, entropy $s \Rightarrow s_V$ increases under electron motion within the free area determined by the electric potential *V*.

Assume that entropy of the building envelope can be described via a function of the three independent macroscopic state—volume of the control area (V_0), temperature (T), and electric potential (V):

$$S_{Env} = f\left(\nu_0, T, N_{EI}\right).$$

Then, its total potential ds_{Env} will read as follows:

$$ds_{Env} = \left(\frac{\partial s_{Env}}{\partial \nu_0}\right)_{T,N_{EI}} d\nu_0 + \left(\frac{\partial s_{Env}}{\partial T}\right)_{\nu_0,N_{EI}} dT = \left(\frac{\partial s_{Env}}{\partial N_{EI}}\right)_{\nu_0,T} dN_{EI}.$$

Consider entropy variation in a component of the building envelope with cross-area $A_0 = 1$ m^2 and thickness δ_W and account for energy scatter (entropy increase) within the control area volume (V_0)—$ds_{\nu_0} = \left(\frac{\partial s_{Env}}{\partial \nu_0}\right)_{T,N_{EI}} d\nu_0$. Assume also that the envelope structural material is in equilibrium with the surroundings, and its initial parameters are temperature T_0 and entropy S_0 while the width of the banned area is $E_g \approx 2.5$ eV (as ZnSe).

Assume that a *flux* of photons *typical** for the solar radiation and belonging to the visible part of the spectrum ($350 \le \lambda_{Solar} < 760$ nm) is initiated along the

* The photons belonging to the solar flux have a distribution direction specified by the angle of attack on the respective building component (see Section 7.2. Solar shading devices (shield) calculation, direction cosines (λ, μ, γ)).

normal of the wall element (axis Ox) at a moment τ_0. Assume also that photon wavelength is $\lambda_{typical}^{Solar} = 650$ nm (the orange color of the solar spectrum, frequency is $\nu_{typical}$ = 461 THz) and photon energy is E_{Ph} = 1.9 eV. Then, the current barrier coefficient K_{FZ} will be larger than 1.

The eigenvalue of the energy of a typical photon* prior to its interaction with the atoms is found as

$$\langle e_{Ph} \rangle_{typical} = hx\nu_{typical}\left(n_x^2 + n_y^2 + n_z^2\right)$$

$$= 6.6260755 * 10^{-34} * 461 * 10^{12} * (1^2 + 0 + 0) = 3.056 * 10^{-19} \text{ eV},$$

where $\nu = \left(\vec{n}_x, \vec{n}_y, \vec{n}_z\right) = (1^2 + 0 + 0)$ is the index of the eigenstate of the solar flux photons. Entropy of that initial microstate will be equal to that of the equilibrium state or to the so-called residual entropy—$S_{No0} = S_0 + k_B \ln 1 = S_0$.

When the typical photons interact with the front-row atomic structures of an envelope component, as stated in Section 2.3.3 (see Figure 2.15), they will *polarize* and start vibrating with frequency ν_E, giving birth to Einstein standing waves, which would propagate along the *three axes* of the Cartesian coordinate system. Hence, *first-generation* phonons will emerge (due to the oscillation of the valence electrons remaining within the atom structure).

In their turn, *phonons, newly generated* during internal polarization, emit isotropic photons in all directions (see Figure 2.10). Their eigen energies will be

$$\langle e_{Ph} \rangle_{\nu:=(1^2+0+0)} = h * \nu_{Sec}(1^2 + 0 + 0) = 2.4448 \times 10^{-19} \text{ eV}$$

$$\langle e_{Ph} \rangle_{\nu:=(0+1^2+0)} = h * \nu_{Sec}(0 + 1^2 + 0) = 2.4448 \times 10^{-19} \text{ eV}$$

$$\langle e_{Ph} \rangle_{\nu:=(0+0+1^2)} = h * \nu_{Sec}(0 + 0 + 1^2) = 2.4448 \times 10^{-19} \text{ eV},$$

where index $\nu = \left(\vec{n}_x, \vec{n}_y, \vec{n}_z\right)$ has the following values ⇒ $(1^2 + 0 + 0)$, $(0 + 1^2 + 0)$, and $(0 + 0 + 1^2)$, while ν_{Sec} is frequency of the secondary *emission*.

Although the secondary *emitted* photons have identical eigen energy, three different eigenstates of the system will emerge after photon emission, due to different values of the index of spatial distribution $\nu = \left(\vec{n}_x, \vec{n}_y, \vec{n}_z\right)$. They are called degenerates (Lee 2002).

The entropy of the system after the emission of *second-generation photons* increases due to the increased number of microstates ($s_{\nu 0}^{i} = s_{N\underline{o}=0} + i s_{N\underline{o}=1} = s_0 + k_B \ln 3$). At any *subsequent* ($i^{Ta}$) emission of *next-generation* photons in depth of the control volume of the envelope component, entropy will increase in an arithmetic sequence with common difference $k_B \ln 3$. Entropy extreme

* Momentum $\langle p \rangle$ of the typical photon amounts to $\langle p \rangle = h/\lambda_{Typical} = 1019 * 10^{-27}$ Js/m.

(equilibrium) value will be proportional to the number of phonon degenerations N_{Degen} or $s_{v0}^{Max} = N_{Degen} * k_B * \ln 3 + S_0$.

The model of photon gas degeneration within the structure of a solid body is copied by *phonon*-excited motion. In their turn, degeneration of phonons having emerged within front-row atom structures and degeneration of next-generation phonons specify the local temperature of the atom lattice and the value of the local Lagrange multiplier.

One can assess the dynamics of phonon degeneration considering the variation of the local Lagrange multiplier $\beta_{Loc} = 1/(k_B T_{Loc})$, which increases till the satisfaction of condition $\beta_{Loc} = \beta_{Bound} = 1/(k_B T_{Bound})$ (theoretically, the local Lagrange multiplier tends to infinity ($\beta_{Bound} \Rightarrow \infty$), entropy attains maximum, and the local temperature T_{Loc} tends to the absolute zero (T_{Loc}). The described process depends on the overall dimensions of the control area, and its volume ν_0 is specified as a functional argument of entropy.

The second mechanism of entropy increase within the control area is that of the scatter of the phonon flux, generated within the solid materials and described in Section 2.3.1. It physically presents a variation of the local temperature and the Lagrange multiplier—$ds_T = \left(\frac{\partial s_{Env}}{\partial T}\right)_{v0,N_{EI}} dT$. As a result of phonon energy release, the radiation spectrum within the control area redistributes, following a scatter mechanism described in Section 2.3. Radiation intensity comprises all frequencies in a model, which can be assessed by the variation of the local Lagrange multiplier β_{Loc} only or by the local temperature value, respectively (see Section 2.4.2.1). The large radiation intensity supposes generation of a large number of phonons in a certain subarea of the control volume. They copy photon degeneration scheme. Moreover, their dominant frequency form or dominant energy orbital specifies respective local temperature T_{Loc}.

Conversely, a decrease in temperature T_{Loc} in depth of the control volume (linked with an increase in the Lagrange multiplier) is an indication that the

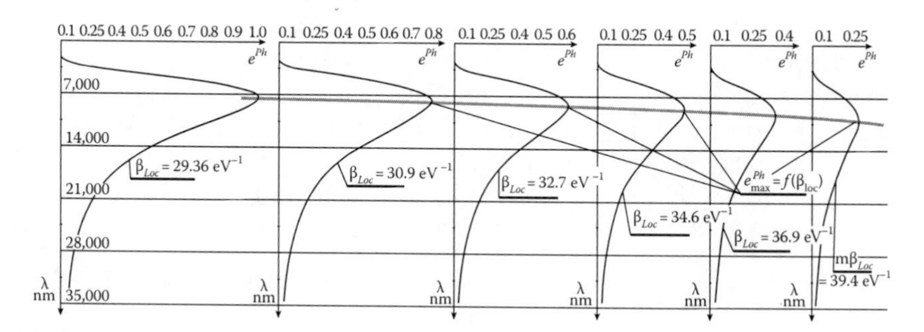

FIGURE 2.19
Reducing the energy of the phonons by increase in the local Lagrange multiplier in the range $29 \geq \beta \geq 40$ eV^{-1}, typical for normal operational conditions in the building envelope.

thermal phonons energetically degenerate in such areas. Figure 2.19 illustrates the predicted distribution of the radiation energy of thermal phonons within the control volume at an increase in the Lagrange multiplier within limits $29 \geq \beta \geq 40$ eV^{-1}, corresponding to the variation of local temperature T_{Loc} within limits (393 ÷ 293 K).

Performing those predictive calculations, we assume linear distribution of the Lagrange multiplier β_{Loc}, which has the form $\beta_{Loc} = 1/[(k_B \left(T_{SUN} - \frac{X}{\delta_W}(T_{SUN} - T_{Indoor} \right)$. The *energy function* corresponding to a certain microstate reads $Z_i = \exp(-\beta_{Loc} \cdot \bar{n}(\upsilon)x\hbar\upsilon)$, while *the total energy* of the entire *macrostate* Z_{Pn}, resulting from phonon radiation is* $Z_{Pn} = \sum_{s=0}^{\infty} \exp\left(-\beta_{Loc} \cdot \bar{n}(\upsilon)x\hbar\upsilon\right) = \frac{1}{1-\exp(-\beta_{Loc}\hbar\upsilon)}$.

The estimations performed hereby prove that the increase in phonon penetration depth within the control volume yields shift of the *dominant energy orbital* toward the long wavelength end of the spectrum—see Figure 2.20.

The wavelength of the energy-carrying phonons in the vicinity of a physical point is linked with the values of the local Lagrange multiplier through the relation

$$\lambda_{max} = 2.51 * 10^{-28} * \beta_{Loc}, \text{nm} \quad (\text{curve 2, see Figure 2.20b}),$$

and the value of the energy charge linked with the dominant radiation length decreases as

$$e_{max}^{Ph} = \frac{4.94 * 10^{30}}{\beta_{Loc}}, \text{eV} \quad (\text{curve 3, see Figure 2.20b}).$$

This is a proof that *new energy forms emerge* within the phonon system of the solid material when the phonon penetration depth increases. They comprise *new long-wave low-intensity vibrations,* whose occurrence is deduced from both the increase in the wavelength of the dominant energy orbital λ_{max} and the increase in the local Lagrange multiplier β_{Loc} (corresponding to lower local temperature T_{Loc}). Thus, the number of microstates $\tilde{\Gamma}_T$ increases resulting in *entropy increase* in the control volume ($s\uparrow \Rightarrow s_T$).

The third mechanism of entropy increase is rooted in electron thermal motion, subsequent execution of electrical work, and electron interaction with the lattice—$ds_{El} = \left(\frac{\partial s_{Env}}{\partial N_{El}}\right)_{\upsilon ol} dN_{El}$. As already clarified, the electrons perform thermal (nonorganized) motion within solids if they do not undergo impacts of an external magnetic field. Besides, free electrons contribute to the increase in

* The total internal energy of the same microstate of the phonon system is assessed via the expression (Lee 2002) $U = (Z_{Pn})^{-1}\, \partial Z_{Pn}/\partial(-\beta_{Loc}) = (\hbar\omega)/[\exp(\beta_{Loc}\hbar\nu) - 1]$.

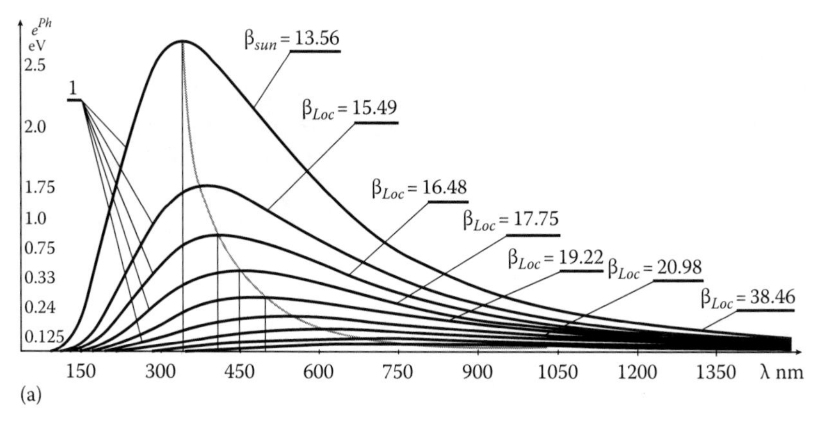

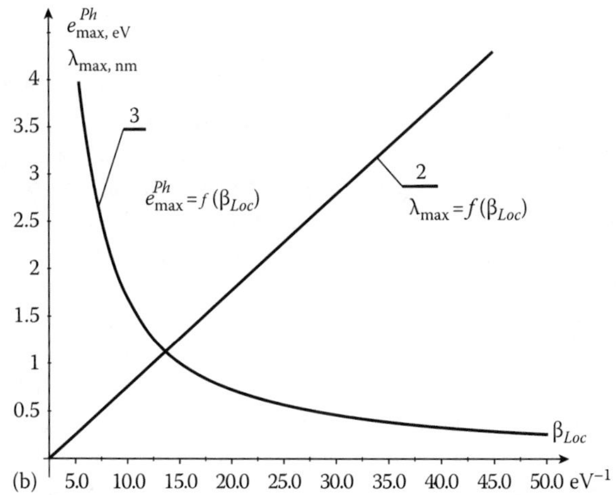

FIGURE 2.20
Reducing the energy of the typical phonons by an increase in the local Lagrange multiplier: (1) in (a) the distribution of energy, corresponding to the local Lagrange multiplier, (2) in (b) length of the dominant energy mode, (3) in (b) power of the dominant energy mode. (a) Evolution of the Spectrum of phonon's energy as a function of the local Lagrange multiplier; (b) evolution of the energy of the dominating phonon as a function of the local Lagrange multiplier.

internal energy $\langle U \rangle$ and temperature T within the control area. In our previous considerations, we noted that despite their minoritarian tribute to the total energy capacity of solids, electrons (mainly those in metals and semiconductors) play an important role in the transfer of energy to the atom lattice. If *electron gas* generated during material internal ionization is present in the control volume, it undergoes two types of degeneration. As a result, the number of its microstates increases, and thus the total entropy in the control volume also increases.

The *first mechanism of degeneration* of free electron energy is owing to a spatial factor. It is similar to that concerning the *photon gas*, and it is described at the beginning of the present section. Since the process of internal ionization runs under stochastic conditions, the number of allowed energy levels (orbitals) occupied by the *electron gas* n_x, n_y, and n_z {1,2,3,...} *increases within the interval* $\varepsilon \leq E_i \leq \varepsilon + d\varepsilon$ when electrons move along the three axes of the Cartesian coordinate system (energy level $\varepsilon = E_i$ can be attained via various combinations of n_x, n_y, and n_z) (Van Carey 1999).

The *second mechanism* of degeneration of the *electron gas* consists of release of its energy ΔE (via special T-waves) in the form of phonons to the vibrating lattice (Van Carey 1999). To assess entropy increase at the expense of phonon emergence, it is convenient to adopt a modified expression—Equation 2.5:

$$ds''_{EI} = \frac{\Delta E}{T_I} = \left(1 - \frac{T_I}{T_e}\right)\frac{G}{T_I}.$$

The analysis should include the electron local concentration within the control volume (respectively, relation between the electron numbers [N_{EI}] and size of the control volume [v_0]) as a primary of factor affecting the amount of energy, which the electron gas releases to the lattice. The second factor is the barrier coefficient ($\varepsilon_F = (K_{FZ})$) or Fermi level (ε_F). Since the coefficient of electron and atom coalition G is a constant quantity for a certain structural material (see Table 8.11) and the condition $\lim(T_I/T_e) \Rightarrow 0$ is fulfilled, increase in entropy s''_{EI} is determined by temperature T_I of the atom lattice, which decreases in depth of the control volume. This phenomenon results from the emergence of new energy forms during the interaction between electrons and atoms. Thus, the number of microstates in the canonical ensemble increases yielding increase in the system total entropy.

Hence, we may *employ* change of entropy s as a measure of the *degree of photon degeneration,* while entropy maximal value s_{max} can be taken as a representative characteristic of the amount of *scattered* (dissipated) energy at equilibrium attainment. Ultimately, the combined effect of the three independent mechanisms of entropy increase in the control volume can be expressed by their sum

$$s_{Env} = s_{v0} + S_T + S_{N_{EI}}.$$

Hence, we assume the following form of entropy change:

$$ds_{Env} = \left(\frac{\partial s_{Env}}{\partial v_0}\right)_{T,N_{EI}} dv_0 + \left(\frac{\partial s_{Env}}{\partial T}\right)_{v_0,N_{EI}} dT + \left(\frac{\partial s_{Env}}{\partial N_{EI}}\right)_{v_0,T} dN_{EI},$$

(we introduce notation $ds_{Env} = ds_0$), which should be interpreted as a *generalized coordinate* of variation of the *thermal charge (phonon system)* resulting from the *thermal work done* for thermal charge (phonon) transfer. We use the known expressions*

- $\delta L_T = \delta Q = TdS_0$ for the whole control volume
- $\delta l_T = \delta q = Tds_0$ for unit mass.

(Heat transferred through the control volume of the envelope is *equivalent to the work done by the thermal forces* and respectively to the energy transferred by the phonons. It is calculated as a product of the generalized force $\boldsymbol{T}$ and the variation of the generalized coordinate $s_0 \Rightarrow ds_0$.)

2.4.3 Conclusions on the General Methodological Approaches to the Study of an Electrothermomechanical System

Surveying the *discussed literature in the field*, we made the following conclusions concerning the general problems formulated and discussed in the present study:

1. Regarding the control volume of a building envelope, we outline the following phenomena—transformation of the external energy (solar radiation, wind, gravitational energy, etc.) into heat and electricity *e* and mass transfer in the form of infiltrated/ex-filtrated air and water vapor. Thus, the building envelope can be treated as an electrothermomechanical system (*ETS*).
2. There is generally no valid physical model describing the mechanism of interaction between the physical environment (the solar radiation) and wind at microlevel. The proposed *hypothetical model of energy transfer through the facade of the solid components—the so-called model of a lagging temperature gradient* (see Figure 2.13)—has been accepted as a methodological basis of further studies.
3. There are no studies outlining the link between processes running at microlevel *and macroscopic* state parameters.
4. The survey of literature in the field shows that unilateral separatist approach is adopted to study the behavior of building components by assessing environmental effects. In this respect, only part of the state parameters are considered, that is, the studies are performed at macrolevel.
5. There is no evidence to formulate a *generalized* macroscopic approach enabling one to describe the simultaneous effect of different phenomena occurring in the building envelope: transfer of entropy (heat), electricity, and mass.

* Those forms are introduced by Clausius (1850, 1868).

6. The survey of literature did not find available methodological ideas and trends of modeling and development of the *intelligent* response of the building envelope to environmental impacts, involving real-time active regulation of the physical properties of structural materials.

Those conclusions stimulated our efforts to follow the sequence of general problems set forth in Section 2.2 in an attempt to propose a generalized theoretical explanation of energy and mass transfer through the building envelope components. In this respect, *we assumed the existence of macroscopic generalized force, work, and potential in an electrothermodynamic field generated by environmental forces.**

* In the absence of suitable levels of energy *charge* of the building surroundings (building exterior), the role of ***ETS*** high-temperature *source* is played by the internal environment (building interior), that is, by the systems providing building internal comfort, respectively (Димитров 2008a). Then, exchange processes invert their direction (we consider here the case of unilateral steady transfer only).

3

Design of a Model of Energy Exchange Running between the Building Envelope and the Surroundings: Free Energy Potential

3.1 Energy-Exchange Models of the Building Envelope

As is known in classical physics and in thermodynamics, in particular, these models are classified with respect to whether a studied system *exchanges mass or energy with the surroundings*. Since modern physics considers mass and energy to be *different forms of matter* being individually uniquely identical, it is assumed that a more general classification of the energy-exchange models should account for the type of the studied system and the number of its interactions with the surroundings. Hence, we may assume the following general division of energy-exchange (energy) systems and the respective models of the running processes:

- Isolated systems (no energy exchange with the surroundings).
- Open systems (one or more interactions with the surroundings—exchange of entropy, momentum, mass, electric charges, or moments). According to the number of possible energy interactions between the studied system and the surroundings, the systems have one or more degrees of freedom.

Another criterion of classifying the energy-exchange systems and models employed to describe the processes running in the building envelope is the number of *energy impacts* that the surroundings exert upon it. The analysis of literature in Chapter 1 shows that the processes running in the components of the building envelope (facade walls, roofs, floors, attics, light collectors, and other integrated and polyfunctional components) have been analyzed in *open systems*, having *one degree of freedom*, only, that is, in monofunctional systems.

These processes comprise transfer of heat, moisture, and ions, mechanical deformation, etc. Some related studies in the field are given as follows:

- Studies on heat transfer; entropy transfer only (Clausius 1868; Adkins 1908; Bosworth 1952; Fast 1962; Obert and Young 1962; Vild 1964; Boelter et al. 1965; Luikov 1968; Klemens 1969; Welty 1974; Kreith and Black 1980; Chapman 1981; Zemansky and Dittman 1981; Ларинков 1985; Incropera and DeWitt 1985; Craig 1992; Thomas 1992; Bejan 1993; Bevensee 1993; Kakas and Yener 1993; Fuchs 1996; Holman 1997; Cengel 1998; Patterson 1999; Richet 2001; Lee 2002; Grevin et al. 2003)
- Studies on moisture transfer; mass transfer only (Hirschfelder et al. 1954; Sissom and Pitts 1972; Treyba 1980, Bennet and Myers 1984; Gebhart 1993; Wilkinson 2000)
- Studies on electric charges; ions transfer only (cations and anions) or electrons (Maxwell 1900; Dampier 1905; Schelkunoff 1963; Von Ness 1964; Edminister1965; Anisimov et al. 1974; Duffin 1980; Hammond 1981; Naidu and Kamataru 1982; Roald 1986; Wangsness 1986; Morely and Hughes 1994; АнаниевЛ and Мавров 1995; Benjamin 1998; Griffith 1999, Kraus and Fleisch 1999; Schwarz 2001)
- Studies on mechanical deformation; momentum or displacements (Galerkin 1915; Fagan 1984; Hang 1987; Бъчваров 2006)

Work done at macrolevel in monofunctional systems can be interpreted in terms of power potential fields of specific physical forces:

- For electric force $\vec{F}_E$: Field of electric potential, (ψ_E)
- For thermodynamic force $\vec{F}_T$: Field of thermal potential, (ψ_T)
- For mechanical force $\vec{F}_M$: Field of mechanical potential, (ψ_M)

Real physical phenomena observed in the vicinity of a point of a building envelope component are the variations $\Delta\vec{r}_{m(m=E,T,M)}$ of three types of *generalized coordinates* (Димитров 2006; Dimitrov 2013):

1. Electric: Change of electrical quantity, $(e_0, C) \rightarrow \Delta\vec{r}_{m=E} = \Delta e_0$
2. Thermal: Entropy increase,* $(s_0, \mathrm{J/°K}) \rightarrow \Delta\vec{r}_{m=T} = \Delta s_0$
3. Volume/mass, $(v_0, \mathrm{m^3/kg–kg/m^3}) \rightarrow \Delta r_{m=M} = \Delta v_0$

It is convenient *to present here* the small work $\Delta I_{m(m=E,T,M)}$ as a scalar product between the vector–force $\vec{F}_{m(m=E,T,M)}$ (electric, thermal, and mechanical) and the generalized displacement $\Delta\vec{r}_{m(m=E,T,M)}$: $\Delta l_m = \vec{F}_m \Delta\vec{r}_m, (m = E, T, M)$.

* For instance, degeneration of a phonon with frequency ν_E.

Theoretical physics and mechanics assume that the conservative *monofunctional potential fields* have a total differential of their potential functions (Бъчваров 2006). It can be *presented here* in the form $(\Psi_m)_{m=E,T,M}$:

$$d\Psi_{m\, m=E,T,M} = \frac{d}{dr_m}(\Psi_m)_{m=E,T,M}\, dr_{m(m=E,T,M)}.$$

The satisfaction of condition $(-d\Psi_m = dI_m)_{m=E,T,M}$ yields that the derivative of Ψ_m with respect to the generalized displacement $d\vec{r}_m \Rightarrow -\frac{d}{dr_m}(\Psi_m)_{m=E,T,M}$ is equal to the generalized force $\vec{F}_{m(m=E,T,M)}$, that is, the following equality holds:

$$|F_m|_{m=E,T,M} = -\frac{d}{dr_m}(\Psi_m)_{m=E,T,M}.$$

Here, $|F_E| = -d/de_0(\psi_E) = V$, $|F_T| = -d/ds_0(\psi_T) = T$, and $|F_M| = -d/dv_0(\psi_M) = p$ are the respective generalized electric, thermal, and mechanical forces applied by the surroundings on the building envelope.

The theory of monofunctional potential fields (Бъчваров 2006) assumes that the variation of the three generalized coordinates starts at a current physical space point with potential values $[(\psi_E)_A = (Ve_0)_A, (\psi_T)_A = (Ts_0)$, and $(\psi_M)_A = (pv_0)_A]$ and proceeds to an infinitely distant point, at which the potential function is nullified and the generalized forces are neglected, that is, $(V)_\infty = 0$, $(T)_\infty = 0$, and $(p)_\infty = 0$ and, respectively, $(\psi_E)_\infty = (Ve_0)_\infty = 0$, $(\psi_T)_\infty = (Ts_0)_\infty = 0$, and $(\psi_M)_\infty = (pv_0)_\infty = 0$. Hence, the work to be done to change the three generalized coordinates, starting from a current point and ending at an *infinitely distant* point, will have the form

$$I_{m_{A,\infty}} = \int_{r_{mA}}^{r_{m\infty}} \frac{d}{dr_m}(-\Psi_m)dr_m = F_m \int_{r_{mA}}^{r_{m\infty}} dr_m = F_m(r_{m_\infty} - r_{m_A}) = F_m \Delta r_m, \quad (m = E,T,M).$$

It will be assessed via the following expressions:

$$I_{E_{A,\infty}} = \int_A^\infty V de_0 = -(V.e_0)_A, \qquad J\text{-for electrical work;}$$

$$I_{T_{A,\infty}} = \int_A^\infty T ds_0 = -(T.s_0)_A, \qquad J\text{-for thermal work;}$$

$$I_{M_{A,\infty}} = \int_A^\infty p.dv_0 = -(p.v_0)_A, \qquad J\text{-for mechanical work.}$$

Since the work in these potential fields is done over finite distances or finite limit states (between two close Points A and B of state 1 and state 2) and depends on the *displacement trajectory* and on the potential of the respective fields at Point A ($\psi_E = V\cdot e_0$, $\psi_T = T\cdot s_0$, or $\psi_M = p\cdot v_0$)A and at Point B ($\psi_E = V\cdot e_0$, $\psi_T = T\cdot s_0$, or $\psi_M = p\cdot v_0$)B, we use the following relations:

$$I_{E_{A,B}} = I_{E_{A,\infty}} - I_{E_{B,\infty}} = \int_A^\infty V\cdot de_0 - \int_B^\infty V\cdot de_0$$

$$= (V\cdot e_0)_B - (V\cdot e_0)_A, \quad J \text{ for electrical work;} \tag{3.1}$$

$$I_{T_{A,B}} = I_{T_{A,\infty}} - I_{T_{B,\infty}} = \int_A^\infty T ds_0 - \int_B^\infty T ds_0$$

$$= (T\cdot s_0)_B - (T\cdot s_0)_A, \quad J \text{ for thermal work;} \tag{3.2}$$

$$I_{M_{A,B}} = I_{M_{A,\infty}} - I_{M_{B,\infty}} = \int_A^\infty p\cdot dv_0 - \int_B^\infty p\cdot dv_0$$

$$= (p\cdot v_0)_B - (p\cdot v_0)_A, \quad J \text{ for mechanical work.} \tag{3.3}$$

Accounting for the physical result of each monofunctional system and the infinitesimally small variations of the generalized coordinates for $\Delta\vec{r}_m \to 0$ and ($\Delta e_0 \to de_0$; $\Delta s_0 \to ds_0$; and $\Delta v_0 \to dv_0$), the differential form of the work done to change e_0, C; s_0, J/°K; and v_0, m³/kg will be

$$\delta I_m \Rightarrow -d\Psi_m = -\frac{d\psi_m}{dr_m} dr_m = |F_m| dr_m \quad (m = E, T, M), \tag{3.4}$$

$$\delta I_E \Rightarrow -d\Psi_E = -\frac{d\psi_E}{de_0} de_0 = |F_E| de_0 = V de_0 \quad (m = E), \tag{3.5}$$

$$\delta I_T \Rightarrow -d\Psi_T = -\frac{d\psi_T}{ds_0} ds_0 = |F_T| ds_0 = t ds_0 \quad (m = T), \tag{3.6}$$

and

$$\delta I_M \Rightarrow -d\Psi_M = -\frac{d\psi_M}{dv_0} dv_0 = |F_M| dv_0 = p dv_0 \quad (m = M). \tag{3.7}$$

Notations are as follows:

- δl_E, δl_T, and δl_M: Infinitesimally small electrical, thermal, and mechanical works and total differentials of the respective potential fields
- $-\frac{d\psi_E}{de_0} = |F_E|$, $-\frac{d\psi_T}{ds_0} = |F_T|$, and $-\frac{d\psi_M}{dv_0} = |F_M|$: Differentials of the potential functions of the electric and thermal fields and the generalized forces, doing work within the respective fields

Table 3.1 synthesizes the types of specific energy impacts on a monofunctional *ETS*: generalized forces (F_m), resulting effect (change of the generalized coordinates $-r_m$), and quantitative estimation of the impact power (or the elementary generalized work done within the *ETS*, $\delta l_m = F_m \cdot dr_m$). The generalized work done in a system is assumed positive if it opposes the external forces.

The analysis of the available literature (Sections 3.2.1 and 3.2.2) shows that *three types of work* (electrical, thermal, and mechanical) *are simultaneously done* within the components of the building envelope. They are interpreted as external energy impacts or impacts introduced in the building system. Depending on the specific problem, the energy-exchange models employed herein comprise two or three interactions (or degrees of freedom) between the system and the surroundings. One can only get correct quantitative estimations of the work done in the components of the building envelope by integrally approaching the building. The latter should be treated as an autonomous energy-exchange *three-functional* medium of open type. In what follows, it will be referred to as *building electrothermodynamic system (ETS) and envelope*. Hence, the description of real processes running in the building envelope at macrolevel and reflecting the basic interior comfort (thermal and luminous) should employ *physical models* of a trifunctional *ETS*. They should be classified with respect to the energy interactions running between the *ETS* and the surroundings, which

TABLE 3.1

Types of Energy Impact Exerted by the Surroundings

Energy Impact	Generalized Force $F_{m(m=M,E,T)}$	Generalized Coordinate $r_{m(m=M,E,T)}$	Field Potential	Small Energy Impact (Elementary Work, Monofunctional System) $\delta l_m = F_m \cdot dr_{m(m=M,E,T)}$
Mechanical	Pressure: p	Volume: v_0	$\psi_M = p \cdot v_0$	Mechanical work: $\delta l_M = p dv_0$
Electrical	Electric potential: V	Electric charge: e_0	$\psi_E = V \cdot e_0$	Electrical work: $\delta l_E = V de_0$
Thermal	Temperature: T	Entropy: s_0	$\psi_T = T \cdot s_0$	Thermal work: $\delta l_T = T ds_0$

specify the admissible ranges of comfort parameters. We discuss here the characteristics of solar radiation (considering direct and diffusive flux) or wind impacts resulting from global solar activity.

Heat and moisture transfer takes place in both directions—from and to the building interior. The following conceptual theoretical questions arise in the development of calculation design of building facades:

- What is the relation between the different types of work?
- What are the factors controlling that relation?
- How can we select a criterion concerning structural materials in order to apply appropriate strategies in organizing the envelope energy-related functions?

These questions will be answered by designing an appropriate energy-exchange model and applying the *law of energy conservation* (the so-called first law of thermodynamics) to the considered system of the building envelope. To study the processes running in a building, we choose a *type* of an energy-exchange model regarding the particular case under consideration. For instance, it seems appropriate to choose a model accounting for *two types* of interactions—thermal and mechanical (Димитров 2006)—in assessing heat and moisture transfer through solid structural elements. A bifunctional system (thermal and electrical) can also do the job in analyzing processes running in integrated photovoltaic cells. A model incorporating *three types of energy impacts* (thermal, mechanical, and electrical) works in the investigation of the so-called hybrid illumination systems, integrated with building envelopes and comprise components of a natural illumination system. The employment of this approach is based on the theory of potential field and respective classical mathematical instrumentation, with capabilities and advantages that will be discussed further on. In what follows, we shall ground the necessity of introducing the term *gross potential* of the studied three-functional system. We shall also provide its mathematical interpretation, considering the system state parameters: temperature, pressure, stress, entropy, enthalpy, and free energy function.*

* Massieu was the first to introduce the term (1869) (see Guggenheim 1957), which is why it was known as Massieu function. Eight years later (1867), Gibbs introduced the terms *free energy function* ψ and *Gibbs function* ξ. These terms were later used by Zemansky (1998) and Munster (1956). They have since been used by Fast (1962), Spanner (1964), Jocky (1965), Saad (1966), Morse (1969), Silver (1971), Sychev and Shier (1978), Ziegler (1983), Wark (1983), Martin (1986), Lawden (1986), Stöcker (1996), Russell and Adebeliyi (1993), Lee (2002), Димитров (2006), and Kondeppud (2008).

3.2 Work Done in the Building Envelope and Energy-Exchange Models

We assume that each building envelope is an open system specified by control volume and surfaces. The control volume in its turn is specified by the building stereometry, while the control surfaces coincide with the building envelope (external and internal walls, roof, and floor). The *effective bounding surface* A_0 *of the building* is assumed to represent the control surface, while the *building volume* v_0 is assumed to represent the control volume. In contrast to the monofunctional systems discussed in Section 3.3.1, the envelope components undergo three types of impacts originating from the surroundings.

Moreover, three types of work are done in the trifunctional power field, with *potential* that has the following structural form:

$$\Psi = \Psi(F_m \cdot r_m) = \Psi(-F_E \cdot r_E, -F_T \cdot r_T, -F_M \cdot r_M), \quad (m = E, T, M).$$

We call this function *free energy potential of the* ETS in what follows, and it has a total differential $d\Psi$:

$$d\Psi = \frac{\partial \Psi}{\partial r_E} dr_E + \frac{\partial \Psi}{\partial r_T} dr_T \frac{\partial \Psi}{\partial r_M} dr_M, \tag{3.8}$$

which is the sum of three elementary works of the type

$$\sum_E^M \delta I_m = \sum_E^M F_m dr_m = -\sum_E^M \left(-\frac{\partial \Psi}{\partial r_m} \right) dr_m, \quad (m = E, T, M). \tag{3.9}$$

Substitution of the partial derivatives in Equation 3.8 for expression (Equation 3.9) and integration within appropriate limits of the coordinate $r_{m_1} \le r_m \le r_{m_2 (m=E,T,E)}$ yields the following generalized form of the integral:

$$\Psi = \int_1^2 d\Psi = -\left(\int_{r_{E1}}^{r_{E2}} F_m dr_E + \int_{r_{T1}}^{r_{T2}} F_T dr_T + \int_{r_{E1}}^{r_{E1}} F_M dr_M \right). \tag{3.10}$$

We give its physical interpretation in Section 3.2.1. Here, F_m, r_m, and $F_m \cdot r_m$ are generalized forces, generalized coordinates, and generalized potentials characterizing the *ETS* of the building envelope. As known from the general theory of potential fields (Бъчваров 2006), these quantities are also partial derivatives of the functional $\Psi = \Psi(F_m \cdot r_m)$ with respect to the generalized displacements (coordinates, $r_{m(m = E,T,M)}$) of the *ETS*. We employ similar characteristics of the physical fields as generalized forces F_m, which emerge as a result of various interactions between the surroundings and the envelope components. These are pressure *p*, electric potential *V*, and temperature *T*, which are already used in Subsection 3.3.1 to describe monofunctional systems (see Table 3.1). The larger the difference between outside and inside generalized forces, the more intensive is the interaction between the surroundings and the envelope and the larger the work (energy exchange) done.

3.2.1 Law of Conservation of the Energy Interactions between the Envelope Components and the Building Surroundings

It is known from physics that energy is not *produced* and does not *vanish*, but only changes its forms at macrolevel. An illustration of this statement is the mechanical displacement of macrobodies, when either entropy increases within the system or motion of electric charges takes place or both effects run simultaneously. Note that the amount of total energy (*TE*) remains constant within an *isolated system* (*TE* = const). In other words, *the TE remains constant* regardless of the various forms of energy exchange running within a closed system. Mathematically, this general physical law is written in terms of *small* energy variations:

$$\Delta TE = \Delta E + \Delta U + \Delta PE = 0. \tag{3.11}$$

The quantities participating in Equation 3.11 represent the variation of the kinetic energy of mechanical motion *E*, the potential energy *PE*, and the internal kinetic energy *U* of the system.

For *nonisolated systems*, the law of energy conservation takes the form

$$\Delta TE = \Delta E + \Delta U + \Delta PE + \sum_{1}^{n} \Delta \overline{W} = 0. \tag{3.12}$$

The energy interactions $\sum_{1}^{n} \Delta \overline{W}_i$ between a nonisolated system, comprising other systems and bodies, and the surroundings are in the form of thermal impacts ($\Delta L_T = \Delta Q = T \cdot \Delta S_0$), mechanical work ($\Delta L_M = p\Delta v_0$), and electrical ($\Delta L_E = V \cdot \Delta e_0$) work or others. Formally, these interactions can be written as (Hopkinson 1913)

$$\sum_{1}^{n} \Delta \overline{W} = (\Delta L_M + \Delta L_E - \Delta Q),$$

while expression (Equation 3.12) is transformed into the following equation of energy balance within the building envelope, written in finite difference form Wark (1983):

$$\Delta Q - \Delta L_E - \Delta U - \Delta L_M = 0. \tag{3.13}$$

Its differential form is

$$dQ - dL_E - dU - dL_M = 0, \tag{3.14}$$

and the integral form of Equations 3.13 and 3.14 is as follows:

$$U_{1-2} - Q_{1-2} + L_{M_{1-2}} + L_{E_{1-2}} = U_{1-2} + L_{1-2}^{Gross} = \text{Const} = -\Omega_{Env}. \tag{3.15}$$

Here,

- Ω_{Env} is an integration constant called *gross potential* (Lee 2002)
- L_{1-2}^{Gross} is work done within the potential field by generalized force F_{Env} originating from the surroundings ($L_{1-2}^{Gross} = -Q_{1-2} + L_{M_{1-2}} + L_{E_{1-2}}$ are the final works done during system transition from state 1 to state 2 characterized by state parameters (V_1, T_1, p_1) and (V_2, T_2, p_2), respectively)
- U_{1-2} is the change of the system internal energy at the process initiation and conclusion
- In terms of the terminology of classical thermodynamics, the difference between the values of the *gross potential* Ω_{Env} (Equation 3.15) and the *internal energy* U_{1-2} is called *potential difference of the free energy* (Ψ_{1-2}) or *free energy function** of the envelope control volume:

$$-\Omega_{Eve} - U_{1-2} = L_{1-2}^{Gross} = -\Psi_{1-2}. \tag{3.16}$$

The potential difference Ψ_{1-2} is interpreted as *work of the field forces* $\left(L_{1-2}^{Gross}\right)$, while U_{1-2} is treated as a measure of the internal energy accumulated by the control volume. This is seen comparing the structure of the integral of the total differential of Ψ (Equation 3.10) and the structure of the integral form of the law of energy conservation (Equation 3.16).

To assess the characteristic features of transfer, we use the free energy potential difference Ψ_{1-2} *in a substantial form* ($\Psi = \Psi_{1-2}$) *called free energy function.* Yet *generalized work* L_{1-2}^{Gross} comprises various components. For instance, while some studies (Fermi 1938; Carrington 1944; Lee 2002) include only the thermal work in the expression of free energy, a study by Spanner (1964) adds also mechanical work to heat. Studies by Fast (1962) and Kelly (1973) consider the tribute of the electromagnetic work. Our survey of literature proves that most of the analytical studies account for the chemical work done (Saad 1966; Bailyn 2002), while one study (Sychev and Shier 1978) *only* incorporates *electrical work* in the potential function.

* The term *free energy* is used in the works of Helmhoitz (1882), Plank (1945), Zemansky (1998), and Munster (1956) and in some recent studies (Fast 1962; Spanner 1964; Jocky 1965; Saad 1966; Morse 1969; Silver 1971; Sychev and Shier 1978; Ziegler 1983; Wark 1983; Martin 1986; Lawden 1986; Stocker 1993; Russell and Adebeliyi 1993; Lee 2002)

Free energy potentials Ψ_i of points located on the physical boundaries of the envelope define the largest potential difference Ψ_{1-2} and the largest work, respectively, that such a generalized force would do. The maximal potential difference is defined by both the magnitude of the surroundings' energy charge and the degree of degeneration of the phonon system within the envelope. When the final state, point 2 of the *ETS*, specified by the value of the potential function of the free energy (Ψ_2) is at infinity, that is, $(r_{m_2} \to \infty)$, and condition $\lim(\Psi_2)_{r_{m2}\to\infty} = 0$ is fulfilled, the work done by the generalized force is equal to the value of the potential function (Ψ_1), assessed at the physical start of interaction between the *ETS* and the surroundings:

$$(\Psi_1) = \sum_{m=T,M,E} \sum_{r_1}^{r_2\to\infty} (L_m) = -Q_{1-\infty} + L_{M_{1-\infty}} + L_{E_{1-\infty}} = \text{Const} = \Psi_{Enter}.$$

Consider the potential function (Ψ_1) found using the *TE function* $Z_{Ph} = \Omega_{En}$ (see footnote 36) and the internal energy U_{En}. Hence, its value at the input control surface is considered to be a boundary (input $(\Psi_{En})_{\beta=\beta_{SUN}}$) condition, and that value is equal to the potential of the surroundings (Ψ_{Sur}), that is,

$$\Psi_1 = (\psi_{En})_{\beta=\beta_{SUN}} = \Omega_{En} - U_{En}.$$

The difference between the free energy potentials Ψ_{1-2} (or the *negative work done by the generalized force* L_{1-2}^{Gross} *of the field*) is used to assess the transfer properties of the physical fields, considering, for instance, mechanical interactions, volume change $\Delta_{rM} = \Delta v_0 = v_2 - v_1$ or shift of macrobodies at a distance $\Delta\vec{r}_M = \vec{r}_2 - \vec{r}_1$; electrical interactions, change of the electric charge $\Delta r_E = \Delta e_0 = e_2 - e_1$; thermal interactions, change of entropy $\Delta r_T = \Delta s_0 = s_2 - s_1$; etc. It is traditionally assumed that *the larger the difference between the free energy potentials* Ψ_{1-2}, *the more intensive is the envelope–surroundings interaction and the more intensive is the energy exchange (phonon degeneration).* That difference is assumed to be positive if the local Lagrange multiplier of the entrance surface of the control volume β_{Enter} is larger than the local value β_{Exit} of the exit surface. The working capability of the primary energy carriers—photons—decreases in that process (the exchanged heat is positive if the direction of its flux is from the surroundings to the system, with increases in entropy).

Reducing Equations 3.13 and 3.14 to unit mass of the system,

$$\delta q = \delta l_e - du - \delta l_M = 0, \tag{3.17}$$

and using infinitesimally small increments of the internal energy du, heat δq, electrical work δl_e, and mechanical work δl_M expressed by the variation of the state parameters of *ETS* ($du = C_v dT$, $\delta q = T \cdot ds$, $\delta l_M = pdv$, and $\delta l_e = Vde_0$), we find the parametric link between temperature T, electric potential V, and pressure p:

$$Tds_0 - Vde_0 - c_v dT - pdv = 0. \tag{3.18}$$

Equation 3.18 yields the following relations between the variations of the three transferred charges:

$$ds_0 = \frac{V}{T}de_0 + c_v\frac{dT}{T} + \frac{p}{T}dv = \frac{V}{T}de_0 + c_P\frac{dT}{T}; \quad de_0 = \frac{T}{V}ds_0 - c_P\frac{dT}{V},$$

and

$$dv = \frac{T}{p}ds_0 - \frac{V}{p}de_0 - c_V\frac{dT}{p}. \tag{3.19}$$

Using the mathematical expressions of the elementary works done to shift the elementary energy carriers and to perform integration, we find the following expressions of the specific works $I_{T_{1-2}}$, $I_{E_{1-2}}$, and $I_{M_{1-2}}$ done to realize shift from point 1 to point 2:

$$\begin{aligned}
&\delta I_T = -T \cdot ds_0 = -(Vde_0 + c_P dT) \Rightarrow I_{T_{1-2}} = (V_1 e_1 - V_2 e_2) - c_P(T_1 - T_2);\\
&\delta I_E = -V \cdot de_0 = -\lfloor (c_v - c_p)dT \rfloor \Rightarrow I_{E_{1-2}} = (c_v - c_p)(T_1 - T_2);\\
&\delta I_M = -p \cdot dv = -(Tds_0 - c_v dT - Vde_0);\\
&\quad \Rightarrow I_{M_{1-2}} = (s_1 T_1 - s_2 T_2) - c_v(T_1 - T_2) - (V_1 e_1 - V_2 e_2).
\end{aligned}$$

The integration of relations (Equation 3.19), considering specified boundary values of the ETS state parameters at point 1 and point 2, yields expressions of entropy, electric charge, or specific volume:

$$S_2 = S_1 + \frac{(V_2 e_2 - V_1 e_1)}{(T_2 - T_1)} + C_V \ln\frac{T_2}{T_1} + \frac{(p_2 v_2 - p_1 v_1)}{(T_2 - T_1)};$$

$$\left(\text{or} \quad S_2 = S_1 + \frac{(V_2 e_2 - V_1 e_1)}{(T_2 - T_1)} + C_P \ln\frac{T_2}{T_1}\right),$$

$$e_2 = e_1 + \frac{(q_2 - q_1)}{(V_2 - V_1)} - C_P\frac{(T_2 - T_1)}{(V_2 - V_1)} \quad \text{and}$$

$$V_2 = V_1 + \frac{(q_2 - q_1)}{(p_2 - p_1)} - \frac{(V_2 e_2 - V_1 e_1)}{(p_2 - p_1)} - C_V\frac{(T_2 - T_1)}{(p_2 - p_1)}.$$

Together with the processes running in a trifunctional ETS and related to the simultaneous transfer of entropy (heat), electric charge (electric current), and mass (infiltration/exfiltration), special energy interactions occur between the surroundings and the envelope components. They are analyzed and described as follows.

3.2.2 Special Cases of Energy Interactions

It is often found that *only two types* of energy interactions are dominate in some cases of energy transfer in solid walls (exchange of mass and heat [entropy] and exchange of heat and electric charges). Yet in other cases, one can observe *only one type* of energy interactions. Then, the law of energy conservation adopts a traditional three- or two-component form, accounting for the variation of system internal energy du and other small increments, including those of entropy, electric charge, and mass.

3.2.2.1 Energy Model of Transfer of Entropy and Electric Charges

The building facade serving as a physical medium, which absorbs the solar energy flux, can also be interpreted as a medium where *electrical and thermal* work is done (bifunctional medium) without mass transfer ($dv \cong 0$ and $dl_M = 0$). Hence, Equation 3.17 takes the form

$$dq - dl_e - du = 0,$$

or

$$T \cdot ds - Vde_0 - C_V dT = 0. \tag{3.20}$$

Equation 3.19 yields the following relation:

$$(c_p - c_v)dT + Vde_0 = 0, \tag{3.21}$$

for such a bifunctional *ETS*, if one uses the link between the variations of the thermal work dq and enthalpy dh and the temperature variation dT, known from thermodynamics. Rewriting relation (Equation 3.21), we find an expression of the electric charge variation (de_0):

$$de_0 = \frac{(c_v - c_p)}{V} dT = -\frac{R}{V} dT, \tag{3.22}$$

since $(c_p - c_v) = R$ (Meyer equation).

The transformation of Equation 3.19 results in entropy variation:

$$ds_0 = \frac{C_P}{T}dT + \frac{V}{T}de_0. \tag{3.23}$$

The analysis of Equation 3.23 proves that the variation of system entropy s_0 consists of two components. The first one is related to the variation of medium temperature $-(C_P/T)dT$. The second component ($(V/T)de_0$) reflects the variation of the electric charge in the system.

Once found, the expressions of *ETS* electric charge and entropy variation (Equations 3.22) and (3.23) can be used to assess the electrical and thermal work done in the *ETS*. Putting them in Equations 3.5 and 3.6, we get

$$\delta I_E = -V \cdot \frac{(c_v - c_p)}{V}dT = -(c_v - c_p)dT; \tag{3.24}$$

$$\delta I_T = -T\left(\frac{C_P}{T}dT + \frac{V}{T}de_0\right) = -C_P dT + V de_0. \tag{3.25}$$

The integration of Equation 3.25, accounting for the two limit states of the system characterized by fixed values of temperature T and electric charge e, T_1, e_1, s_1, and T_2, e_2, produces the entropy s_2 at the second limit state:

$$S_2 = S_1 + C_V \quad \ln\frac{T_1}{T_2} + \frac{(V_1e_1 - V_2e_2)}{(T_1 - T_2)}. \tag{3.26}$$

In contrast to a purely thermodynamic system (*ETS*), entropy increases here from s_1 to s_2 owing to the electrical work normalized via the temperature difference, as seen in Equation 3.26 (the third term). Consider variation of the system state parameters from state 1 (T_1,V_1,e_1) to state 2 (T_2,V_2,e_2). Then, Equations 3.24 and 3.25 yield the following expressions of the final work done in the system:

$$I_{E_{1-2}} = (c_v - c_p)(T_2 - T_1) = R(T_1 - T_2); \tag{3.27}$$

$$I_{T_{1-2}} = c_p(T_1 - T_2) + (V_1e_1 - V_2e_2). \tag{3.28}$$

The interpretation of these two expressions leads to the following conclusions concerning the work done in a bifunctional *ETS*:

- The electrical work can be estimated via the temperature variation only. It is the difference between the variations of system internal energy u_{1-2} and enthalpy $\bar{h}_{1-2}$, since $I_{E_{1-2}} = u_{1-2} - \bar{h}_{1-2}$ (here, $u_{1-2} = c_v$ $(T_1 - T_2)$ and $\bar{h}_{1-2} = c_p(T_1 - T_2)$).

- Hence, it follows that the variation of the total internal energy of the *ETS* (enthalpy) $\bar{h}_{1-2}$ takes place at the expense of the variation of internal energy u_{1-2} and electrical work $I_{E_{1-2}}$:

$$\bar{h}_{1-2} = u_{1-2} - I_{E_{1-2}}.$$

- The thermal work $I_{T_{1-2}}$ is the sum of enthalpy $\bar{h}_{1-2}$ and electrical work $I_{E_{1-2}}$ (product of the electric potential drop $\Delta V = V_1 - V_2$ and the variation of the electric charge $\Delta e = e_1 - e_2$; see Equation 3.1):

$$I_{T_{1-2}} = \bar{h}_{1-2} + I_{E_{1-2}}.$$

Comparing expressions $I_{E_{1-2}} = u_{1-2} - h_{1-2}$ and $I_{E_{1-2}} = I_{T_{1-2}} - h_{1-2}$, found after the transformation of Equation 3.27, we may conclude that the variation of the internal energy $u_1 \to u_2$ occurs at the expense of the thermal work done over the *ETS* (the introduced heat $I_{T_{1-2}}$):

$$u_{1-2} = I_{T_{1-2}}.$$

This means that the variation of the internal temperature and the local Lagrange factor (β_{Loc}), respectively, takes place at the expense of the thermal work introduced in the system through its control surface.

3.2.2.2 Energy Model of Entropy Transfer with or without Mass Transfer

Energy transfer through the walls of the wall components accompanied or not by mass transfer (infiltration–exfiltration) is a basic mechanism that runs in various building and building elements. In case of heat transfer via *conductivity and convection* (mass transfer of mechanical work δl_M), the law of energy conservation attains a form known from thermodynamics (the first law of thermodynamics):

$$\delta q = du + \delta I_M. \tag{3.29}$$

Putting the parametric links in expression (Equation 3.29), we obtain the following equations:

$$Tds_0 = c_V dT + pdv = c_P dT - vdp, \tag{3.30}$$

$$Tds_0 = c_V dT + wdw = d\bar{h} - wdw. \tag{3.31}$$

Here, $wdw = d(w^2)/2 = -vdp$ is the variation of the kinetic energy of the mass flux $0.5\, w^2$, assessed under ideal conditions by the Bernoulli differential. The measure of system deenergization is found from relation (Equations 3.30 or 3.31):

$$ds_0 = c_v \frac{dt}{T} - w\frac{dw}{T} \quad \text{or} \quad ds_0 = c_p \frac{dT}{T} + \frac{v}{T} dp. \tag{3.32}$$

Integrating Equation 3.32 regarding two limit states of the system having fixed temperature T, pressure p, specific volume v, and velocity w of the convective flux—T_1, p_1, v_1, w_1, s_1 and T_2, p_2, v_2, w_2—we find the entropy value at the second limit state s_2:

$$s_2 = s_1 + c_v \ln \frac{T_1}{T_2} - \frac{\left(w_1^2 - w_2^2\right)}{2(T_1 - T_2)} \quad \text{or} \quad s_2 = s_1 + c_v \ \ln \frac{T_1}{T_2} - \frac{\left(p_1 v_1 - p_2 v_2\right)}{(T_1 - T_2)}. \tag{3.33}$$

As seen in Equation 3.33 and in contrast to a pure *ETS*, entropy increases here from s_1 to s_2 due not only to the temperature difference but also to fluid mass motion within the system control boundaries (the third term of Equation 3.33).

Equations 3.17, 3.20, 3.30, and 3.31 provide the links between the variations of the system internal energy, thermal work, and electrical work and the corresponding variations of the system state parameters. Yet we must note that the relation between the *electrical work and the thermal work*, both done within the building envelope components under identical conditions, will depend (1) on the physical properties of the building materials (on the energy barrier of the banned area) and (2) on the order of material deposition following the direction of energy transfer within the envelope. However, some general considerations concerning this matter will be set forward in Section 7.8.

In conclusion, we can say that the discussed model of energy interactions between the components of the building envelope and the environment (Equation 3.17) is a special case of the general law of energy conservation regarding a *trifunctional physical medium*. We will use this model in the solution of engineering problems, tackling the energy design of building facades.

3.3 Specification of the Structure of the Free Energy in the Components of the Building Envelope (Electrothermodynamic Potential of the System)

According to the definition, state parameters are quantities that uniquely characterize the state of an *ETS* and do not depend on its prehistory. Temperature (T), electric potential (V), and pressure (p) are assumed to be the three independent state parameters of an *ETS*. Their choice is based on the circumstances that they have clear physics and that they can be measured using simple measuring instruments. Processes running in an *open ETS* as is the building envelope urge the *ETS* to equilibrium under conditions imposed by energy reservoirs (Lee 2002)* located at the boundaries of the *ETS* control surface.

Transfer fluxes are directed from areas with larger generalized forces (temperature T, (resp $\cdot \beta\downarrow$), electric potential V, and pressure p) to areas with smaller ones. This is in accordance with L. Boltzmann's statement that *nature tends to transition from less probable to more probable states* (Dimitrov 2014).

Studying various energy interactions between an *ETS* and the surroundings, it seems useful to look for a generalized representative state parameter of the *ETS*.

Potential field mathematics proves the convenience of assuming a total differential of the parameter. Since temperature T, potential V, and pressure p are independent state parameters of the studied *ETS*, the generalized parameter (function of the system state) will be functionally bound up with those quantities. The problem of its definition reads as follows: assume that T, p, and V are three independent state parameters and the *free energy function* $\psi = f(T, p, V)$ is a dependent parameter (state function). The total differential of the *free energy functional* reads as

$$d\psi = \left(\partial\psi/\partial s_0\right)_{V,p} \cdot ds_0 + \left(\partial\psi/\partial v_0\right)_{T,V} \cdot dv_0 + \left(\partial\psi/\partial e_0\right)_{T,p} \cdot de_0 \tag{3.34}$$

Put

- $M = (\partial\psi / \partial s_0)_{V,p}$, partial derivative of $\psi = f(T, p, V)$ with respect to entropy (s_0), generalized coordinate of the thermal work at V = const and p = const
- $N = (\partial\psi / \partial v_0)_{T,V}$, partial derivative of $\psi = f(T, p, V)$ with respect to volume (v_0), generalized coordinate of the mechanical work at T = const and V = const
- $P = (\partial\psi / \partial e_0)_{T,p}$, partial derivative of $\psi = f(T, p, V)$ with respect to the electric charge (e_0), generalized coordinate of the electrical work at T = const and p = const

Then, Equation 3.34 will read as $d\psi = M \cdot ds_0 + N \cdot dv_0 + P \cdot de_0$.

Requirement $d\psi = 0$ is satisfied by the equipotential surfaces of the free energy $[\psi = f(T, p, V) = \text{const}]$, and their equations will have the form

$$M \cdot ds_0 + N \cdot dv_0 + P \cdot de_0 = 0, \tag{3.35}$$

* The second law of thermodynamics for open and nonisolated systems requires the *free energy potential to tend to minimum*.

or

$$\frac{ds_0}{N \cdot P} + \frac{dv_0}{M \cdot P} + \frac{de_0}{M \cdot N} = 0.$$

Using Equation 3.35, the increments of the respective parameters can be written as

$$ds_0 = -\left(\frac{N}{M}dv_0 + \frac{P}{M}de_0\right) \text{ and } dv_0 = -\left(\frac{M}{N}ds_0 + \frac{P}{M}de_0\right) \tag{3.36}$$

The following substitutions hold for differentials dv_0, de_0, and ds_0:

$$\frac{dv_0}{de_0} = q \, and \, \frac{dv_0}{de_0} = n \tag{3.37}$$

Putting Equation 3.36 and Equation 3.37 in Equation 3.35, we get

$$Nq + Mn + P = 0 \quad \text{or} \quad \frac{Nq + Mn}{P} = -1 \tag{3.38}$$

To satisfy the requirement for the existence of a total differential of the free energy $\psi = f(V, T, p)$, it is necessary that

$$\frac{\left(\frac{\partial \psi}{\partial v_0}\right)_{T,V}\left(\frac{\partial v_0}{\partial e_0}\right)_T + \left(\frac{\partial \psi}{\partial s_0}\right)_{V,p}\left(\frac{\partial s_0}{\partial e_0}\right)_p}{\left(\frac{\partial \psi}{\partial e_0}\right)_{T,p}} = -1 \tag{3.39}$$

Expression 3.39 sets the link between the three gradients at a point of the 4D surface, and it is called differential equation of the state of the *ETS* work medium.

Regarding the special case of a bifunctional system, Equation 3.39 gets a structure similar to that of Maxwell's equation. For instance, for a system subjected to thermal and electric impacts, it will have the form

$$\frac{M}{P}\frac{ds_0}{de_0} = -1 \quad \text{or} \quad \frac{de_0}{\left(\frac{\partial \psi}{\partial s_0}\right)_{V,p}} + \frac{ds_0}{\left(\frac{\partial \psi}{\partial V_0}\right)_{T,p}} = 0. \tag{3.40}$$

Yet for a system subjected to electric and mechanical impacts, only, Equation 3.39 will be modified as

$$\frac{N}{P}\frac{dv_0}{de_0} = -1 \quad \text{or} \quad \frac{de_0}{\left(\dfrac{\partial \psi}{\partial v_0}\right)_{V,T}} + \frac{dv_0}{\left(\dfrac{\partial \psi}{\partial e_0}\right)_{T,p}} = 0. \tag{3.41}$$

Equations 3.40 and 3.41 are differential equations of the two types of bifunctional system. It is worth studying them in finer detail after having defined the free energy functional $\psi = \psi(F_m \cdot r_m)_{m=T,M,E}$. The differential equations (Equation 3.41)—the last bifunctional *ETS* (undergoing thermal and mechanical interactions only)—can be derived assuming that $P = \left(\dfrac{\partial \psi}{\partial e_0}\right)_{T,p} = 0$ and rewriting Equation 3.38:

$$\left(\frac{\partial \psi}{\partial v_0}\right)_{V,T} dv_0 + \left(\frac{\partial \psi}{\partial s_0}\right)_{V,p} ds_0 = 0.$$

The conditions of *ETS* potential existence in these two cases are transformed into second-order differential equations of the free energy function $\psi = \psi(F_m \cdot r_m)_{m=T,M,E}$:

$$\frac{\partial \psi^2}{\partial s_0 \partial e_0} = \frac{\partial \psi^2}{\partial e_0 \partial s_0} \quad \text{and respectively} \quad \frac{\partial \psi^2}{\partial v_0 \partial e_0} = \frac{\partial \psi^2}{\partial e_0 \partial v_0} \tag{3.42}$$

3.3.1 Finding the Structure of the Free Energy Function

We propose here the following operational form of the structure of the analytical function of the *free energy* ψ, which will be used to characterize the *ETS* state:

$$\psi_{ETS} = p.v_0 - T \cdot s_0 + V \cdot e_0 \tag{3.43}$$

Six state parameters of the system participate in this relation: T, temperature; p, pressure; v, specific volume; s_0, entropy; V, electric potential; and e_0, charge of the *ETS*.* The analytical function (Equation 3.43) is representative for the *ETS* state, and it can be used instead of all six particular parameters. That form of expressing the characteristics of the free energy called *function of the ETS free energy* ψ displays properties of a total differential:

$$d\psi = \nabla\psi = d(p \cdot v - T \cdot s_0 + V \cdot e_0), \tag{3.44}$$

since it meets the *potentiality requirements*. Then, the estimation of the work done does not depend on the law of charge motion (i.e., on charge trajectories), but only on charge limit generalized coordinates.

The partial derivatives of the free energy function ψ (Equation 3.43) with respect to the *ETS* generalized coordinates are of practical interest. They are needed to assess the generalized or partial work done in the generalized force field. These partial derivatives are employed to estimate the variation of the system volume v_0, electric charge e_0, and entropy s_0 (see Section 5.2):

$$N = (\nabla\psi)_V = \left(\frac{\partial\psi}{\partial V_0}\right)_{T,V} = p, \quad P = (\nabla\psi)_{e0} = \left(\frac{\partial\psi}{\partial e_0}\right)_{T,p} = V, \tag{3.45}$$

and

$$M = (\nabla\psi)_{s0} = \left(\frac{\partial\psi}{\partial s_0}\right)_{p,V} = -T.$$

* We also found the other forms of the *free energy function* are possible:

- Those forms show the meaning of the term free energy, using *Helmholtz function* $H = u - Ts_0$ (Fermi 1938; Carrington 1944; Parker 1950; Sears 1950; Von Ness 1964; Morse 1969; Andrews 1971; Silver 1971; Martin 1986; Russell and Adebeliyi 1993; Stöcker 1996; Cheng 2006; Kondeppud 2008); the suggested *form 1* $\rightarrow \psi''_{ETS} = H - u + p.v_0 + V.e_0$.
- Using *Gibbs function* $G = h - Ts_0$ (Fast 1962; Spanner 1964; Kelly 1973; Zeigler 1983; Lawden 1986; Martin 1986; Russell and Adebeliyi 1993; Cheng 2006; Kondeppud 2008); the suggested *form 2* $\rightarrow \psi^{III}_{ETC} = G - h + pv_0 + V \cdot e_0$; here, h is enthalpy $\rightarrow$ ($h = u + p \cdot v_0$).
- Using *Fermi function* $\Phi = A + p \cdot v_0$ (Fermi 1938; Guggenheim 1957; Morse 1969; Wark 1983; Stocker 1988; Bailyn 2002); the suggested *form 3* $\rightarrow \psi^{IV}_{ETC} = \Phi + u + V \cdot e_0$ *energy function* is variative.

They also illustrate *the difference* between quantities ψ; H; G; and Φ.

Putting Equation 3.45 in Equation 3.39, we get

$$\frac{p\left(\dfrac{dv_0}{de_0}\right)_T + (-T)\left(\dfrac{ds_0}{de_0}\right)_p}{V} = -1,$$

and it follows that $d\psi = pdv_0 - Tds_0 + Vde_0 = 0$. This however is the equation of the equipotential surface of the free energy (ψ = const) with integral that is similar to Equation 3.43.

Systems performing *electric and thermal interactions*, only, are described by means of the equation of the free energy equipotential surface:

$$\frac{M}{P}\frac{ds_0}{de_0} = -1 \quad \text{or} \quad Vde_0 - Tds_0 = 0,$$

while the potential function of the free energy takes the form

$$\psi_{E,T} = -Ts_0 + Ve_0 = \text{const.}$$

For systems performing *electric and mechanical interactions*, only, the type of Equation 3.38 will be modified as

$$\frac{N}{P}\frac{dv_0}{de_0} = -1 \quad \text{or} \quad Vde_0 + pdv_0 = 0,$$

and we find after integration that the equation of the potential function of the free energy is $\psi_{E,M} = -Ve_0 + pv_0 = \text{const}$.

We can derive the differential equations of the last bifunctional system (the one *thermally and mechanically interacting with the surroundings or the* ETS) by rewriting Equation 3.38 and assuming that $P = 0$:

$$Nq + Mn = 0 \rightarrow d\psi = pdv_0 - Tds_0 = 0,$$

whereas we find the function of the potential of that type of bifunctional system as

$$\psi_{M,T} = \int pdv_0 - \int Tds_0 + \text{Const} = pv_0 - Ts_0.$$

It is known as Gibbs potential or Gibbs formula.

The condition of potentiality existence, Equation 3.42 and expressions (Equation 3.45), yields the following relations, which may prove useful:

$$\frac{\partial T}{\partial e_0} - \frac{\partial V}{\partial s_0} = 0 \quad \text{and} \quad \frac{\partial p}{\partial e_0} - \frac{\partial V}{\partial v_0} = 0.$$

The *ETS free energy functional* $\psi = f(T, VИp)$ specified in the present paragraph via its operational form, Equation 3.43, can be taken as a *generalized characteristic of the system state* instead of temperature T, electric potential V, and pressure p. That consideration is accounted for in Section 5.2 in order to find a generalized parameter participating in the *laws of transfer*.*

3.3.1.1 Links between Entropy and the System Basic Parameters

We discussed in Section 2.4.2.4 entropy as a macroscopic state parameter depending on volume v_0, Lagrange multiplier β_{Loc}, and Fermi energy level ε_F or barrier coefficient K_{FZ}, as well as its importance for the energy processes running in the control volume. As already established, entropy possesses a total differential:

$$dS_{Env} = \left(\frac{\partial S_{Env}}{\partial V_0}\right)_{T,N_{El}} dv_0 + \left(\frac{\partial S_{Env}}{\partial T}\right)_{v_0,N_{El}} dT + \left(\frac{\partial S_{Env}}{\partial N_{El}}\right)_{v_0,T} dN_{El},$$

and its distribution can be described via analytical functions.

To find the partial derivatives, one needs to transform and differentiate Equation 3.16:

$$-d\Omega_{Eve} = du - d\Psi = c_v dT - T ds_0 + p dv_0 + \overline{\varepsilon}_{Eff} dN_{El} = 0,$$

whereas entropy increase can be assessed via the formula

$$ds_0 = \frac{P}{T} dv_0 + \frac{c_v}{T} dT + \frac{\overline{\varepsilon}_{Eff}}{T} dN_{El}.$$

We use here the *following relation* between the electric potential and the Fermi energy level:

$$V de_0 = \frac{\overline{e}_{Eff}}{e_0} d(e^0 * N_{El}) = \overline{e}_{Eff} * dN_{El},$$

* The form of the free energy function found here differs from the so-called gross potential used in the treatment of chemical–physical heat–mass exchange processes, where the chemical potential μ_0 is involved and variation of the substance mass serves as a generalized coordinate (Smith 1952; Stocker 1993; Tassion 1993; Bailyn 2002; Kondeppud 2008): $G = p \cdot v - Ts_0 + \sum \mu_0 \cdot n$.

while the efficient electric potential is defined as

$$\bar{\varepsilon}_{Eff} = \frac{1}{N_{El}} \sum_{N_{El}} (E_i - \varepsilon_F).$$

A comparison between the two expressions of entropy increase shows that the partial derivatives of the first equation are equal to $\left(\frac{\partial s_{Env}}{\partial V_0}\right) T, N_{El} = \frac{p}{T}$, $\left(\frac{\partial S_{Env}}{\partial T}\right)_{v_0, N_{El}} = \frac{c_v}{T}$, and $\left(\frac{\partial S_{Env}}{\partial N_{El}}\right)_{v_0 t} = \frac{\bar{\varepsilon}_{Eff}}{T}$, respectively.

Then, the equation of entropy increase will take the following final form:

$$ds_{Env} = \frac{p}{T} dv_0 + \frac{c_v}{T} dt + \frac{\bar{\varepsilon}_{Eff}}{T} dN_{El}.$$

We propose here an alternative form to calculate entropy increase. The assumed definition of the free energy potential Ψ and its increase $d\Psi$ is our starting point, that is,

$$d\psi = -T ds_0 + p dv_0 + \bar{\varepsilon}_{Eff} dN_{El}.$$

Transforming the total differential with respect to ds_0 (changing the notations $ds_0 \Rightarrow ds_{Env}$), we find the following expression:

$$ds_{Env} = \frac{p}{T} dv_0 + \frac{\varepsilon_{Eff}}{T} dN_{El} - \frac{1}{T} d\psi,$$

which formalizes the physical considerations in Section 2.4 of entropy increase.

The novelty of the expression of ds_{Env} *consists in* the use of the variation of the free energy function $d\psi$, accounting for temperature T. Hence, we may conclude that entropy will be at its maximum when the *free energy of the control volume is minimized* ($-d\psi \Rightarrow 0$). *Moreover,* the processes will run in the control volume to *produce extremum of the partial works done.* Such will be the process development in the control volume when one type of work is done (only electrical or mechanical), as seen in the following expressions:

$$ds_{Env} = \frac{\varepsilon_{Eff}}{T} dN_{El} - \frac{1}{T} d\psi \quad \text{and} \quad dS_{Env} = \frac{p}{T} dv_0 - \frac{1}{T} d\psi.$$

It stands clear that knowing the distribution of the free energy function Ψ_{Env} in the control volume is as important as knowing the distribution of entropy s_{Env}. The physical difference between these two quantities consists in determining the direction of the processes running in the control area (their progress is such that entropy tends to a local maximum when *the free energy tends to a local minimum*).

3.4 Distribution of the Free Energy within the Building Envelope

We will show in this book that the calculation of the distribution of the free energy function (ψ_{Env}) in the control volume can be performed using a procedure that is significantly simpler than that finding the entropy distribution (s_{Env}). On the other hand, it is fully capable of defining the behavior of all other *ETS* state parameters of interest and the energy fluxes flowing in the entire control area.

Generally, the variation of the free energy function of the envelope system in the time–space continuum can be described by means of a continuous scalar function in Cartesian coordinates:

$$\psi = \psi_1(x, y, z, \tau), \tag{3.46}$$

$$\psi = \psi_2(r, \theta, z, \tau), \tag{3.47}$$

$$\psi = \psi_3(r, \theta, \lambda, \tau). \tag{3.48}$$

Figure 3.1 shows a plot of the free energy potential obtained using the isopotential surfaces ψ = const plotted within the three main coordinate systems. The normal vector $\vec{n}_\psi$, also shown there, specifies the direction of gradψ defined by the expression

$$\text{grad}\psi = \lim_{\Delta\vec{n}_\psi \to 0}\left(\Delta\psi/\Delta\vec{n}_\psi\right) = \frac{\partial\psi}{\partial\vec{n}}. \tag{3.49}$$

Similarly to grad T, grad V, and grad p, the gradient of ψ is a vector directed along $\vec{n}_\psi$. The assessment of diffusion quantitative characteristics, which will be treated in Section 4.2, directly needs finding the distribution of the vector

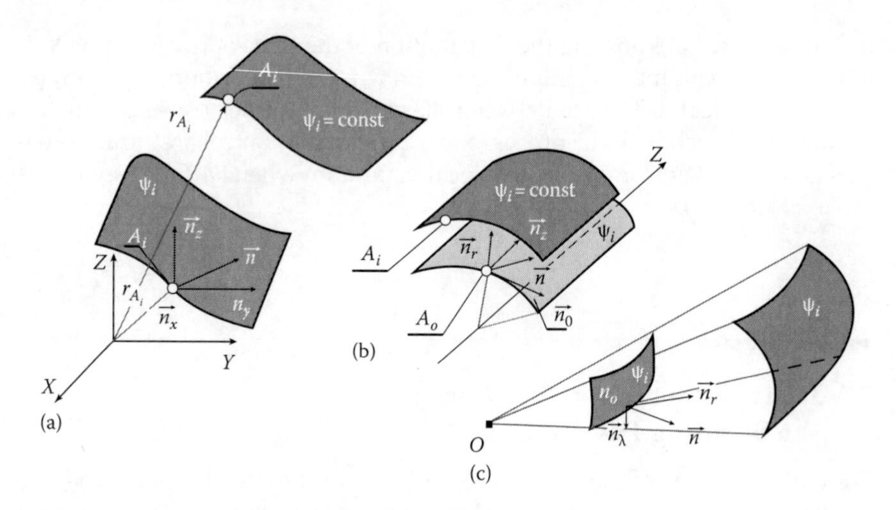

FIGURE 3.1
Visualization of the free energy potential field characteristics in (a) Cartesian, (b) cylindrical, and (c) spherical coordinate systems.

function gradψ = $F(x, y, z, \tau)$. Its form is different in different coordinate systems:

- In Cartesian coordinates,

$$\text{grad}\psi = \frac{\partial \psi}{\partial \vec{n}_\psi} = \nabla_D \psi = \vec{i} \cdot \frac{\partial}{\partial x}\left(p \cdot v - T \cdot s_0 + V \cdot e_0\right)$$
$$+ \vec{j} \cdot \frac{\partial}{\partial y}\left(p \cdot v - T \cdot s_0 + V \cdot e_0\right) + \vec{k} \cdot \frac{\partial}{\partial z}\left(p \cdot v - T \cdot s_0 + V \cdot e_0\right).$$

- In cylindrical coordinates,

$$\text{grad}\psi = \frac{\partial \psi}{\partial \vec{n}_\psi} = \nabla_C \psi = \vec{n}_{\psi_r} \cdot \frac{\partial}{\partial r}\left(p \cdot v - T \cdot s_0 + V \cdot e_0\right)$$
$$+ \frac{1}{r}\vec{n}_{\psi\theta} \cdot \frac{\partial}{\partial \theta}\left(p \cdot v - T \cdot s_0 + V \cdot e_0\right) + \vec{n}_{\psi_z} \cdot \frac{\partial}{\partial z}\left(p \cdot v - T \cdot s_0 + V \cdot e_0\right).$$

- In spherical coordinates,

$$\text{grad}\psi = \frac{\partial \psi}{\partial \vec{n}_\psi} = \nabla_{Sp} \psi = \vec{n}_{\psi_r} \cdot \frac{\partial}{\partial r}\left(p \cdot v - T \cdot s_0 + V \cdot e_0\right)$$
$$+ \frac{1}{r}\vec{n}_{\psi\theta} \cdot \frac{\partial}{\partial \theta}\left(p \cdot v - T \cdot s_0 + V \cdot e_0\right) + \frac{1}{r \cdot \sin\theta}\vec{n}_{\psi\lambda} \cdot \frac{\partial}{\partial \lambda}\left(p \cdot v - T \cdot s_0 + V \cdot e_0\right).$$

Obviously, there are some difficulties in handling the mathematical forms of the vector functions and in their analytical determination. Yet it is worth extracting the entire useful information out of the *ETS* free energy function.

3.4.1 State Parameters Subject to Determination via the Free Energy Function

As already noted, the free energy function is qualitatively conformable to the *ETS* entropy (s_{Env}). Yet in contrast to entropy that increases by itself until attaining local maximum (the second law of thermodynamics), the free energy function decreases in the control volume $\psi_{Env} = -TS_{Env} + pV_{Env} + \varepsilon_{Eff}N_{El}$ until attaining local equilibrium and minimum. Note the purely theoretical assumption that the variation of the free energy function $d\psi_{Env}$ specifies the direction of the thermodynamic process via the signs of the partial derivatives $\left(\frac{\partial \psi_{Env}}{\partial s_{Env}}\right)_{P,\bar{\varepsilon}}$, $\left(\frac{\partial \psi_{Env}}{\partial V_{Env}}\right)_{T,\bar{\varepsilon}}$, and $\left(\frac{\partial \psi_{Env}}{\partial N_{El}}\right)_{P,T}$. Yet their absolute values enable one to assess the basic macroscopic parameters of the system:

- Temperature: $T = -\left(\frac{\partial \psi_{Env}}{\partial s_{Env}}\right)_{P,\bar{\varepsilon}}$.
- Pressure: $p = \left(\frac{\partial \psi_{Env}}{\partial V_{Env}}\right)_{T,\bar{\varepsilon}}$, as well as $p = T\frac{ds_{Env}}{dv_0}$.
- Effective electric potential: $\bar{\varepsilon}_{Eff} = \left(\frac{\partial \psi_{Env}}{\partial N_{El}}\right)_{P,T}$ or $\bar{\varepsilon}_{Eff} = T\left(\frac{ds_{Env}}{dN_{El}}\right)$.

From a microscopic point of view, the assessment of the free energy amount can throw additional light on energy degeneration of the phonon system in particular areas of the building envelope.

As is known (Saad 1966; Lee 2002), the total thermodynamic energy (a synonym of the gross thermodynamic potential (Ω_{Env})) is the sum of the products of the mathematical expectations for the microstate energy E_j and the probabilities P_j with which the system would get into that microstate, that is,

$$\begin{aligned}\langle E_{Env}\rangle &= \sum_j E_j P_j = \frac{1}{Z}\sum E_j \cdot e^{-\beta E_j} = -\frac{1}{Z}\frac{\partial}{\partial \beta} Z(\beta, E_1, E_2, \ldots) \\ &= -\frac{\partial}{\partial \beta}(\ln Z) = k_B T^2 \frac{\partial}{\partial T}(\ln Z).\end{aligned}$$

Since the internal energy reads $\langle U \rangle = c_{v0}T = \frac{1}{k_B T}\frac{\partial^2}{\partial \beta^2}(\ln Z)$, the free energy function can be calculated by definition via the expression (Equation 3.20) modified as follows:

$$\psi = \langle E_{Env} \rangle - \langle U_{Env} \rangle = k_B T^2 \frac{\partial}{\partial T}(\ln Z) - \frac{1}{k_B T}\frac{\partial^2}{\partial \beta^2}(\ln Z).$$

Hence, the following equation is obtained, providing an assessment of the *gross sum of the energy of macrostate Z*:

$$\frac{T}{\beta}\frac{\partial}{\partial T}(\ln Z) - \beta\frac{\partial^2}{\partial \beta^2}(\ln Z) = \psi_{Env},$$

and it can serve as assessment of the processes of energy degeneration in the control volume. Note the specific difficulties in handling the mathematical forms of the vector functions gradψ and the scalar function $(\psi)_{Env}$, introduced by Equations 3.46, 3.47, and 3.48. Nevertheless, we shall use those forms in Chapters 4 and 5 to design differential and integral equations of transfer in building envelopes, instead of gradient functions of temperature (Equation 2.9), pressure (Equation 2.10), and electric potential (Equation 2.11). The approach is expected to provide significant calculation advantages in applying modern numerical methods and, particularly, the finite element method discussed in Chapter 6.

4

Definition of the Macroscopic Characteristics of Transfer

The physical results of the energy interactions running in electrothermodynamic systems (ETSs), such as building envelopes comprising facades, roofs, shading devices, etc., consist in transfer of entropy, electric charge, or mass (change of the generalized coordinates) and energy exchange (performance of work).

Suppose that a designer faces the problem of *calculating the building envelope components*, especially stressing on the organization and control of their energy-related functioning. The major issue in engineering practice is to provide quantitative assessment of motion of molecules, ions, or electrons in solid envelope components, directed from areas with more free energy (temperature, pressure, or electric potential, respectively) to areas with less free energy.

Once having established the integral character of the running processes and the existence of free energy in free energy potential in the building envelope components, we pose herein the following problems:

- Formulate a generalized transfer* law valid under all physical conditions imposed on the building envelope.
- Design new differential and integral equations of transfer.
- Apply the generalized law to classical numerical methods.
- Design an algorithm of solving integral equations.
- Development of engineering methods applicable in the design of components of the building envelope and obtain integral assessment of envelope efficiency.
- Perform numerical simulations of *1D, 2D,* and *3D ETS* using the newly designed equations of transfer.

The assessment of these *transfer processes* presents a basis for all *engineering algorithms* concerning the organization of the building envelope operation

* More than 10 specific laws are known in physics, including those of Fourier (1822), Thomson (1856), Reynolds (1879), Sorel (1879), Benedicks (1918), Ohm (1826), Reuss (1808), Jaul/Thomson (1852), Quincke (1850), Poiseuille (1846), Dufour (1873), Dorn (1879) and Nolet (1748), and Fick (1855), *which employ gradient forms* (Macaulay and Hopkinson 1913).

TABLE 4.1

Quantitative Characteristics of Transfer Rates

Heat (Heat Flux, $\vec{Q}$)	Electrical (Current, $\vec{I}_E$)	Fluid (Mass Rate, $\vec{G}$)	Matter (Matter Rate, $\vec{I}$)
$\vec{Q} = \lim\left(\frac{\Delta\vec{H}}{\Delta\tau}\right)_{\Delta\tau\to 0} = \frac{d\vec{H}}{d\tau}, (\text{J/s} = \text{W})$	$\vec{I}_E = \lim\left(\frac{\Delta\vec{\Xi}_E}{\Delta\tau}\right)_{\Delta\tau\to 0} = \frac{d\vec{\Xi}_E}{d\tau}, (Q/s = A)$	$\vec{G} = \lim\left(\frac{\Delta\vec{M}}{\Delta\tau}\right)_{\Delta\tau\to 0} = \frac{d\vec{M}}{d\tau}, (\text{kg/s})$	$\vec{I} = \frac{d\vec{\Xi}}{d\tau}$

already discussed. Moreover, the envelope should be treated as a barrier, filter, and intelligent membrane or it should serve as a photoionization medium (where photovoltaic recuperation [conversion] of solar energy in electricity takes place).

As a result of the analysis performed, we assume that to model the quantitative characteristics of transfer, one should use its vector characteristics called *vectors of transfer*. We present in what follows the basic quantitative characteristics of these vectors and propose a *generalized rule** of performing appropriate calculations, regardless of the physical nature of the object of transfer (mass, heat, electricity, etc.).

The quantitative characteristics of transfer can be modeled by means of vectors generally called *transfer vectors*. These vectors are

- For energy (heat or electrical†), transferred energy quantity ($\vec{H}$ or $\vec{\Xi}_E$), [J]‡
- Transferred quantity of fluid (mass) $\vec{M}$, (kg)

They will be presented by means of a generalized characteristic of the amount of transferred matter.

Other characteristics of transfer are *rate* ($\vec{Q}$, $\vec{I}_E$, $\vec{G}$, and $\vec{\Xi}$ —Table 4.1) and *density* ($\vec{q}$, $\vec{\delta}$, $\vec{g}$, and $\vec{h}$ —Table 4.2).

* The term is known from the book by *Numerical Methods of Transfer* (Patankar 1980). Yet, *it does not propose a form that would be adequate to the introduced term*. The forms used are "≈ −grad *T*," "≈ −grad *p*," which other authors use, too; see, for instance, Bosworth (1952), Schneider (1955), Carslaw and Jaeger (1959), Bird et al. (1960), Kutateladze (1963), Boelter et al. (1965), Luikov (1968), Klemens (1969), Eckert and Darke (1972), Welty (1974), Kreith and Black (1980), Chapman (1981), Todd and Ellis (1982), Wolf (1983), Lucas (1991), Ларинков (1985), Kakas (1985), Thomas (1992), Ashcroft (1992), Taine and Petit (1993), Poulikakos (1994), Holman (1997), Rohsenow et al. (1998), Cengel (1998), Pitts and Sisson (1998), Wilkinson (2000), Kreith and Bohn (2001), Bailyn (2002), Incropera and DeWitt (1996, 2002), Sucec (2002), Димитров (2006), and Atkins (2007).

† $1\ \text{Q} = 6.24 \cdot 10^{18}$ electrons.

‡ $1\ \text{J} = 1\ \text{Nm}$.

TABLE 4.2

Quantitative Characteristics of Transfer Densities

Heat (Specific Heat Flux)	Electrical (Density of Current)	Fluid (Velocity)	Matter (Density Rate, $\vec{h}$)
$\vec{q} = \lim\left(\frac{\Delta\vec{Q}}{\Delta A_0}\right)$	$\vec{\delta} = \lim\left(\frac{\Delta I_E}{\Delta A_0}\right)$	$\vec{g} = \lim\left(\frac{\Delta\vec{G}}{\Delta A_0}\right)$	$\vec{h} = \frac{d\vec{I}}{dA_0}$
$= \frac{d\vec{Q}}{dA_0}$, (W/m^2)	$= \frac{d\vec{I}_E}{dA_0}$, (A/m^2)	$= \frac{d\vec{G}}{dA_0}$, (kg/m^2)	

Here,

$$\Delta\vec{H} = \vec{H}(\tau+\Delta\tau) - \vec{H}(\tau); \quad \Delta\vec{M} = \vec{M}(\tau+\Delta\tau) - \vec{M}(\tau); \quad \Delta\vec{\Xi}_E = \vec{\Xi}_E(\tau+\Delta\tau) - \vec{\Xi}_E(\tau);$$

$$\Delta\vec{\Xi} = \vec{\Xi}(\tau+\Delta\tau) - \vec{\Xi}(\tau)$$

are variations of the vectors of transfer of heat and moisture in the interval $\Delta\tau$: $\vec{Q}$, rate of heat transfer (*heat flux*); $\vec{I}_E$, rate of transfer of electricity flux (*current*); $\vec{G}$, rate of fluid transfer (*mass rate of fluid*); and $\vec{I}$, transferred matter rate (*current*).

Transfer density is defined accounting for the area of the collecting surface (ΔA_0): $\vec{q}$, density of heat transfer (specific heat flux); $\vec{i}$, density of current (specific current); $\vec{v}$, density of transferred fluid (velocity); and $\vec{h}$, density of transferred matter (specific matter flux).

The definitions of Table 4.1 yield the following relations after performing the subsequent integration:

$$\vec{H} = \int \vec{Q} d\tau; \quad \vec{\Xi}_E = \int \vec{I}_E d\tau; \quad \text{and} \quad \vec{M} = \int \vec{G} d\tau, \tag{4.1}$$

and it follows from the definitions of Table 4.1 that $d\vec{Q} = \vec{q} \cdot dA_0$ or $\vec{Q} = \int_{A_0} \vec{q}\, dA_0$, $d\vec{I}_E = \vec{\delta} \cdot dA_0$ or $\vec{I}_E = \int_{A_0} \vec{\delta} \cdot dA_0$, and $d\vec{G} = \vec{g} \cdot dA_0$ or $\vec{G} = \int_{A_0} \vec{g}\, dA_0$.

Putting relations of Table 4.2 in relations of Table 4.1, we find

- The amount of energy (heat):

$$\vec{H} = \int_{\tau_1}^{\tau_2} \left(\int_{A_0} \vec{q} \cdot dA_0 \right) \cdot d\tau$$

or

$$\vec{H} = \sum_{\tau_1}^{\tau_2} \vec{Q} \cdot \Delta\tau = \sum_{\tau_1}^{\tau_2} \left(\sum_{A_0} \vec{q} \cdot \Delta A_0 \right) \cdot \Delta\tau$$

- The amount of electricity:

$$\vec{\Xi} = \int_{\tau_1}^{\tau_2} \left(\int_{A_0} \vec{\delta} \cdot dA_0 \right) \cdot d\tau$$

or

$$\vec{\Xi} = \sum_{\tau_1}^{\tau_2} \vec{I}_E \cdot \Delta\tau = \sum_{\tau_1}^{\tau_2} \left(\sum_{A_0} \vec{\delta} \cdot \Delta A_0 \right) \Delta\tau$$

- The amount of fluid:

$$\vec{M} = \int_{\tau_1}^{\tau_2} \left(\int_{A_0} \vec{g} \cdot dA_0 \right) \cdot d\tau$$

or

$$\vec{M} = \sum_{\tau_1}^{\tau_2} \vec{G} \cdot \Delta\tau = \sum_{\tau_1}^{\tau_2} \left(\sum_{A_0} \vec{g} \cdot \Delta A_0 \right) \Delta\tau$$

(A_0 is the total area of the control surface—note that estimations are made regarding that surface).

If transfer in the *studied object* proceeds in a 3D spatial continuum, it is useful to specify the vectors of transfer and their characteristics by means of their components in a Cartesian coordinate system connected to the wall element:

- For the *generalized characteristic* of the transferred quantity of matter $\vec{\Xi}$ ($\vec{H}$, $\vec{M}$, or Ξ_E):

$$\vec{\Xi} = \Xi_x \vec{i} + \Xi_y \vec{j} + \Xi_z \vec{k}.$$

- For *transfer rate* $\vec{I}$ of energy $\vec{Q}\left(\vec{Q}_T, \vec{Q}_M, \vec{Q}_E\right)$ and matter (of mass $\vec{G}$ and electricity $\vec{I}_E$):

$$\vec{I} = \vec{I}_x\vec{i} + \vec{I}_y\vec{j} + \vec{I}_z\vec{k};$$

$$(\vec{Q} = Q_x\vec{i} + Q_y\vec{j} + Q_z\vec{k}; \quad \vec{G} = G_x\vec{i} + G_y\vec{j} + G_z\vec{k}; \quad \text{and} \quad \vec{I}_E = I_{E_x}\vec{i} + I_{E_y}\vec{j} + I_{E_z}\vec{k}).$$

- For transfer density:

$$\vec{h} = h_x\vec{i} + h_y\vec{j} + h_z\vec{k};$$

$$(\vec{q} = q_x\vec{i} + q_y\vec{j} + q_z\vec{k}; \quad \vec{g} = g_x\vec{i} + g_y\vec{j} + g_z\vec{k}; \quad \text{and} \quad \vec{\delta} = \delta_x\vec{i} + \delta_y\vec{j} + \delta_z\vec{k}).$$

The vectors of the generalized quantity of matter $\vec{\Xi}$ and transfer rate $\vec{I}$ are related as follows:

$$\vec{\Xi} = \int_{\tau_1}^{\tau_2} \left(\vec{i} \int_{A_x} \vec{h}_x dA_x + \vec{j} \int_{A_w} \vec{h}_y dA_y + \vec{k} \int_{A_z} \vec{h}_z dA_z \right) d\tau, \tag{4.2}$$

$$\vec{I} = \vec{i} \int_{A_x} \vec{h}_x dA_x + \vec{j} \int_{A_w} \vec{h}_y dA_y + \vec{k} \int_{A_z} \vec{h}_z dA_z, \tag{4.3}$$

where dA_x, dA_y, and dA_z are control areas, normal to the axes Ox, Oy, and Oz, respectively.

For finite values of the variations

$$dA_x = \Delta A_x; dA_y = \Delta A_y; dA_z = \Delta A_z;$$

$$d\tau = \Delta\tau = (\tau_2 - \tau_1),$$

other generalized characteristic of transferred matter $\vec{\Xi}$ acquires the following form:

$$\vec{\Xi} = (h_x \Delta A_x \Delta\tau)\vec{i} + (h_y \Delta A_y \Delta\tau)\vec{j} + (h_y \Delta A_z \Delta\tau)\vec{k}.$$

These equations serve as estimation of the energy interactions between an *ETS* and the surroundings. They are incorporated in all engineering approaches discussed in Chapter 5.

4.1 General Law of Transfer

Based on the analysis of the significance of the *ETS* state parameters in Section 3.1, the *free energy function* $\psi = f(T,V$ and $p)$ defined by expressions (Equations 3.42 and 3.43) is treated as a *representative characteristic of the transfer processes. We assume here* that the quantitative characteristics of transfer (shown in Table 4.1 equations) are better described via a *linear functional dependence between the vectors of transfer*_density (shown in Table 4.2 equations) *and the gradient of the ETS free energy function, grad* ψ (Equation 3.48), using the form

$$\vec{h} = -\bar{\Gamma} \cdot \text{grad}\,\psi, \tag{4.4}$$

where $\bar{\Gamma}$ is a proportionality coefficient called general coefficient of transfer.

We shall call this functional link *general law of transfer,** and coefficient $\bar{\Gamma}$ will be dimensionalized depending on the physical nature of the object of transfer. Substituting grad ψ in Equation 4.4 for its expressions (Equation 3.48), *we find here* the following new forms of the generalized law of transfer:

$$\vec{h} = -\bar{\Gamma} \cdot \text{grad}\;\psi = -\bar{\Gamma} \cdot \vec{n}_{\varphi} \cdot \nabla(p \cdot \text{v} - s \cdot T + e_0 V). \tag{4.5}$$

The generalized transfer coefficient $\bar{\Gamma}$ depends on the physical nature of the transferred matter. However, its dimension in energy transfer is $1/\text{m} \cdot \text{s}$, in mass transfer s/m^3, and in transfer of electric charges $\text{C/N} \cdot \text{s} \cdot \text{m}^2$.

The coefficient of transfer $\bar{\Gamma}$ in practical engineering calculations is in the form of a diagonal matrix of rank 3, and the members of its main diagonal are *coefficients of heat and mass transfer and electroconductivity* $\bar{\Gamma}_{\lambda}$, $\bar{\Gamma}_{\beta}$, $u\bar{\Gamma}_{\bar{\Omega}}$:

$$[\Gamma] = \begin{bmatrix} \Gamma_{\lambda} & 0 & 0 \\ 0 & \Gamma_{\beta} & 0 \\ 0 & 0 & \Gamma_{\bar{\Omega}} \end{bmatrix}. \tag{4.6}$$

The matrix form of the transfer law (Equation 4.4) is

$$h = -[\Gamma]\{\nabla\psi\} = \begin{bmatrix} \Gamma_{\lambda} & 0 & 0 \\ 0 & \Gamma_{\beta} & 0 \\ 0 & 0 & \Gamma_{\bar{\Omega}} \end{bmatrix} \begin{Bmatrix} (\nabla\psi)_s \\ (\nabla\psi)_v \\ (\nabla\psi)_e \end{Bmatrix}. \tag{4.7}$$

Here, $\{\nabla\psi\}$ is a matrix column with rank 3, with components that are the partial derivatives of the function ψ with respect to the *ETS* generalized

* It retains the gradient shape but sets qualitatively different content.

coordinates—entropy s_0, volume ν, and electric charge e_0 (see Equation 3.43). Besides, gradients of the quantities not participating in transfer are zero.

The laws of Fourier and Fick known in the theory of heat and mass transfer and that of Ohm known in electrotechnics can be expressed by the generalized law of transfer written in a matrix form:

Equation of Fourier

$$q = -[\Gamma]\{\nabla\psi\} = -\begin{bmatrix} \Gamma_\lambda & 0 & 0 \\ 0 & \Gamma_\beta & 0 \\ 0 & 0 & \Gamma_{\bar{\Omega}} \end{bmatrix} \begin{Bmatrix} \{\nabla\psi\}_s \\ 0 \\ 0 \end{Bmatrix} = -\Gamma_\lambda \cdot \nabla T = -\lambda \cdot \mathrm{grad}(T), \ (\mathrm{W/m^2}).$$

Equation of Fick

$$g = [-\Gamma](\nabla\psi) = -\begin{bmatrix} \Gamma_\lambda & 0 & 0 \\ 0 & \Gamma_\beta & 0 \\ 0 & 0 & \Gamma_{\bar{\Omega}} \end{bmatrix} \begin{Bmatrix} 0 \\ \{\nabla\psi\}_v \\ 0 \end{Bmatrix} = -\Gamma_\beta \cdot \nabla p = -\lambda \cdot \mathrm{grad}(p), \quad \mathrm{kg/m^2}.$$

Equation of Ohm

$$\delta = -[\Gamma](\nabla\psi) = -\begin{bmatrix} \Gamma_\lambda & 0 & 0 \\ 0 & \Gamma_\beta & 0 \\ 0 & 0 & \Gamma_{\bar{\Omega}} \end{bmatrix} \begin{Bmatrix} 0 \\ 0 \\ \nabla V \end{Bmatrix} = -\Gamma_{\bar{\Omega}} \cdot \nabla V = -\bar{\Omega} \cdot \mathrm{grad}(V), \quad \mathrm{A/m^2}.$$

Here, the coefficients of heat and mass transfer and electroconductivity are traditionally denoted, respectively, $\Gamma_\lambda = \lambda$, W/mK, $\Gamma_\beta = \beta$, kg/m·Pa, and $\Gamma_{\bar{\Omega}} = \bar{\Omega}$, A/mV.

Equations describing transfer of the three forms of energy in the control volume, in the vicinity of an *ETS* point, can be written in a form similar to that of the general law of transfer. The proportionality coefficient $\bar{\Gamma}$ reads as a diagonal matrix with rank 3:

$$[\Gamma] = \begin{bmatrix} \Gamma_M & 0 & 0 \\ 0 & \Gamma_E & 0 \\ 0 & 0 & \Gamma_T \end{bmatrix},$$

where Γ_M, m/s is the coefficient of transfer of mechanical energy; Γ_E, m/s is the coefficient of transfer of electrical energy; and Γ_T, m/s is the coefficient of transfer of thermal energy.

Then, using Equation 4.5 derived herein, we avoid the use of a set of algorithms, and the engineering calculations *are reduced to the solution of the transfer equation written in a general form.*

4.2 Physical Picture of the Transmission Phenomena

Transfer can be quantitatively and qualitatively assessed by visualization of the its current surfaces and lines. Essentially, *visualization* of the vector field of the transfer density is performed by means of the so-called stream surfaces. These are surfaces formed by streamlines. Being basic building elements of the stream surfaces, streamlines are defined by the transfer vectors $\vec{h}$, $\vec{q}$, and $\vec{g}$ (see Figure 4.1). They are spatial curves with tangential vectors $\vec{\tau}$ that are collinear with the *transfer density* vectors $\vec{h}$ (or $\vec{q}$ and $\vec{g}$) (see Figure 4.1).

The equation of streamlines is found from the condition of colinearity between the vectors of transfer density $\vec{h}$ (or $\vec{q}$ and $\vec{g}$) and the elementary displacement along the streamline $d\vec{r}$ (see Figure 4.1).

Then, their vector product is identical to zero:

$$d\vec{r} \times \vec{h} = \begin{vmatrix} \vec{i} & \vec{j} & \vec{k} \\ dx & dy & dz \\ h_x & h_y & h_z \end{vmatrix} = 0. \tag{4.8}$$

It follows from Equation 4.8 that the differential equation of the streamlines will have the form

$$\frac{dx}{h_x} = \frac{dy}{h_y} = \frac{dz}{h_z}. \tag{4.9}$$

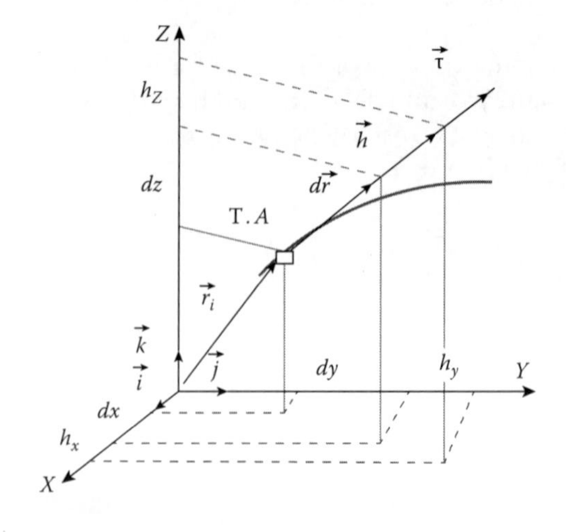

FIGURE 4.1
Streamline.

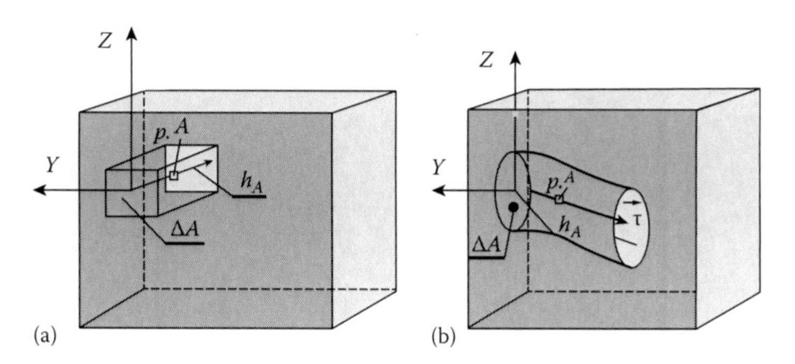

FIGURE 4.2
Stream pipe with (a) plane symmetry and (b) axial symmetry.

Note that only one streamline will pass through a fixed point A of space (x, y, z) at a moment τ. If vector $\vec{h} = f(x, y, z, \tau)$ changes its direction in time, then streamlines of unsteady transfers will change their form in time. For a steady transfer, streamlines do not vary in time.

Surfaces, formed by streamlines, display interesting properties; for instance, no transversal transfer of heat or mass takes place through them, and they serve as membranes. If these surfaces form closed contours, they are called *stream pipes* (see Figure 4.2a, b).

Figure 4.2a shows a stream tubule with transversal face ΔA corresponding to 1D steady transfer through a structural element of a flat wall, while Figure 4.2b is a pipe formed during 3D transfer in a similar object. Considering the second case, the pipe has a complex spatial topology, while in the first case, it has a rectangular or axisymmetrical cross section. It is oriented along the direction of transfer. The set of all streamlines and surfaces forms the stream pattern of transfer.

The identification of the stream pattern in a wall component is one of the first tasks of a component study. It can be practically conducted in two ways: experimental and theoretical.

The experimental method consists in extracting appropriate information about the change of the basic state parameters of the system, followed by a mathematical processing and adequate presentation of the result found. The theoretical method for identification of the stream pattern is based on the use of two important characteristics of the vector field of transfer $\vec{h} = h(x, y, z, \tau)$:

- A potential function $\Phi(x, y, z, \tau)$* defined by

$$\phi(x,y,z,\tau) = \int_s \vec{h} \cdot d\vec{r} = \int \left(h_x \cdot \vec{i} + h_y \cdot \vec{j} + h_z \cdot \vec{k}\right)\left(dx \cdot \vec{i} + dy \cdot \vec{j} + dz \cdot \vec{k}\right)$$

* The definition equation of $\Phi(x, y, z, \tau)$ yields $\phi(x,y,z,\tau) = \int_s \vec{h} \cdot d\vec{r} = \int_s h \cdot ds$ and hence $\vec{h} = \nabla\phi$.

along an arc element of the streamline ($\vec{\tau} = d\vec{r}/ds$, unit vector of the tangent to the streamline)

- A stream function $\bar{\Psi}(x,y,z,\tau)$* defined by

$$\bar{\Psi}(x,y,z,\tau) = \int_{A_0} h\, dA$$

The potential function $\Phi(x, y, z, \tau)$† replaces the vector function $\vec{h} = h(x,y,z,\tau)$ and facilitates the mathematical analysis of transfer. If potential function $\Phi(x, y, z, \tau)$ of the density vector is known, the determination of vector $\vec{h} = h_x\vec{i} + h_y\vec{j} + h_z\vec{k}$ is quite an easy task, finding its partial derivatives $\partial\phi/\partial x$, $\partial\phi/\partial y$, and $\partial\phi/\partial z$ only:

$$\vec{h} = \frac{\partial\phi}{\partial x}\vec{i} + \frac{\partial\phi}{\partial y}\vec{j} + \frac{\partial\phi}{\partial z}\vec{k}.$$

Then, we get

$$h_x = \frac{\partial\phi}{\partial x}, \quad h_y = \frac{\partial\phi}{\partial y}, \quad h_z = \frac{\partial\phi}{\partial z}.$$

This operation is significantly simpler if the streamline $s(x, y, z, \tau)$ is analytically set forth in advance. Vector $\vec{h} = h(x,y,z,\tau)$ is always directed along the tangent to the streamline s (this specificity determines its orientation), and the vector magnitude $|h|$ is calculated by means of the derivative $\partial\phi/\partial s$ (i.e., $|h| = \partial\phi/\partial s$).

Potential function $\Phi(x, y, z, \tau)$ can be visualized by means of the so-called equipotential surfaces where $\Phi(x, y, z, \tau) = \text{const}$ (see Figure 4.3). The potential of transfer density is constant for each of them. Vector $\vec{h} = \text{grad}\,\phi$ is normal to the element ΔA_i belonging to each plane element of the equipotential surface. Hence, it is collinear with the tangent to the streamline (see Figure 4.3).

Another important characteristic of the vector field of transfer density is that $\bar{\Psi}(x,y)$ possesses complete differential $d\bar{\Psi}$, that is,

$$d\bar{\Psi} = \frac{\partial\bar{\Psi}}{\partial x}dx + \frac{\partial\bar{\Psi}}{\partial y}dy = -h_y dx + h_x dy \tag{4.10}$$

(by definition $\bar{\Psi}(x,y)$ $h_x = \partial\bar{\psi}/\partial y$ and $h_y = -(\partial\bar{\Psi}/\partial x)$) and that its isolines $(\bar{\Psi}(x,y) = \text{const})$ coincide with the streamlines, that is, the streamline

* The stream function is used in 2D fields only.

† It follows from the definition equation for $\Phi(x, y, z, \tau)$ that $\phi(x,y,z,\tau) = \int_s h\, ds$. Hence, $\vec{h} = \nabla\phi$.

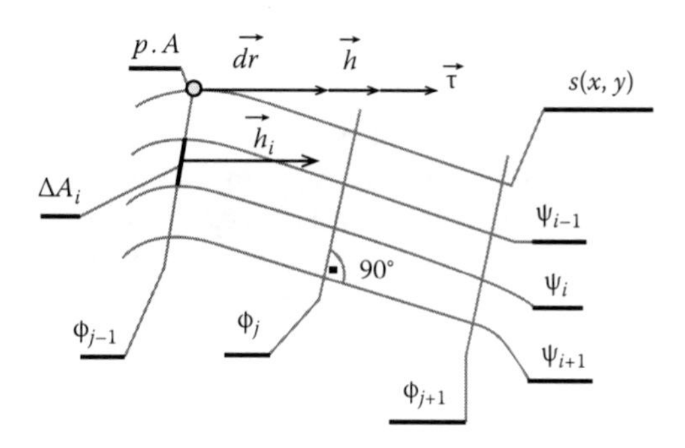

FIGURE 4.3
Stream pattern.

differential equation (Equation 4.9) is identical to that of lines $\bar{\Psi}(x,y) = \text{const}$. This means that the two curve families are equidistant. Moreover, the isolines of the stream function and those of the potential function are orthogonal (see Figure 4.3).

The stream function $\bar{\Psi}(x,y)$ has a specific physical interpretation if the components of the vector of transfer density $h_x = f_1(x, y)$ and $h_y = f_2(x, y)$ are known; it can be easily calculated by integrating (Equation 4.10)

$$\bar{\Psi} = \int d\bar{\psi} = \int -f_2(x,y)\cdot dx + f_1(x,y)\cdot dy = \int_{A_0} h\cdot dA.$$

The value of the stream function $\bar{\Psi} = \int_{A_0} h\cdot dA$ is the magnitude of the resultant flux (heat flux or mass debit) transferred through the control volume bounded by lines $\bar{\Psi} = \text{const}$ ($dz = 1$, conditionally), since the streamlines and surfaces display property characteristics for impermeable membranes.

The method of stream surfaces has been continuously developing in time in accordance with the development of engineering methods and means of study. Considering the study of transfer in solid wall elements, various analogues and numerical techniques of its application are known. However, the most accurate ones are visual methods employing infrared cameras.

4.3 Conclusions

The newly found law of transfer (Equations 4.4, 4.5, and 4.7), involving the gradient of the free energy function, can be used either in direct engineering methods via expressions Equations 4.2 and 4.3 or in more complex numerical methods, such as the methods of differential relations.

The used generalized form of the equations of transfer significantly facilitates the numerical calculations, especially in the treatment of objects with complex geometry such as building envelopes. Calculation comfort can be provided by upgrading some of the existing software packages, such as COMSOL, SolidWorks, Unigraphics, and Cosmos, adding libraries with data on the physical properties of structural materials (coefficients of heat and mass, electroconductivity characteristics, etc.). The advantages consist in the reduction of the technical operations needed to design the graphical model of the physical medium and specify the boundary conditions. All other procedures described in Section 5.4 are automated and running simultaneously. Together with the distribution of the generalized potential within the envelope components, we will also find the transfer vector characteristics.

5

Numerical Study of Transfer in Building Envelope Components

In engineering practice, apart from visualization via streamlines and surfaces, two fundamentally different methods of transport evaluation are currently employed:

1. Method of the differential relations
2. Method of the integral forms

In what follows, we will discuss the fundamental principles and relationships of those methods in light of their applicability in the study of transfer of entropy, electrical charges, and mass through a solid wall being a component of facade structures. At the same time, detailed basic equations will be designed to treat one- (1D), two- (2D), and three-dimensional (3D) envelope components (Dimitrov 2013).

5.1 Method of the Differential Relations

This method is based on the principle of matter conservation at *local level*, assuming that the area under consideration is infinitesimally small and its volume Δv tends to zero without violating space continuity. The final result is a differential equation whose dependent variable is any of the quantities characterizing the state of an electrothermodynamic system (ETS).

A dependent variable can be, for instance, the function of free energy (ψ), temperature (T), pressure (p), or electrical potential (V) at a point A of space. On the other hand, independent variables are spatial coordinates of (x_i, y_i, and z_i) of the space enveloping the ETS as well as time τ.

The derived differential equation is written in two forms:

1. A gradient one
2. A divergent one

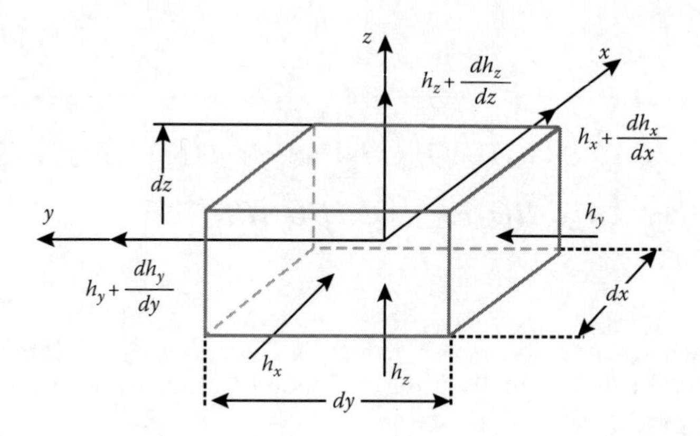

FIGURE 5.1
Geometry of the control volume.

Those forms are applied to each point of an appropriately designed time–space continuum. An infinitesimally small spatial element with dimensions *dx*, *dy*, and *dz* in a Cartesian coordinate system is considered (see Figure 5.1), and the principle of matter conservation is applied to a volume $\Delta v = dx\, dy\, dz$ whose boundary surface has area A_0. In our case, *to transfer active substances* means to carry entropy, electric charges, and mass across the border of the control surface.

The following relation is valid: *quantity of inflowing matter* Ξ_{in} – *quantity of outflowing matter* Ξ_{out} = *quantity of recuperated matter* Ξ_g – *quantity of accumulated matter* Ξ_a. It formally reads

$$\vec{\Xi}_{in} - \vec{\Xi}_{out} = \left|\vec{\Xi}_g - \vec{\Xi}_a\right|, \tag{5.1}$$

where the participating terms are expressed by the following equalities:

$$\vec{\Xi}_{in} = (h_x \cdot \Delta A_x \cdot \Delta\tau)\cdot\vec{i} + (h_y \Delta A_y \Delta\tau)\cdot\vec{j} + (h_y \Delta A_z \Delta\tau)\cdot\vec{k}, \tag{5.2}$$

$$\vec{\Xi}_{out} = \left(h_x + \frac{\partial h_x}{\partial x}\cdot\Delta x\right)\cdot\Delta A_x \cdot d\tau\cdot\vec{i} + \left(h_y + \frac{\partial h_y}{\partial y}\cdot\Delta y\right)\cdot\Delta A_y \cdot d\tau\cdot\vec{j}$$
$$+ \left(h_z + \frac{\partial h_z}{\partial z}\cdot\Delta z\right)\cdot\Delta A_z \cdot d\tau\cdot\vec{i}, \tag{5.3}$$

$$\Xi_g = G_h \cdot \Delta v \cdot \Delta\tau, \tag{5.4}$$

$$\Xi_a = \overline{\Xi}\cdot\Delta v\cdot\Delta\tau. \tag{5.5}$$

(The specific capabilities of the medium enclosed within the control volume, that is, the accumulating $\bar{\Xi}_a$ capability and the recuperating G_h capability, are reduced to a volume unit and a unit of time, and $\Delta A_x = \Delta y \Delta z$, $\Delta A_y = \Delta x \Delta z$, and $\Delta A_z = \Delta x \Delta y$.)

Based on the analysis of the inflows and outflows along the respective coordinate axes, we get expressions for axis $Ox: \left(\frac{\partial h_x}{\partial x}\Delta x\right)\Delta y \Delta z \Delta \tau$, axis $Oy: \left(\frac{\partial h_y}{\partial y}\Delta y\right)\Delta x \Delta z \Delta \tau$, and axis $Oz: \left(\frac{\partial h_z}{\partial z}\Delta z\right)\Delta x \Delta y \Delta \tau$.

In conformity with the notations in Figure 5.1, the left-hand side of relation (Equation 5.1) takes the following form:

$$\vec{\Xi}_{in} - \vec{\Xi}_{out} = \left(\frac{\partial h_x}{\partial x}\right)\Delta v \cdot \Delta \tau \cdot \vec{i} + \left(\frac{\partial h_y}{\partial y}\right)\Delta v \cdot \Delta \tau \cdot \overleftarrow{j} + \left(\frac{\partial h_z}{\partial z}\right)\Delta v \cdot \Delta \tau \cdot \vec{k}. \quad (5.6)$$

To balance inflowing and outflowing quantities at a moment τ and at point A, Δv, and $\Delta \tau$ should tend to zero—$\Delta v \to 0$ and $\Delta \tau \to 0$. Then, the balance value of the two quantities in relation (Equation 5.6) will be equal to the following limit:

$$\lim \left(\vec{\Xi}_{in} - \vec{\Xi}_{out}\right)_{\substack{\Delta v \to 0, \ \Delta\tau \to 0 \\ A}} = \frac{\partial h_x}{\partial x} + \frac{\partial h_y}{\partial y} + \frac{\partial h_z}{\partial z} = \text{div } \vec{h}_A = \nabla \vec{h}_A, \quad (5.7)$$

which is interpreted as *source density* at point A, and it is the divergence of the field of vector *transfer density* $\vec{h}_A$ at this point.

In short,

$$\text{div } \vec{h}_A = \nabla \vec{h}_A = \lim_{\Delta v \to 0} \frac{1}{\Delta v} \oiint_{A_0} \left(\vec{\Xi}_{in} - \vec{\Xi}_{out}\right)\Delta \tau \Delta A = \lim_{\Delta v \to 0} \frac{1}{\Delta v} \oiint_{A_0} \left(\vec{I}_{in} - \vec{I}_{out}\right)\Delta A$$

is the balance value of the inflowing and outflowing matter in the vicinity of a point and in the absence of recuperation and accumulation.

The expression at the right-hand side of relation (Equation 5.1) is found after the substitution of the two quantities by relations (Equations 5.6 and 5.7), that is,

$$\left|\Xi_g - \Xi_a\right| = \sum_{\tau}\sum_{V}(G_h \cdot \Delta v)\Delta \tau - \sum_{\tau}\sum_{V}\left(\bar{\Xi} \cdot \Delta v\right)\Delta \tau,$$

where, after limit transition for $\Delta v \to 0$ and $\Delta \tau \to 0$, we find

$$\lim \left|\Xi_g - \Xi_a\right| = G_h - \frac{\partial \bar{\Xi}}{\partial \tau}. \quad (5.8)$$

Based on relations (Equations 5.7 and 5.8), we get the divergent form of the differential equation from relation (Equation 5.1):

$$\frac{\partial \Xi}{\partial \tau} + \text{div}\,\vec{h} - G_h = 0 \quad (5.9)$$

or

$$\frac{\partial \Xi}{\partial \tau} + \nabla \vec{h} - G_h = 0. \quad (5.10)$$

The first term $\partial \Xi / \partial \tau$ is called *accumulation* term, the second term $\Delta \vec{h}$—*divergent* term, and the third term G_h—*recuperation* term. Those terms express the three components of conservation low. From a mathematical point of view, Equations 5.9 and 5.10 show that the algebraic sum of the accumulation, divergent, and recuperation components should be uniquely equal to zero in the vicinity of a point. From a physical point of view, the equations state a balanced equality between the accumulated, transferred, and recuperated matter in the vicinity of any point of the time–space continuum where transfer takes place.

The gradient form of the differential equation is obtained by substituting the members of Equation 5.9 or 5.10 as follows:

- Term $\partial \Xi / \partial \tau$ denoting *accumulation rate* at point A should be substituted by the following equation:

$$\frac{\partial \Xi}{\partial \tau} = \frac{\partial}{\partial \tau}(C_\psi \cdot \rho \cdot \psi), \quad (5.11)$$

- (where C_ψ is the specific recuperated energy and ρ—medium density).
- The recuperation term G_h[quantity/m^{3}s] should be substituted by the value of the specific recuperated energy–C_ψ (Dimitrov 2014a).
- Putting (Equation 5.11) and (Equation 5.1) or (Equation 5.2) in (Equation 5.9), the differential equation of transfer takes the form

$$\frac{\partial}{\partial \tau}(C_\psi \cdot \rho \cdot \psi) - \nabla\left(\left[\Gamma\right] \cdot \text{grad}\,\psi\right) - G_h = 0, \quad (5.12)$$

and the substitution of the free energy function $\psi = f(T, p, V)$ in Equation 3.43 yields the *following new differential equation* comprising the state parameters (T, p, V, s_0, e_0, v):

$$\frac{\partial}{\partial \tau}\left[C_\psi \cdot \rho\left(pv - s_0T + e_0V\right)\right] - \nabla\left[\left[\Gamma\right] \cdot \vec{n}_\psi \nabla\left(pv - s_0T + e_0V\right)\right] - G_h = 0. \quad (5.13)$$

The equation represents the gradient form of Equation 5.1. Under steady transfer, relation (Equation 5.1) is converted to

$$\nabla^2\psi - \frac{G_h}{[\Gamma]} = 0 \quad \text{or} \quad \text{div}\left(\frac{\partial\psi}{\partial\vec{n}_\psi}\right) - \frac{G_h}{[\Gamma]} = 0. \tag{5.14}$$

It is called Poisson's condition, where the term $\nabla^2\psi = \Delta\psi = \nabla\text{grad } \psi$ is the Laplace operator. If there is no recuperation of matter in the medium, the last equation degenerates into Laplace equation under steady conditions, then we have

$$\text{div}\left(\frac{\partial\psi}{\partial\vec{n}_\psi}\right) = 0 \quad \text{or} \quad \Delta\psi = 0,$$

and in a Cartesian coordinate system,

$$\frac{\partial^2\psi}{\partial x^2} + \frac{\partial^2\psi}{\partial y^2} + \frac{\partial^2\psi}{\partial z^2} = 0.$$

Respectively, the general equation of transfer is written employing Cartesian, cylindrical, and spherical coordinates, that is,

$$\frac{\partial\psi}{\partial\tau} - \frac{[\Gamma]}{C_\psi\rho}\left(\frac{\partial^2\psi}{\partial x^2} + \frac{\partial^2\psi}{\partial y^2} + \frac{\partial^2\psi}{\partial z^2}\right) - \frac{G_h}{C_\psi\rho} = 0;$$

$$\frac{\partial\psi}{\partial\tau} - \frac{[\Gamma]}{C_\psi\rho}\left[\frac{1}{r}\frac{\partial}{\partial r}\left(r\frac{\partial\psi}{\partial r}\right) + \frac{1}{r^2}\frac{\partial^2\psi}{\partial\theta^2} + \frac{\partial^2\psi}{\partial z^2}\right] - \frac{G_h}{C_\psi\rho} = 0;$$

$$\frac{\partial\psi}{\partial\tau} - \frac{[\Gamma]}{C_\psi\rho}\left[\frac{1}{r^2}\frac{\partial}{\partial r}\left(r^2\frac{\partial\psi}{\partial r}\right) + \frac{1}{r^2\sin\theta}\frac{\partial}{\partial\theta}\left(\sin\theta\frac{\partial\psi}{\partial\theta}\right) + \frac{1}{r^2\sin^2\theta}\frac{\partial^2\psi}{\partial\lambda^2}\right] - \frac{G_h}{C_\psi\rho} = 0. \tag{5.15}$$

For a 1D unsteady transfer, these equations take the following form:

$$\frac{\partial}{\partial\tau}(\psi) - \frac{1}{r^n}\frac{d}{dr}\left(r^n\frac{[\Gamma]}{C_\psi\rho}\frac{d\psi}{dr}\right) - \frac{G_h}{C_\psi\rho} = 0,$$

where

- For a Cartesian coordinate system, we have $r \to x$, and $n = 0$.
- For a cylindrical coordinate system, we have $n = 1$.
- And for a spherical coordinate system, we have $n = 2$.

If the transfer is 3D, but steady, Equation 5.15 reads as follows:

$$\frac{\partial^2\psi}{\partial x^2}+\frac{\partial^2\psi}{\partial y^2}+\frac{\partial^2\psi}{\partial z^2}+\frac{G_h}{[\Gamma]}=0. \tag{5.16}$$

$$\frac{\partial}{\partial r}\left(r\frac{\partial\psi}{\partial r}\right)+\frac{1}{r}\frac{\partial^2\psi}{\partial\theta^2}+r\frac{\partial^2\psi}{\partial z^2}+\frac{G_h r}{[\Gamma]}=0. \tag{5.17}$$

$$\frac{\partial}{\partial r}\left(r^2\frac{\partial\psi}{\partial r}\right)+\frac{1}{\sin\theta}\frac{\partial}{\partial\theta}\left(\sin\theta\frac{\partial\psi}{\partial\theta}\right)+\frac{1}{\sin^2\theta}\frac{\partial^2\psi}{\partial\lambda^2}+\frac{G_h r^2}{[\Gamma]}=0. \tag{5.18}$$

From a mathematical point of view, the partial differential equations (Equations 5.16, 5.17, and 5.18) are of hyperbolic type (with respect to time τ), and they display elliptic behavior with respect to the spatial coordinates x, y, z.

If equations display a hyperbolic behavior, this means that the values of the ***ETS*** state parameters found as varying in time depend on the ***ETS*** pre-history only, that is, on their initial values. If the partial differential equations are elliptic, it is assumed that the system state parameters depend on their values at the continuum boundary.

The method of differential relations discussed in Section 5.1 and used to find the transfer equation (Equation 4.5) serves as the starting point in developing a large and popular group of methods for numerical study of transfer. It is generally known as *the method of finite differences* (Jaluria and Torrance 2003; Dimitrov 2013).

The method provides an approximation of the differential equations expressing transfer balance in the vicinity of a point. Although simple and easily programmable, the numerical technique based on the *finite differences* is not universally applied. Regarding some cases where a study of transfer in bodies with complex geometry is to be performed, researchers employ other numerical techniques, which enable them to plausibly model the body shape. These techniques are united in a single numerical method generally known as *the finite element method* (FEM) (Clough 1960; Fagan 1984; Pepper and Heinrich 1992; Dimitrov 2013). They are described in more detail in the next section, noting their growing importance and role in building envelope design and transport assessment. Several dozens of commercial FEM-based packages were developed in the last 50–60 years using *FEM*, some of them supplied with their own graphical editors as the already mentioned Comsol, SolidWork, Unigraphic, and Cosmos. Some of the existing and widespread FEM software packages are open to programming codes embedding libraries of physical characteristics, such as thermal conductivity ($\Gamma_\lambda = \lambda$, W/m^{0}K), permeability ($\Gamma_\beta = \beta$, kg/m·Pa), and conductivity ($\Gamma_{\bar{\Omega}} = \bar{\Omega}$, A/m·V). Hence, in what follows, we will make an authorized, creative, and critical review of the *FEM*. Next, FEM equations will be proposed to tackle transfer in 1D, 2D, and 3D areas of the building envelope.

5.2 Method of the Integral Forms

The integral equations of transfer serve as a theoretical basis of the FEM, and they incorporate the idea of designing a balance of transfer fluxes within an entire segment of an area with finite dimensions. The balance equations found concerning the entire spatial area serve as the starting point of the development of various techniques for numerical analysis of transfer. The most popular of them are discussed in the exposé that follows, deriving the integral equations of transfer and formulating the boundary conditions of wall element performance. As emphasized, the differential equations express the change of the ***ETS*** state variables at each point of the time–space continuum ***R*** occupied by the wall element. They are derived by applying the law of conservation of matter within an elementary volume Δv.

Consider a larger area of the time–space continuum ***R*** and note that the conservation law is also valid in a global sense (see Figure 5.2). This enables one to derive a new type of equations of transfer called *integral equations*. If ***R*** is a nondeformable 3D area as shown in Figure 5.2, one can integrate equation (Equation 5.9) over that area.

The theoretical basis of the method of integral forms is presented by the integral equations of transfer. They reflect the idea of balance of transfer flows within a finite segment of the area. The balance equations concerning the entire spatial area serve as the starting point of the employment of various techniques in the numerical analysis of transport. The most relevant ones are derived in the following paragraphs: integral equations of transfer (Section 5.3) and description and systematization of the boundary conditions of a wall element operation (Chapter 6).

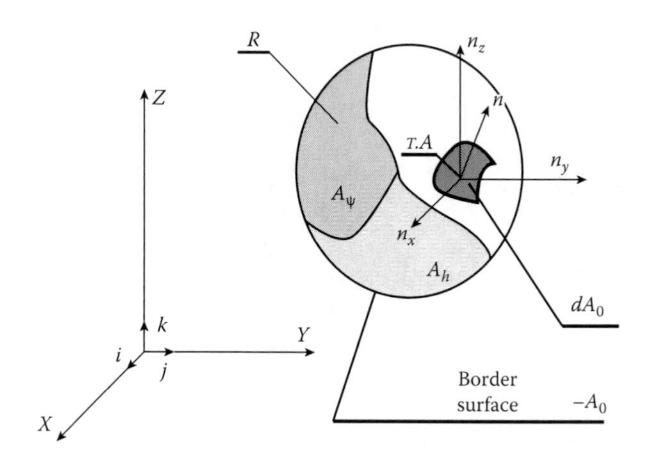

FIGURE 5.2
Border surfaces of the global area.

As already noted, the differential equations concern the ***ETS*** state parameters at any point of the space–time continuum ***R*** of the wall element. They are based on the law of conservation of matter within an elementary volume and on the general law of transfer (Equations 4.5 and 4.7). The conservation law is valid in a broader area of the space–time continuum ***R*** (see Figure 5.2) thus specifying transfer equations called *generalized integral equations of transfer*.

If ***R*** is a nondeformable 3D area that can be seen in shown in Figure 5.2, Equation 5.9 can be integrated within that area:

$$\int_v \left(\frac{\partial \Xi}{\partial \tau}\right) dv + \int_v \operatorname{div} \vec{h}\, dv - \int_v G_h\, dv = 0. \tag{5.19}$$

It follows from Gauss theorem that

$$\int_v \operatorname{div} \vec{h}\, dv = \int_{A_0} \vec{n} \cdot \vec{h}\, dA_0,$$

where

- $\operatorname{div} \vec{h}$ is the divergence of the field of the transfer density
- $dv = dx\,dy\,dz$ is an elementary volume belonging to ***R***
- dA_0 is a segment of the boundary (control) surface A_0 with an elementary area

If $\vec{h}$ is the density of the transferred flux, Equation 5.19 proves that the volume integral of the divergence of vector $\vec{h} - \int_V \operatorname{div} \vec{h}\, dV$ equals the pure outflowing flux $\int_{A_0} \vec{h}_n\, ds$ passing through the boundary surfaces A_0 of area ***R***. Vector $\vec{h}_n$ is a component of $\vec{h}$ along the normal of the area element dA (see Figure 5.2), and it can be presented as follows:

$$\vec{h}_n = \vec{n}_D \cdot \vec{h} = \vec{i} \cdot [\Gamma]\frac{\partial \psi}{\partial x} + \vec{j} \cdot [\Gamma]\frac{\partial \psi}{\partial y} + \vec{k} \cdot [\Gamma]\frac{\partial \psi}{\partial z} = \left|\vec{h}\right| \cdot \cos\alpha$$

in a Cartesian coordinate system.

$$\vec{h}_n = \vec{n}_C \cdot \vec{h} = \vec{n}_r \cdot [\Gamma]\frac{\partial \psi}{\partial r} + \vec{n}_\theta \cdot \frac{[\Gamma]}{r}\frac{\partial \psi}{\partial \theta} + \vec{n}_z \cdot [\Gamma]\frac{\partial \psi}{\partial z}$$

in a cylindrical coordinate system.

$$\vec{h}_n = \vec{n}_{Sp} \cdot \vec{h} = \vec{n}_r \cdot [\Gamma]\frac{\partial \psi}{\partial r} + \vec{n}_\theta \cdot \frac{[\Gamma]}{r}\frac{\partial \psi}{\partial \theta} + \vec{n}_\lambda \cdot \frac{[\Gamma]}{r \cdot \sin\theta}\frac{\partial \psi}{\partial \lambda}$$

in a spherical coordinate system.

Here, h_n is the component of $\vec{h}$ along the outer-pointing normal to the element dA_0 of $\boldsymbol{R}$ (see Figure 5.2). Thus, Equation 5.19 yields the following integral equation:

$$\int_{v} \frac{\partial \Xi}{\partial \tau} dv + \int_{A_0} \vec{n} \cdot \vec{h}\, dA - \int_{v} G_h\, dv = 0,$$

which is the balance equation of the spatial area $\boldsymbol{R}$. It shows that the sum of the *accumulating flux*, the flux exported through boundary area A_0 and the flux *recuperated* by an internal source, should be zero.

If density of the transfer flux $\vec{h}$ is proportional to the gradient of field $\bar{\psi}$, in compliance with the generalized flow of transfer (Equation 5.12), and the function subject to integration is expressed by relations (Equations 4.4, 4.5, and 4.7), the earlier equation takes the following form:

$$\int_{v} \frac{\partial \psi}{\partial \tau} dv - \frac{[\Gamma]}{C_\psi \cdot \rho} \int_{A_0} \vec{n}_\psi \cdot \nabla \psi dA - \frac{1}{C_\psi \cdot \rho} \int_{v} G_h\, dv = 0, \tag{5.20}$$

since $\int_{v} \nabla \vec{h}\, dv = \int_{v} \nabla\left(-[\Gamma] \cdot \text{grad }\psi\right) dv = \int_{A_0} \vec{n}_\psi \cdot \left(-[\Gamma] \nabla\left(Ve_0 - Ts_0 + pv\right)\right) dA.$

Note the more popular form of Equation 5.20:

$$\frac{\partial}{\partial \tau} \int_{v} \psi\, dv - \frac{[\Gamma]}{C_\psi \cdot \rho} \int_{A_0} \frac{\partial \psi}{\partial \vec{n}_\psi} dA - \frac{1}{C_\psi \cdot \rho} \int_{v} G_h\, dv = 0. \tag{5.21}$$

and the new form of the integral equation of transfer

$$\frac{\partial}{\partial \tau} \int_{v} \left(Ve_0 - Ts_0 + pv\right) dv - \frac{[\Gamma]}{C_\psi \cdot \rho} \int_{A_0} \frac{\partial}{\partial \vec{n}_\psi} \left(Ve_0 - Ts_0 + pv\right) dA - \frac{1}{C_\psi \cdot \rho} \int_{v} G_h\, dv = 0,$$

are found for the purpose of the present study.

Putting the different coordinate forms of (Equation 3.49) in relation (Equation 5.21), we find the following integral equations:

$$\frac{\partial}{\partial \tau} \int_{v} \psi\, dv - \frac{[\Gamma]}{C_\psi \cdot \rho} \int_{A_0} \left(n_x \cdot \frac{\partial \psi}{\partial x} + n_y \cdot \frac{\partial \psi}{\partial y} + n_z \cdot \frac{\partial \psi}{\partial z} \right) dA - \frac{1}{C_\psi \cdot \rho} \int_{v} G_h\, dv = 0; \tag{5.22}$$

$$\frac{\partial}{\partial \tau} \int_{v} \psi\, dv - \frac{[\Gamma]}{C_\psi \cdot \rho} \int_{A_0} \left(n_r \cdot \frac{\partial \psi}{\partial r} + n_\theta \cdot \frac{1}{r} \cdot \frac{\partial \psi}{\partial \theta} + n_z \cdot \frac{\partial \psi}{\partial z} \right) dA$$
$$- \frac{1}{C_\psi \cdot \rho} \int_{v} G_h\, dv = 0;$$

and

$$\frac{\partial}{\partial\tau}\int_v \psi\, dv - \frac{[\Gamma]}{C_\psi\cdot\rho}\int_{A_0}\left(n_r\cdot\frac{\partial\psi}{\partial r}+n_\theta\cdot\frac{1}{r}\frac{\partial\psi}{\partial\theta}+n_\lambda\cdot\frac{1}{r\cdot\sin\theta}\cdot\frac{\partial\psi}{\partial\lambda}\right)dA$$
$$-\frac{1}{C_\psi\cdot\rho}\int_v G_h\, dv = 0.$$

written in Cartesian, cylindrical, and spherical coordinates, respectively.

Equations 5.20 ÷ 5.22 are different mathematical forms of the balance equation of transfer in a finite area ***R***. If transfer is steady, the integral analogues of *Poisson* and *Laplace* equations are read as follows:

$$-[\Gamma]\int_{A_0}\vec{n}_\psi\cdot(\text{grad }\psi)dA+\int_v G_h\, dv=0 \quad\text{or}\quad -[\Gamma]\int_{A_o}\frac{\partial\psi}{\partial\vec{n}_\psi}dA+\int_v G_h\, dv=0$$

$$\int_{A_0}\vec{n}_\psi\cdot(\text{grad }\psi)dA=0 \quad\text{or}\quad \int_{A_0}\frac{\partial\psi}{\partial\vec{n}_\psi}dA=0.$$

Although the form of the integral equations differs from that of the differential ones, the solutions found keep their parabolic character with respect to time and are elliptic with respect to the spatial coordinates. From a physical point of view, the differential and integral equations present a local and global balance of the quantities that participate in the conversional low. Those equations offer alternative starting points of the development of numerical methods.

5.3 Weighted Residuals Methodology Employed to Assess the ETS Free Energy Function

The weighted residuals methodology (*WRM*) was successfully developed during the 1950s of the last century thanks to the rapid development of computers. It became popular as a universal method of solving linear and nonlinear differential equations. The use of *WRM* to estimate the parameters of an ***ETS*** and the characteristics of transfer in a physical domain v with boundary A_0 seems logical and rationally well sustained from a physicomethodological point of view, since the method is based on the following hypotheses:

Hypothesis 1: Integral relations (Equations 5.20 through 5.22), specified in Section 5.2, are applicable in the global domain v and in its subdomains $v_{(e)}$ called finite elements.

Hypothesis 2: Errors $R_A^k, k=\psi,T,p,V$ in estimating the ***ETS*** parameters (sought at point $A\in V_{(e)}$) occur during the numerical solution of the differential equations. Those errors are in fact differences between the real values of ψ, T, V, or p and their approximate values ($\bar{\psi}, \bar{T}, \bar{V}$, or $\bar{p}$. They are due to the approximation model chosen, and one can express them as

$$R_A^\psi = |\psi_A - \bar{\psi}_A|, \quad R_A^t = |T_A - \bar{T}_A|, \quad R_A^V = |V_A - \bar{V}_A|, \text{ and } R_A^P = |p_A - \bar{p}_A|.$$

Hypothesis 3: The total error in the global domain V, as well as that in each FE, can be eliminated by appropriately choosing the weighting functions $W_A^k = W_m^k$, that is,

$$\int_{v_e} W_m^k \cdot R_m^k \, dV = 0, k = \psi, T, V, p, m = i, j, k, l\ldots, \quad e = 0, 1, \ldots, M$$ or in a matrix form

$$\int_{v_e} \{W^e\}_k [R^e]_k \, dV = 0, k = \psi, T, V, p, m = i, j, k, l\ldots, \quad e = 0, 1, \ldots, M$$

where

- v_e is the volume of the eth FE ($v_0 = v$ is the area volume)
- $m = i, j, k, l,\ldots$ is the index of the current point A (or node) of v_e
- $e = 0, 1,\ldots, M$ is the index of the current FE in the global domain (M is the number of FE)
- $\{W^e\}_k$ is the matrix-column containing the weighting function of the nodes of the discrete analogue of the eth finite element
- $[R^e]_k$ is the matrix-row containing the errors of $\bar{\psi}, \bar{T}, \bar{V}$, or $\bar{p}$ of the nodes of the discrete analogue of the eth finite element

For instance, the error R_A^φ at point A can be estimated using Equation 5.12 in Section 5.1 and substituting ψ for the estimation $\bar{\psi}_A$, which is valid for the function of the free energy at the same point:

$$R_A^\psi = \frac{\partial}{\partial \tau}(c_\psi \cdot \rho \cdot \bar{\psi}_A) - \nabla([\Gamma] \text{grad } \bar{\psi}_A) - G_h.$$

If $\psi_A = \bar{\psi}_A$, then $R_A^\psi = 0$. A coincidence between the estimated and the real values of a quantity at an arbitrary point is accidental, and it is not expected to occur in the process of numerical solving. Hence, the **WRM** requires coincidence within each FE and within the global area only, and not at a specific point. These requirements to an FE (e) with n nodes yield the following condition:

$$\int_{v_e} W_m^e \left(\frac{\partial}{\partial \tau}(c_\psi \cdot \rho \cdot \bar{\psi}_m) - \nabla([\Gamma] \text{grad } \bar{\psi}_m) - G_{h_e} \right) dv = 0, \tag{5.23}$$
$$e = 0, 1, \ldots, M \quad m = i, j, k, l\ldots$$

or in a matrix form

$$\int_{v_e} \{W^e\} \left[\frac{\partial}{\partial \tau}(c_\psi \cdot \rho \cdot \bar{\psi}_m) - \nabla([\Gamma] \text{grad } \bar{\psi}_m) - G_{h_e} \right] dv = 0,$$
$$e = 0,1,\ldots,M \quad m = i,j,k,l\ldots.$$

The earlier equation is basic to the **WMR**, and it expresses the requirement that the total error should be zero for each FE and for the total global area. *Newmann's* boundary condition yields a similar requirement (Equation 52, Dimitrov 2013):

$$\int_{A_{h_e}} \hat{W}_m^e \left[[\Gamma] \cdot \frac{\partial \bar{\psi}_m}{\partial \vec{n}_\psi} + \bar{h}_e \right] dA = 0. \tag{5.24}$$

The summation of Equations 5.23 and 5.24 results in

$$\int_{v_e} W_m^e \left[\frac{\partial}{\partial \tau}(c_\psi \cdot \rho \cdot \bar{\psi}_m) - \nabla([\Gamma] \text{grad } \bar{\psi}_m) - G_{h_e} \right] dv + \int_{A_{he}} \hat{W}_m^e \left[[\Gamma] \cdot \frac{\partial \bar{\psi}_m}{\partial \vec{n}_\psi} + \bar{h}_e \right] dA = 0.$$

Applying Green's lemma, we get

$$\begin{aligned} &\int_{v_e} W_m^e \frac{\partial}{\partial \tau}(c_\psi \cdot \rho \cdot \bar{\psi}_m) dv - \int_{v_e} W_m^e G_{h_e} dv \\ &+ \int_{v_e} \left[\frac{\partial W_m^e}{\partial x} [\Gamma_x] \frac{\partial \bar{\psi}_m}{\partial x} + \frac{\partial W_m^e}{\partial y} [\Gamma_y] \frac{\partial \bar{\psi}_m}{\partial y} + \frac{\partial W_m^e}{\partial z} [\Gamma_z] \frac{\partial \bar{\psi}_m}{\partial z} + \right] dx\, dy\, dz \\ &+ \int_{A_{\psi_e} + A_{h_e}} [\Gamma] \frac{\partial \bar{\psi}_m}{\partial \vec{n}_\psi} W_m^e dA + \int_{A_{h_e}} \hat{W}_m^e \left[[\Gamma] \frac{\partial \bar{\psi}_m}{\partial \vec{n}_\psi} + \bar{h}_e \right] dA = 0. \end{aligned} \tag{5.25}$$

The choice of functions W_m^e and $\hat{W}_m^e$ can be restricted to the choice of a set of functions, which meet the requirements: $W_m^e = 0$ over A_{ψ_e} $\hat{W}_m^e = -W_m^e$ over A_{h_e}.

Then, Equation 5.25 gets a simplified form:

$$\begin{aligned} &\int_{v_e} W_m^e \frac{\partial}{\partial \tau}(c_\psi \cdot \rho \cdot \bar{\psi}_m) dv + \int_{v_e} \left[\frac{\partial W_m^e}{\partial x} [\Gamma_x] \frac{\partial \bar{\psi}_m}{\partial x} + \frac{\partial W_m^e}{\partial y} [\Gamma_y] \frac{\partial \bar{\psi}_m}{\partial y} + \frac{\partial W_m^e}{\partial z} [\Gamma_z] \frac{\partial \bar{\psi}_m}{\partial z} + \right] \\ &dx\, dy\, dz + - \int_{v_e} W_m^e G_{h_e}\, dv + \int_{A_{he}} W_m^e \bar{h}_e\, dA = 0. \end{aligned} \tag{5.26}$$

For a steady volume, it is modified as

$$\int_{v_e}\left[\frac{\partial W_m^e}{\partial x}[\Gamma_x]\frac{\partial\overline{\psi}_m}{\partial x}+\frac{\partial W_m^e}{\partial y}[\Gamma_y]\frac{\partial\overline{\psi}_m}{\partial y}+\frac{\partial W_m^e}{\partial z}[\Gamma_z]\frac{\partial\overline{\psi}_m}{\partial z}+\right]dx\,dy\,dz$$

$$-\int_{v_e} W_m^e G_{h_e}\,dV+\int_{A_{he}} W_m^e \overline{h}_e\,dA=0. \tag{5.27}$$

The last equation is known as a weak formulation of the equation of steady transfer (Dimitrov 2013). Index m of the weighting functions W_m^e in the FE gets values $i, j, k, l, \ldots$ corresponding to the nodes. If n tends to infinity $n \to \infty$, then R_v disappears. Usually, a finite number of weighting functions are used.

5.3.1 Basic Stages of the Application of WRM in Evaluating Transport within the Envelope

Although there are a wide variety of approaches in applying the **WRM**, the following six stages are worth outlining:

Stage 1: Discretization of the continuous physical space into FEs

Stage 2: Design of interpolating functions of the FEs introduced

Stage 3: Modeling of transfer within an FE using a matrix equation

Stage 4: Design of the matrix equation of the global area

Stage 5: Finding the parameter values at the element nodes and outlining the transfer *pattern*

Stage 6: Error estimation and solution specification

To find a solution of an ***ETS***-related engineering-applied problem, one should account for an appropriate physical model, and the domain is to be substituted for a set of FEs. This specific process consists of the following stages:

To analyze the *properties* of a real physical domain, one needs to

- *Divide it into a limited number of subdomains* (choose the type of FEs and adopt a scheme of element numbering)
- *Design a discrete analogue* of the FEs (i.e., specify a node analogue scheme of the FE, element number, and size and node order)

The first stage aims at establishing (if any) domain symmetry, process steadiness, and homogeneity and isotropy of the domain physical properties, as well as outlining domain boundaries. All real objects are *3D*, but when one of the dimensions is significantly larger (or smaller), object modeling in 1D or 2D domains would be an indication of good engineering flair. Experience suggests that when 3D physical domains and their boundaries have planar,

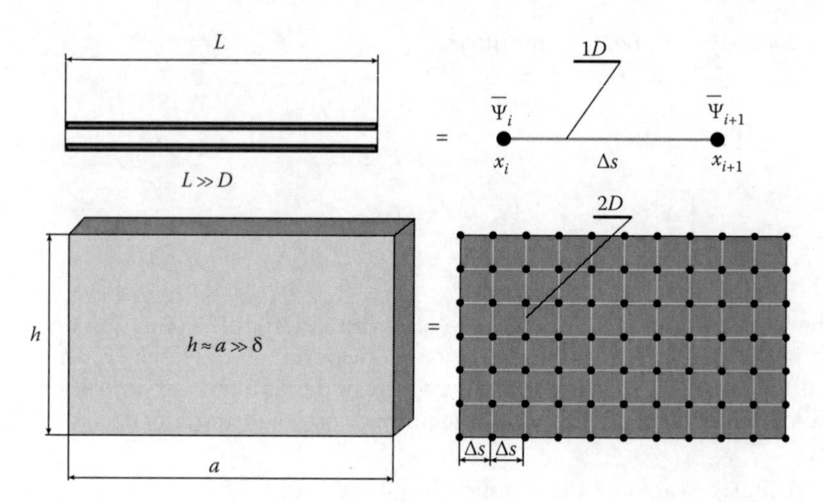

FIGURE 5.3
Physical objects (pipe and wall) modeled as simplified global areas.

axial, or central symmetry, they can be analyzed by using simple spatial models (see Figure 5.3). When a global physical domain consists of subdomains of different materials, the interfaces should bound domains with different physical properties as shown in Figure 5.4.

Perform an analysis of a physical domain. Choose an appropriate geometrical model of the global domain together with a coordinate system. Then, switch to the second stage of dividing the physical domain into a collection of subdomains (FEs).

Figure 5.5 presents in tabular form the most common types of FEs and their application in specific tasks, depending on the complexity of the

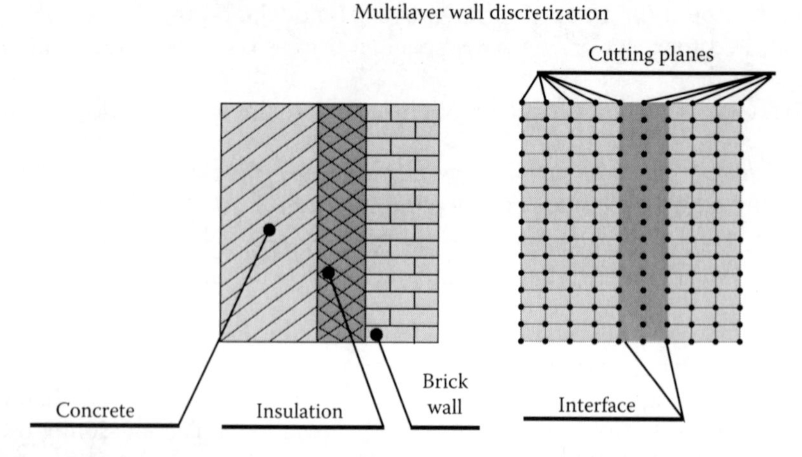

FIGURE 5.4
Discretization of a compound physical area.

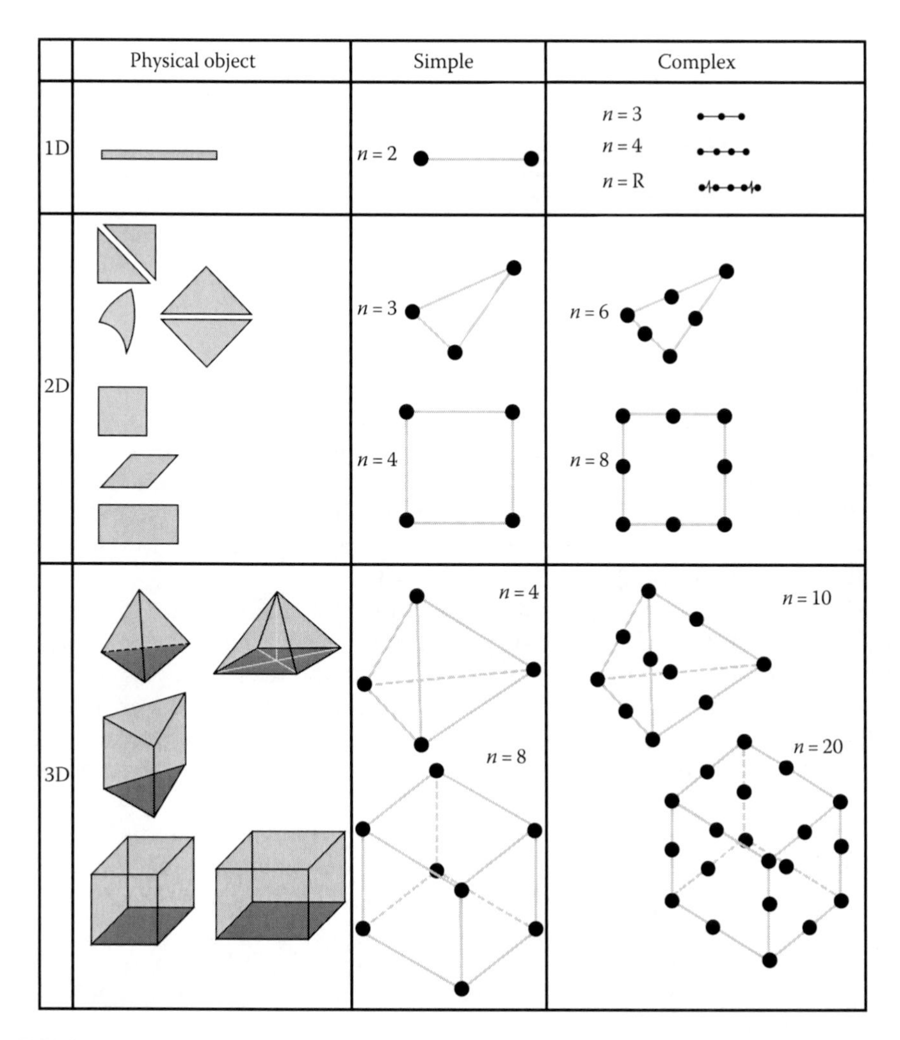

FIGURE 5.5
Types of finite elements used for the discretization of physical objects and classification of the discrete analogues of finite element.

global physical domain, required accuracy, and availability of computational resources. The number of FEs used in domain discretization is not limited. Moreover, to achieve the best match between the real object and the geometric model, one can increase the FE number (mesh the domain) accordingly.

The third (last) stage of modeling the physical domain is the substitution of the FE continuum for a discretely distributed domain, that is, an FE discrete analogue. The latter is a set of nodes distributed along the element boundary. Specification of the scheme type, number, location, step, and order of nodes is needed for that purpose. Depending on the type of the FE discrete analogue, one can use different number of nodes. The FE classification according to their spatial dimension is given in Figure 5.5. Although the FE have a defined geometry, in the current stage it is important to get their discrete analogues, such as

- Node location
- Number of nodes of an FE
- Numbering of nodes and FEs of the global domain

As seen in Figure 5.5, despite the identical geometry, an FE can be presented using different discrete analogues. The aim is to increase the accuracy of the estimates performed using more accurate approximating functions, which will be discussed in this section. FE numbering in the global domain is important, and it affects the method of forming information massifs, as well as the buffer memory needed. The speed of algorithm operation will depend on the width of the matrices processed and on the direction of node overtravel (node indexation) in the global domain.

If the discretization of the continuous physical domain is a significant first step in the **WRM** implementation, the second step—the choice of an interpolation function—is most important in attaining the required accuracy with minimum computational operations. Each FE is part of the global area v_0, but it is defined at the nodes of the discrete analogue only. Hence, Equation 5.27 is valid only for a specific element, and it is a mini-model of the medium where transfer takes place.

The **WRM** interpolation function $\overline{\psi}_A(x,y,z,\tau)$ modeling the distribution of the ***ETS*** field parameters at a current point A is set forth using two methods. Although they yield identical results, their distinction is useful for the task of our study (Dimitrov 2013).

- *First method*: Modeling the density distribution of the domain occupied by the FE

One of the approaches of finding an estimate of the quantity $\overline{\psi}$, at point $A(x, y, z)$ of the FE and at moment τ, consists in the design of an interpolation function $\overline{\psi}_A(x,y,z,\tau)$ of the form

$$\psi_A(x,y,z,\tau) \approx \overline{\psi}_A(x,y,z,\tau) = \sum_{m=i,j,k,l,\ldots} P_m^e(x,y,z)\overline{\psi}_m^e(\tau), \qquad (5.28)$$

where

$\overline{\psi}_m^e(\tau)$ is the amplitude of ψ at the mth node of the FE

$P_m^e(x,y,z)$ is the approximating function of the mth node of the FE, known also as a shape function

Each shape function sets forth the distribution of the solution $\overline{\psi}_A(x,y,z,\tau)$ in the space between nodes. It depends on the spatial coordinates (x, y, z) only. From a methodical point of view, however, $\overline{\psi}$ is interpreted as a function *organizing* density of the spatial continuum around the nodes. Moreover, the values of the integrals in Equation 5.28, that is, the accuracy of the estimation $\overline{\psi}_m^e$ of the studied quantity, depend on the structure of the shape function.

Product $P_m^e(x,y,z)*\overline{\psi}_m^e(\tau)$ in relation (Equation 5.28) presents the tribute of the mth node of the FE (e) to the estimation of quantity $\overline{\psi}_A(x,y,z,\tau)$. The estimation of quantity $\overline{\psi}$ is carried out via equality (Equation 5.28) as a superposition of its amplitudes at the nth nodes of the FE. For $n = 3$, that is, for a three-node complex linear element (pipe/beam) or a simplex (simple) triangular element, this expression is reduced to the following one:

$$\overline{\psi}_A = P_i^e\overline{\psi}_i^e + P_j^e\overline{\psi}_j^e + P_k^e\overline{\psi}_k^e. \tag{5.29}$$

- *Second method*: Modeling the distribution density of the assessed quantity

This is an alternative method employed in a predominant number of ***FEM*** techniques. Unlike the earlier considerations, density of the distribution of the ***ETS*** characteristics within the FE is an object of modeling here, and not the space density. For instance, $\overline{\psi}_A(x,y,z,\tau)$ is sought in the form

$$\overline{\psi}_A(x,y,x,\tau) = C_0^e + \sum_{m=i,j,k,l,\ldots} C_m^e P_m^e(x,y,z), \tag{5.30}$$

considering Dirichlet's boundary conditions

$$\overline{\psi}_0 = C_0^e + \sum_{m=i,j,k,l\ldots} C_m^e P_m^e(x,y,z),$$

over area A_ψ of the boundary surface A_0, and Neumann's boundary conditions

$$\Gamma\frac{\partial}{\partial n}\left[C_0^e + \sum_{m=i,j,k,l\ldots} C_m^e P_m^e(x,y,z)\right] - \overline{h}_0 = 0$$

over area A_h of the boundary surface A_0, where $A_0 = A_\varphi \cup A_h$. Here C_0^e and C_m^e are the analytical functions sought satisfying the boundary conditions. Comparison between (Equation 5.28) and (Equation 5.30) shows that the two approaches formally differ from each other, and they provide identical results.

One can choose various mathematical functions as shape functions $P_m^e(x,y,z)$. Practically, polynomials of the following type are predominantly employed as shape functions $P_m^e(x,y,z)$ in the FEM techniques:

$$p(x) = a + bx + cx^2 + dx^3 + \cdots + qx^n \quad \text{for one-dimensional elements,} \tag{5.31}$$

$$p(x,y) = a + bx + cy + dxy + ex^2 + fy^2 \quad \text{for two-dimensional elements,} \tag{5.32}$$

$$p(x,y,z) = a + bx + cy + dz + exy + fxz + gyz + hx^2 + jy^2 + iz^2$$
$$\text{for three-dimensional elements.} \tag{5.33}$$

Note that polynomials are convenient for computer processing, and this is their main advantage. However, an additional advantage is that the accuracy of their approximation can be easily increased—that is, by increasing the polynomial degree only (see Figure 5.6).

This status quo has been established due to the ability of computers to easily handle polynomials. An additional benefit is that the accuracy

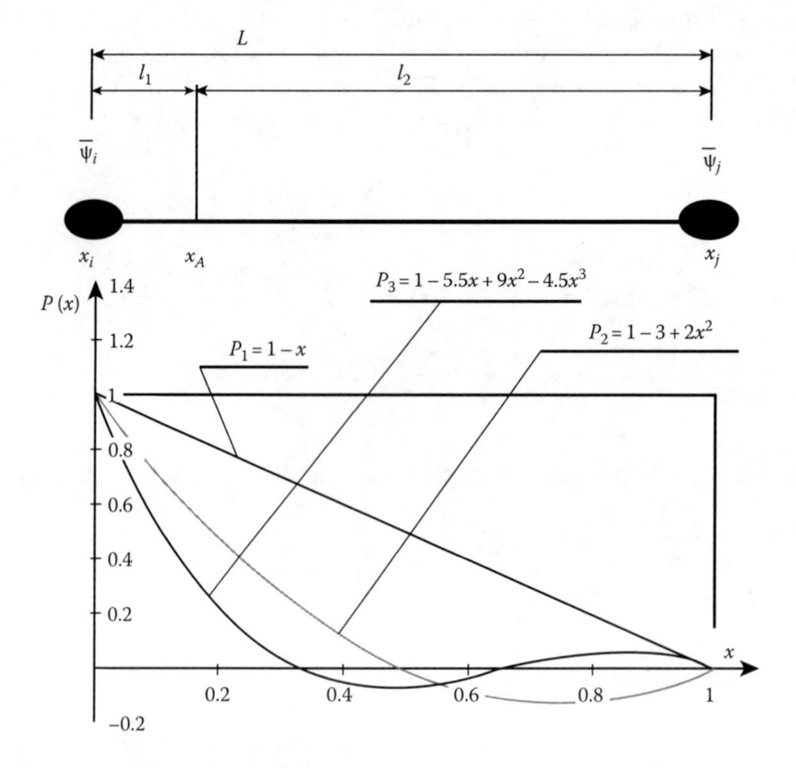

FIGURE 5.6
Linear, quadratic, and cubic shape functions.

of the approximation increases easily—only by raising the degree of the polynomial. It is considered, however, that a very good approximation can be achieved only by a first-degree polynomial, using a finer meshing step (smaller dimensions of the elements). This is proved when using commercial ***FEM*** software packages where the process of discretization in the global domain is fully automated. To penetrate deeper into shape functions, we will discuss in what follows the impact of the polynomial degree on the design of the FE discrete analogue and on the interpolation accuracy.

Consider a 1D linear FE—a pipe (beam) (see Figure 5.5), that is, the use of polynomials of the type (Equation 5.31) illustrated in Figure 5.7. Note that we use polynomials whose degree does not exceed 3. The coefficients in front of variable x are chosen such as to satisfy the most essential requirement to the shape function $P_m^e(x,y,z)$—that is, it should damp very fast in the vicinity of the *m*th node being zero at other FE nodes. The following equalities express that functional condition:

$$P_m^e(x_s,y_s,z_s)=\begin{cases}1, & s=m\\ 0, & s\neq m\end{cases}, \quad \text{i.e.,} \quad \sum_{m=i,j,k} P_m^e=1, \tag{5.34}$$

where (x_m, y_m, z_m) is the absolute coordinate of a node of the 1D element that a shape function P_m^e has been designed for.

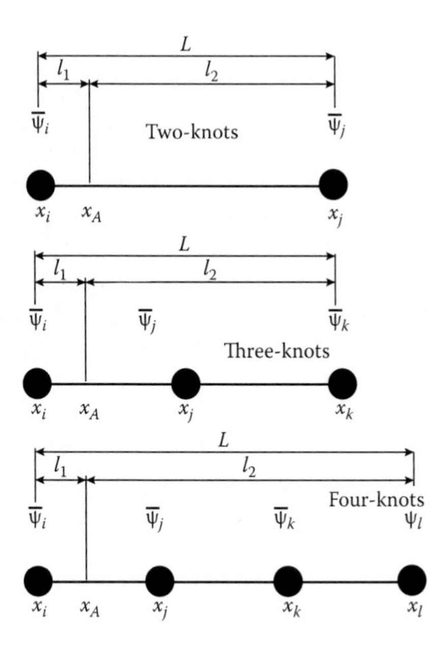

FIGURE 5.7
Linear finite element in absolute and natural coordinates.

The distribution of quantity ψ along a linear simple (simplex) FE (e) reads

$$\overline{\psi}_A = P_i^e \overline{\psi}_i^e + P_j^e \overline{\psi}_j^e, \tag{5.35}$$

where P_i^e, P_j^e are shape functions, and $\overline{\psi}_i^e, \overline{\psi}_j^e$ are estimations of ψ at two nodes i, j building the FE.

In case of a linear shape function, polynomial (Equation 5.31) is reduced to the expression $a + bx$, and obeying condition (Equation 5.34), we find the following values of coefficients a and b—$a = 1$ and $b = -1$ at the ith node (see Figure 5.6). As seen in Figure 5.6, the linear function *damps* slower for $x \to 1$ as compared to quadratic and cubic functions. This is due to the larger values of the derivative at point $x = 0$ and, to be more precise, to the validity of inequalities $\frac{dP_1(x)}{dx} > \frac{dP_2(x)}{dx} > \frac{dP_3(x)}{dx}$ in the vicinity of the first node. Considering the quadratic and cubic shape functions of the FE thus chosen, the integral (Equation 5.31) will take different values—0.1666 and 0.125. As seen, they differ by 66.7% and 75% as compared to the integral values 0.5, found via a linear function.

Thus, the qualitative advantage of higher-degree polynomials is illustrated. Those shape functions however imply another structure of the discrete analogue of the FE. Since the linear shape function requires that the analogue contain two nodes in order to find the two coefficients a and b, the quadratic function requires the availability of three nodes in order to find three coefficients, and the cubic one requires four nodes in order to find four coefficients (see Figure 5.7). Those requirements transform the discrete analogue of the FE from a simplex (a simple) one to a complex one (see Figure 5.5). New nodes should be introduced in the element sections where the approximation function is zero. According to the graph shown in Figure 5.6, the quadratic function requires the introduction of an additional node at point $x = 0$, and the cubic function—requires two additional nodes at points $x = 0.33$ and $x = 0.66$. The new structures of the discrete analogue of a 1D FE will use information coming from three and four nodes, respectively, in order to estimate quantity ψ—in compliance with relation (Equation 5.28).

The interpolating functions in the new *1D* discrete analogues, according to relation (Equation 5.28), use three and four nodes, and they will have the following forms:

- *For the three-noded analogue*

$$\overline{\psi}_A(x, y, z, \tau) = P_i^e \overline{\psi}_i^e + P_j^e \overline{\psi}_j^e + P_k^e \overline{\psi}_k^e, \tag{5.36}$$

- *For the four-noded analogue*

$$\overline{\psi}_A(x, y, z, \tau) = P_i^e \overline{\psi}_i^e + P_j^e \overline{\psi}_j^e + P_k^e \overline{\psi}_k^e + P_l^e \overline{\psi}_l^e, \tag{5.37}$$

where P_i^e, P_j^e, P_k^e, and P_l^e are node shape functions, and $\overline{\psi}_i^e, \overline{\psi}_j^e, \overline{\psi}_k^e$, and $\overline{\psi}_l^e$ are estimates of the quantity ψ at the nodes of the respective scheme of the 1D discrete analogue. In compliance with condition (Equation 5.34), P_i^e, P_j^e, P_k^e, and P_l^e should satisfy the following relations:

$$P_i + P_j = 1 \quad \text{for the two-noded analogue,} \tag{5.38}$$

$$P_i + P_j + P_k = 1 \quad \text{for the three-noded analogue,}$$

$$P_i + P_j + P_k + P_l = 1 \quad \text{for the four-noded analogue.}$$

Thus, the following important problem concerning the design of the type of the approximating functions emerges. One can perform the design in absolute coordinates related to the global domain, but it is significantly more convenient to operate in a coordinate system linked with the FE itself. It is called a natural coordinate system (see Figure 5.7). More details on the use of various types of coordinate systems will be given in (Dimitrov 2013).

In what follows, we present for clarity the approximation functions of a 1D element in an absolute coordinate system. We find the following relations employing the instrumentation of the analytical geometry:

- For a *two-noded analogue* (corresponding to the linear approximation)

$$\bar{\psi}_A = \bar{\psi}_i \frac{x_j - x_A}{x_j - x_i} + \bar{\psi}_j \frac{x_A - x_i}{x_j - x_i},$$

- For a *three-noded analogue* (corresponding to a quadratic shape function)

$$\bar{\psi}_A = \bar{\psi}_i \cdot \left(1 - 3\frac{x_k - x_A}{x_k - x_i} + 2\left(\frac{x_k - x_A}{x_k - x_i}\right)^2\right) + \bar{\psi}_j \cdot 4 \cdot \frac{x_k - x_A}{x_k - x_i}\left(1 - \frac{x_k - x_A}{x_k - x_i}\right)$$
$$+ \bar{\psi}_k \cdot \frac{x_k - x_A}{x_k - x_i}\left(2\frac{x_k - x_A}{x_k - x_i} - 1\right),$$

- For a *four-noded analogue* (corresponding to a cubic shape function)

$$\bar{\psi}_A = \bar{\psi}_i \frac{x_A - x_i}{x_l - x_i}\left(1 - 4.5\frac{(x_A - x_i).(x_l - x_A)}{(x_l - x_i)^2}\right)$$
$$+ 4.5\bar{\psi}_j \frac{(x_A - x_i).(x_l - x_A)}{(x_l - x_i)^2}\left(3\frac{x_A - x_i}{x_l - x_i} - 1\right)$$
$$+ 4.5\bar{\psi}_k \frac{(x_A - x_i).(x_l - x_A)}{(x_l - x_i)^2}\left(3\frac{x_l - x_A}{x_l - x_i} - 1\right)$$
$$+ \bar{\psi}_l \frac{x_l - x_A}{x_l - x_i}\left(1 - 4.5\frac{(x_A - x_i).(x_l - x_A)}{(x_l - x_i)^2}\right).$$

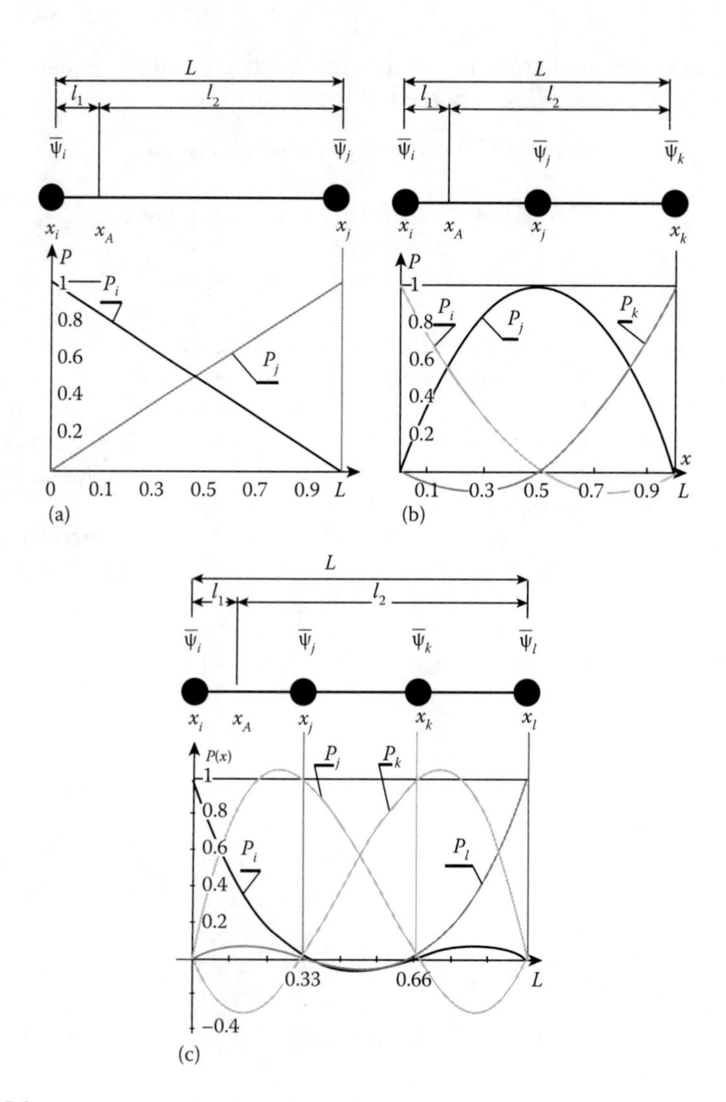

FIGURE 5.8
Plots of the approximating function in discrete analog elements: (a) in a two-noded element; (b) in a three-noded element; (c) in a four-noded element.

An analysis of the structure of these equations shows that it is similar to the structure of Equations 5.35 through 5.37. Figure 5.8 illustrates the development of the approximating functions P_m^e in natural coordinates, along 1D analogue with two, three, and four nodes, respectively.

The visual examination of the internal spatial distribution of P_m^e in the 1D elements shows that the best *filling* is attained in the case shown in Figure 5.8c. This is a plausible proof of the assertion that the increase in the number of nodes of an FE element (i.e., increase in the degree of the approximating polynomial, respectively) yields increase in the approximation accuracy.

If dimensionality of the FEs is larger than 1, the interpolating polynomials are written in a matrix form, since that form is more compact.

Thus, instead of relations (Equations 5.28, 5.35 through 5.37), the following form is used:

$$\bar{\psi} = \left[P^e\right] \cdot \left\{\bar{\psi}^e\right\} \quad m = i, j, k, l, \ldots, \tag{5.39}$$

where

$[P^e]$ is a vector of the approximation functions of the element nodes

$\left\{\bar{\psi}^e\right\}$ is a matrix-column containing the values of the estimates of quantity ψ at the nodes of the FE

Similarly, one can specify the approximating functions using relations (Equations 5.36 and 5.37) for *2D* and *3D* FEs.

To illustrate the earlier arguments, we shall use linear approximating functions only, applied to simple (simplex) *2D* and *3D* elements (see Figure 5.5). In the second part of this book, we shall pay special attention to approximating functions of higher degree applied in complex and multiplex finite elements, respectively.

Quantity $\bar{\psi}$ is interpolated by means of Equation 5.28, which gets the following form in the 2D case:

$$\psi_A(x, y, \tau) \approx \bar{\psi}_A(x, y, \tau) = \sum_{m=i,j,k} P_m^e(x, y)\bar{\psi}_m^e(\tau) = P_i \cdot \bar{\psi}_i + P_j \cdot \bar{\psi}_j + P_k \cdot \bar{\psi}_k$$

$$\text{as } P_i + P_j + P_k = 1. \tag{5.40}$$

As for the simplex FE, its linear part is kept in the approximating function (Equation 5.32), only (the element discrete analogue has three points only, which are sufficient to specify coefficients a_m, b_m, and c_m):

$$p_m^e(x, y) = a_m + b_m.x + c_m \cdot y. \tag{5.41}$$

Figure 5.9 illustrates the schemes of finding coefficients a_m, b_m, and c_m for **m = k**. The simple 2D FE is a three-noded triangle, where its apexes i, j, and k lie in the plane *xOy*, and they have coordinates (x_i, y_i), (x_j, y_j), (x_k, y_k), as shown in Figure 5.9.

Since the approximating function P_k^e for point *k* should satisfy conditions (Equations 5.34 and 5.40), it can be found using the equation of a plane passing through points with coordinates $\lfloor(x_k, y_k, 1), (x_i, y_i, 0), (x_j, y_j, 0)\rfloor$:

$$\det\begin{bmatrix} (x - x_k) & (y - y_k) & (P_k^e - 1) \\ (x_i - x_k) & (y_i - y_k) & (0 - 1) \\ (x_j - x_k) & (y_j - y_k) & (0 - 1) \end{bmatrix} = 0. \tag{5.42}$$

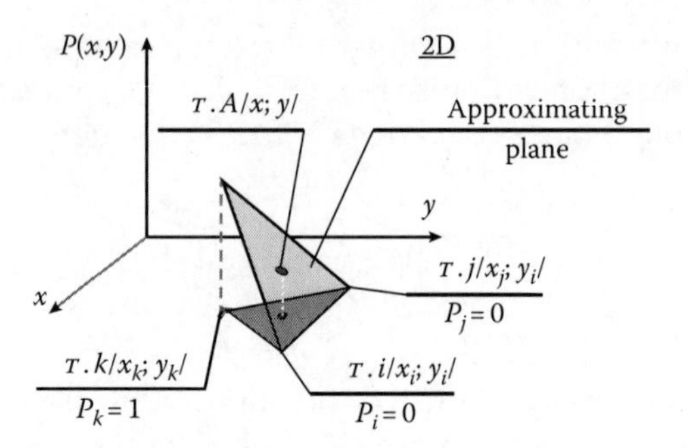

FIGURE 5.9
Distribution of the interpolating function in a 2D simple finite element.

Coefficients a_m, b_m, and c_m in Equation 5.41 are found according to the formulas:

$$a_k = \frac{\Delta_1^k}{\Delta_3^k} x_k - \frac{\Delta_2^k}{\Delta_3^k} y_k + 1, \qquad b_k = -\frac{\Delta_1^k}{\Delta_3^k}, \qquad c_k = \frac{\Delta_2^k}{\Delta_3^k}. \tag{5.43}$$

Here Δ_1^k, Δ_2^k, and Δ_3^k are adjugate matrices of the elements belonging to the first row of matrix (Equation 5.42):

$$\Delta_1^k = \begin{bmatrix} (y_i - y_k) & (-1) \\ (y_j - y_k) & (-1) \end{bmatrix}, \quad \Delta_2^k = \begin{bmatrix} (x_i - x_k) & (-1) \\ (x_j - x_k) & (-1) \end{bmatrix}, \quad \Delta_3^k = \begin{bmatrix} (x_i - x_k) & (y_i - y_k) \\ (x_j - x_k) & (y_j - y_k) \end{bmatrix}. \tag{5.44}$$

Coefficients of the approximating functions P_i and P_j for the other two nodes of the triangular FE are found via cyclic substitution of indexes *i, j, k* in Equations 5.43 and 5.44.

The interpolating function (Equation 5.40) gets the following form:

$$\bar{\psi}_A(x,y) = \sum_{m=i,j,k} (a_m + b_m x + c_m y)\bar{\psi}_m = (a_i + b_i x + c_i y)\bar{\psi}_i$$
$$+ (a_j + b_j x + c_j y)\bar{\psi}_j + (a_k + b_k x + c_k y)\bar{\psi}_k$$

while for $m = i, j, k$,

$$(a_i + b_i x + c_i y) + (a_j + b_j x + c_j y) + (a_k + b_k x + c_k y) = 1.$$

Considering a *3D* simplex/simple element (tetrahedron), ψ is interpolated using relation (Equation 5.33) under conditions specified by relation (Equation 5.34). Polynomials of the following form are chosen as shape functions P_m^e for nodes $m = i, j, k, l$ (see Figure 5.10):

$$P_m^e(x, y, z) = a_m + b_m x + c_m y + d_m z, \tag{5.45}$$

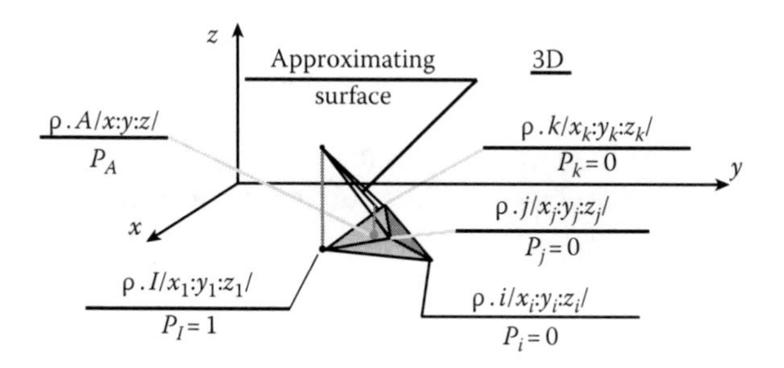

FIGURE 5.10
Interpolating function in a 3D/simple element (tetrahedron).

where only linear terms are present. As in the 2D case, each of those functions can be found from an equation similar to Equation 5.42. It has the following form for the lth node:

$$\det\begin{bmatrix} (x-x_l) & (y-y_l) & (z-z_l) & (P_l^e-1) \\ (x_i-x_l) & (y_i-y_l) & (z_i-z_l) & -1 \\ (x_j-x_l) & (y_j-y_l) & (z_j-z_l) & -1 \\ (x_k-x_l) & (y_k-y_l) & (z_k-z_l) & -1 \end{bmatrix}=0. \quad (5.46)$$

Expanding the determinant with respect to the elements in the first row, we find formulas similar to those of (Equation 5.44) for the coefficients a_l, b_l, c_l, d_l in relation (5.46), that is,

$$a_l=\left(-\frac{\Delta_1^l}{\Delta_4^l}x_l+\frac{\Delta_2^l}{\Delta_4^l}y_l-\frac{\Delta_3^l}{\Delta_4^l}z_l+1\right),\quad b_l=\left(\frac{\Delta_1^l}{\Delta_4^l}\right),\quad c_l=\left(-\frac{\Delta^l}{\Delta_4^l}\right),\quad \text{and}\quad d_l=\left(\frac{\Delta_3^l}{\Delta_4^l}\right). \quad (5.47)$$

Here $\Delta_s^l, s=i,j,k,l$ are adjugate matrices of the elements in the first row of the matrix in (5.46). Putting the coefficients of relation (Equation 5.47) into relation (Equation 5.45), we get the following form of the shape function of the lth node:

$$P_l^e=\left(-\frac{\Delta_1^l}{\Delta_4^l}x_l+\frac{\Delta_2^l}{\Delta_4^l}y_l-\frac{\Delta_3^l}{\Delta_4^l}z_l+1\right)+\frac{\Delta_1^l}{\Delta_4^l}x-\frac{\Delta_2^l}{\Delta_4^l}y+\frac{\Delta_3^l}{\Delta_4^l}z, \quad (5.48)$$

where

- x_l, y_l, z_l are the coordinates of the apex, at which we estimate the approximating function
- x, y, z are the coordinates of the current point A belonging to the tetrahedron interior (see Figure 5.10)

The coefficients of the approximating functions P_i^e, P_j^e, and P_k^e for the other three nodes of the simple 3D FE are found via cyclic substitution of indexes i, j, k, and l in Equation 5.46 through 5.48.

Consider subareas where point A divides the tetrahedron. In what follows, we shall use notations v_i, v_j, v_k, v_l for the subarea volume, where v_e is the tetrahedron volume ($v_e = v_i + v_j + v_k + v_l$). Note that the volume of the tetrahedron is calculated via the known formula:

$$v_e = \frac{1}{6}\det\begin{bmatrix} x_i & y_i & z_i & 1 \\ x_j & y_j & z_j & 1 \\ x_k & y_k & z_k & 1 \\ x_l & y_l & j_l & 1 \end{bmatrix}$$

Volume v_m, $m = i, j, k, l$ is found by substituting the row containing the coordinates of the mth node by row $\begin{bmatrix} x & y & z & 1 \end{bmatrix}$ corresponding to the current point A.

In natural/relative coordinates, the shape functions of *3D* elements are found following the procedure of finding the shape functions of 2D FEs, whereas ratios between the volumes introduced earlier are used instead of ratios between domains (Dimitrov 2013):

$$P_m^e = \frac{v_m}{v_e}, \qquad m = i, j, k, l, \qquad \sum_{m=i,j,k,l} P_m^e = 1.$$

The interpolating function of $\bar{\psi}_A$ at point $A \in v_{(e)}$ gets the following form:

$$\bar{\psi}_A(x, y) = \frac{v_i}{v_e}\bar{\psi}_i + \frac{v_j}{v_e}\bar{\psi}_j + \frac{v_k}{v_e}\bar{\psi}_k + \frac{v_l}{v_e}\bar{\psi}_l,$$

$$\text{where} \quad \frac{v_i}{v_e} + \frac{v_j}{v_e} + \frac{v_k}{v_e} + \frac{v_l}{v_e} = 1, \quad \text{i.e.,} \quad v_i + v_j + v_k + v_l = v_e.$$

In conformity with the method of interpolating the ***ETS*** field characteristics by means of expressions of the form (Equation 5.28), it seems useful to describe the discrete form of the corresponding **vector-gradient** characteristic function. It is an analogue of the vector-gradient function of the continuous medium.

We shall use the matrix form of Equation 5.39 for convenience:

$$\bar{\psi} = \left[P^e\right]\cdot\left\{\bar{\psi}^e\right\} = \begin{bmatrix} P_i^e & P_j^e & P_k^e & \dots & P_m^e \end{bmatrix}\begin{Bmatrix} \bar{\psi}_i \\ \bar{\psi}_j \\ \bar{\psi}_k \\ \vdots \\ \bar{\psi}_m \end{Bmatrix},$$

where

$\{\bar{\psi}^e\}$ is the matrix-column, containing the estimates $\bar{\psi}_m, m = i, j, k, l, \ldots$ at the nodes of the discrete analogue of an arbitrary FE

$\bar{\psi}_m, m = i, j, k, l, \ldots$

$[P^e]$ is the matrix-row of the approximating functions for the nodes of an FE

$P_i^{(e)} \quad P_j^{(e)} \quad P_u^{(e)} \quad \ldots \quad P_n^{(e)}$ are the approximating functions at a node of an FE

$\bar{\psi}_i \quad \bar{\psi}_j \quad \bar{\psi}_k \quad \ldots \quad \bar{\psi}_m$ are the estimates of the field characteristic at the corresponding node

Vector grad $\bar{\psi}_m$ in the corresponding coordinate system can also be presented in a matrix form:

- *In Cartesian coordinates*

$$\text{grad}\ \bar{\psi} = \left\{g_D^{(e)}\right\} = \left[B_D^{(e)}\right]\left\{\psi_i^{(e)}\right\} \tag{5.49}$$

- *In cylindrical coordinates*

$$\text{grad}\ \bar{\psi} = \left\{g_C^{(e)}\right\} = \left[B_C^{(e)}\right]\left\{\psi_i^{(e)}\right\}.$$

- *In spherical coordinates*

$$\text{grad}\ \bar{\psi} = \left\{g_{Sh}^{(e)}\right\} = \left[B_{Sh}^{(e)}\right]\left\{\psi_i^{(e)}\right\}.$$

We use here the following notations:

$$[B^e] = \nabla[P^e] \quad \text{and respectively} \quad [B^e]^T = \nabla[P^e]^T,$$

where $[B^e]$ gets the following form using the corresponding coordinates:

$$\left[B_D^e\right] = \left[\frac{\partial\left[P^e\right]}{\partial x} \quad \frac{\partial\left[P^e\right]}{\partial y} \quad \frac{\partial\left[P^e\right]}{\partial z}\right], \quad \left[B_C^e\right] = \left[\frac{\partial\left[P^e\right]}{\partial r} \quad \frac{\partial\left[P^e\right]}{\partial \theta} \quad \frac{\partial\left[P^e\right]}{\partial z}\right],$$

$$\left[B_{Sh}^e\right] = \left[\frac{\partial\left[P^e\right]}{\partial r} \quad \frac{\partial\left[P^e\right]}{\partial \theta} \quad \frac{\partial\left[P^e\right]}{\partial \lambda}\right].$$

The earlier expressions are valid for matrices containing derivatives of the matrix $[P^e]$ with respect to the coordinates of Cartesian, cylindrical, and spherical coordinate systems, respectively. In what follows, we shall derive the links between vector grad $\bar{\psi}$, matrix **[B]**, and coefficients of the approximating functions, for *1D*, *2D*, *3D* simple FE.

5.3.1.1 One-Dimensional Simple Finite Element

Since the element is composed of two nodes [i and j], we look for an interpolation of the form

$$\bar{\psi} = P_i \cdot \bar{\psi}_j + P_j \cdot \bar{\psi}_j.$$

Here P_i and P_j are equal to $P_i = a_i + b_i \cdot x = (x_j/L) - (1/L) \cdot x$ and $P_j = a_j + b_j \cdot x = -(x_i/L) + (1/L) \cdot x$, where L is the step between two nodes (respectively ($dP_i/dx = b_i = -(1/L)$; $dP_j/dx = b_j = 1/L$) (Dimitrov 2013).

The matrix of the approximating functions has the following form:

$$\left[P^e\right] = \left[\left(\frac{x_j}{L} - \frac{1}{L} \cdot x\right)\left(-\frac{x_i}{L} + \frac{1}{L} \cdot x\right)\right].$$

The matrix of the gradient reads as follows:

$$\text{grad } \bar{\psi} = \left\{g^e\right\} = \left(\left[\frac{dP^e}{dx}\right] \cdot \left\{\bar{\psi}^{(e)}\right\}\right) = \left[\frac{d[P_i]}{dx} \quad \frac{d[P_j]}{dx}\right]\left\{\begin{matrix}\bar{\psi}_i \\ \bar{\psi}_j\end{matrix}\right\} = \left[b_i \quad b_j\right]\left\{\begin{matrix}\bar{\psi}_i \\ \bar{\psi}_j\end{matrix}\right\}$$

$$= \left[-\frac{1}{L} \quad \frac{1}{L}\right]\left\{\begin{matrix}\bar{\psi}_i \\ \bar{\psi}_j\end{matrix}\right\},$$

or in a concise form

$$\text{grad } \bar{\psi} = \left\{g^e\right\} = \left[B^e\right]\left\{\bar{\psi}^e\right\}.$$

The following inequalities hold for matrix $[B]$ and its transposed matrix $[B]^T$:

$$\left[B^e\right] = \frac{1}{L}\left[-1 \quad 1\right] \text{ and } \left[B^e\right]^T = \frac{1}{L}\left\{\begin{matrix}-1 \\ 1\end{matrix}\right\}. \tag{5.50}$$

5.3.1.2 Two-Dimensional Simple Finite Element in Cartesian Coordinates

Following the classification adopted, the FE consists of three nodes, and the interpolating function has the following form:

$$\bar{\psi} = \left[P^e\right]\left\{\bar{\psi}^e\right\} = P_i \cdot \bar{\psi}_i + P_j \cdot \bar{\psi}_j + P_k \cdot \bar{\psi}_k.$$

The matrix of the approximating functions has the following components (Dimitrov 2013):

$$\left[P^e\right] = \begin{bmatrix} a_i & b_i & c_i \\ a_j & b_j & c_j \\ a_k & b_k & c_{ik} \end{bmatrix} \begin{Bmatrix} 1 \\ x \\ y \end{Bmatrix}$$

$$= \left[a_i + b_i \cdot x + c_i \cdot y \quad a_j + b_j \cdot x + c_j \cdot y \quad a_k + b_k \cdot x + c_k \cdot y\right].$$

The gradient function, given in a discrete form, reads as follows:

$$\text{grad } \overline{\psi} = \left\{g^e\right\} = \begin{bmatrix} \left\{\dfrac{dP^e}{dx}\right\}^T \\ \left\{\dfrac{dP^e}{dy}\right\}^T \end{bmatrix} \left\{\overline{\psi}^e\right\} = \frac{1}{2A_0} \begin{bmatrix} b_i & b_j & b_k \\ c_i & c_j & c_k \end{bmatrix} \begin{Bmatrix} \overline{\psi}_i \\ \overline{\psi}_j \\ \overline{\psi}_k \end{Bmatrix} = \left[B^e\right]\left\{\overline{\psi}^e\right\}. \quad (5.51)$$

$$\left[B^e\right] = \frac{1}{2A_0} \begin{bmatrix} b_i & b_j & b_k \\ c_i & c_j & c_k \end{bmatrix} \text{ and } \left[B^e\right]^T = \frac{1}{2A_0} \begin{bmatrix} b_i & c_i \\ b_j & c_j \\ b_k & c_k \end{bmatrix}.$$

5.3.1.3 Two-Dimensional Simple Finite Element in Cylindrical Coordinates

Again, the FE consists of three nodes, and the approximation function has the same general form. All equalities expressed in Cartesian coordinates are valid here after performing the substitution: $x = r$, $y = z$.

5.3.1.4 Three-Dimensional Simple Finite Element

According to the classification adopted, the FE consists of four nodes, and the interpolating function has the following form:

$$\overline{\psi} = \left[P^e\right] \cdot \left\{\overline{\psi}^e\right\} = P_i \cdot \overline{\psi}_i + P_j \cdot \overline{\psi}_j + P_k \overline{\psi}_k + P_l \cdot \overline{\psi}_l$$

$$\left[P^e\right] = \begin{bmatrix} a_i & b_i & c_i & d_i \\ a_j & b_j & c_j & d_j \\ a_k & b_k & c_k & d_k \\ a_l & b_l & c_l & d_l \end{bmatrix} \begin{Bmatrix} 1 \\ x \\ y \\ z \end{Bmatrix} = \begin{Bmatrix} a_i + b_i x + c_i y + d_i z \\ a_j + b_j x + c_j y + d_j z \\ a_k + b_k x + c_k y + d_k z \\ a_l + b_l x + c_l y + d_l z \end{Bmatrix}^T. \quad (5.52)$$

The gradient function contains derivatives with respect to z:

$$\text{grad}\,\bar{\psi} = \{g^e\} = \begin{bmatrix} \left\{\dfrac{dP^e}{dx}\right\}^T \\ \left\{\dfrac{dP^e}{dy}\right\}^T \\ \left\{\dfrac{dP^e}{dz}\right\}^T \end{bmatrix} \{\bar{\psi}^e\} = \frac{1}{6v_0}\begin{bmatrix} b_i & b_j & b_k & b_l \\ c_i & c_j & c_k & c_l \\ d_i & d_j & d_k & d_l \end{bmatrix} \begin{Bmatrix} \bar{\psi}_i \\ \bar{\psi}_j \\ \bar{\psi}_k \\ \bar{\psi}_l \end{Bmatrix} = [B^e]\{\bar{\psi}^e\}$$

$$[B^e] = \frac{1}{6v_0}\begin{bmatrix} b_i & b_j & b_k & b_l \\ c_i & c_j & c_k & c_l \\ d_i & d_j & d_k & d_l \end{bmatrix} \qquad [B^e]^T = \frac{1}{6v_0}\begin{bmatrix} b_i & c_i & d_i \\ b_j & c_j & d_j \\ b_k & c_k & d_k \\ b_l & c_l & d_l \end{bmatrix}. \tag{5.53}$$

The notations thus introduced and equations found will be used to derive formulas for different types of FEs. This will be the subject of Chapter 6.

5.3.2 Modeling of Transfer in a Finite Element Using a Matrix Equation (Galerkin Method)

Assume that the two stages of **FEM** application have been completed if we have managed to design an adequate scheme of FE discrete analogues, with nodes appropriately numbered and shape functions appropriately chosen.

Section 5.3.1 presents a derivation of the basic **WRM** Equations 5.26 and 5.27, which can be used after introducing weighting functions $W_m^e, m = i, j, k, l$. From a mathematical point of view, Galerkin B.G. solved the problem (Galerkin 1915), who assumed that the weighting function for a specific node is identical to the shape function for that node, that is,

$$W_m^e = P_m^e, \quad m = i, j, k, l \quad \text{where} \quad \{W^e\} = [P^e]^T. \tag{5.54}$$

Substituting expression (5.54) into (5.23), we find

$$\int_{v_e} [P^e]^T \left(\frac{\partial}{\partial \tau} \left(c_\phi \cdot \rho \cdot \bar{\psi}^e \right) \right) dv - \int_{v_e} [P^e]^T \cdot \nabla [\Gamma] \text{grad} \{\bar{\psi}^e\} dv - \int_{v_e} [P^e]^T \cdot G_{h_e}\, dv = 0, \tag{5.55}$$

and Newmann's boundary condition (Equation 5.24) gets the following form (Dimitrov 2013):

$$\int_{A_h} \left[P^e\right]^T \left([\Gamma]\text{grad}\left\{\bar{\psi}^e\right\} + \bar{h}_e\right) dA = 0. \tag{5.56}$$

The term-wise summation of expressions (Equations 5.55 and 5.56) yields the matrix form of the FE integral equation of transfer:

$$\int_{v_e} \left[P^e\right]^T \left(\frac{\partial}{\partial \tau}\left(c_{\bar{\psi}} \cdot \rho^e \cdot \bar{\psi}^e\right)\right) dv - \int_{v_e} \left[P^e\right]^T \nabla\left([\Gamma]\text{grad}\left\{\bar{\psi}^e\right\}\right) dv - \int_{v_e} \left[P^e\right]^T G_{he}\, dv$$
$$+ \int_{A_h} \left[P^e\right]^T \left([\Gamma]\text{grad}\left\{\bar{\psi}^e\right\} + \bar{h}_e\right) dA = 0.$$

In case of steady transfer, the first (accumulation) term of that equation is disregarded, and we find

$$\int_{v_e} \left[P^e\right]^T \nabla\left([\Gamma]\text{grad}\left\{\bar{\psi}^e\right\}\right) dv + \int_{v_e} \left[P^e\right]^T G_{he}\, dv$$
$$- \int_{A_h} \left[P^e\right]^T \left([\Gamma]\text{grad}\left\{\bar{\psi}^e\right\} + \bar{h}_e\right) dA = 0. \tag{5.57}$$

Apply *Green's theorem* to the first member of the earlier equation. Then,

$$\int_{v_e} \left[P^e\right]^T \nabla\left([\Gamma]\text{grad}\left\{\bar{\psi}^e\right\}\right) dv = -\int_{v_e} \left(\nabla\left[P^e\right]^T [\Gamma]\text{grad}\left\{\bar{\psi}^e\right\}\right) dv$$
$$+ \int_{A_0 = A_h + A_{\bar{\psi}}} \left[P^e\right]^T [\Gamma]\text{grad}\left\{\bar{\psi}^e\right\} dA.$$

Using relations (Equation 5.49), we find the following form of the integral equation (Equation 5.57):

$$\int_{v_e} \left[P^e\right]^T G_{h_e}\, dv - \int_{v_e} \left[B^e\right]^T [\Gamma]\left[B^e\right] dv \cdot \left\{\bar{\psi}^e\right\} + \int_{A_h} \left[P^e\right]^T \left([\Gamma]\text{grad}\left\{\bar{\psi}^e\right\} + \bar{h}_e\right) dA$$
$$+ \int_{A_0 = A_h + A_{\bar{\psi}}} \left[P^{e(i)}\right]^T [\Gamma]\text{grad}\left\{\bar{\psi}^e\right\} dA = 0. \tag{5.58}$$

The last integral of the earlier expression is reduced to an integral over A_h only, since we have $\operatorname{grad}\{\overline{\psi}^e\} = 0$ in $A_{\overline{\psi}}$ by definition. Unifying the last two terms of the earlier equation, we get

$$\int_{v_e} [P^e]^T G_{h_e}\, dv - \int_{v_e} [B^e]^T [\Gamma][B^e] dv \cdot \{\overline{\psi}^e\} - \int_{A_h} [P^e]^T \overline{h}_e\, dA = 0.$$

The specific flux $\overline{h}_e$ can be assumed as composed of a direct flux ($\overline{h}$) and convective flux ($\overline{h}_c$), that is,

$$\overline{h}_e = \overline{h} + h_c = \overline{h} + \alpha(\overline{\psi}_m - \overline{\psi}_{Out}), \tag{5.59}$$

where

- α is Newton's coefficient of mass release from or transport to the element surface
- $\overline{\psi}_{Out}$ is the value of $\overline{\psi}$ in the fluid area being adjacent to the surface $A_{\overline{\psi}}$ of the element

The insertion of Equation 5.59 into the integral equation (Equation 5.58) yields the following equality:

$$\left[\int_{v_e} [B^e]^T [\Gamma][B^e] dv + \int_{A_\phi} [P^e]^T \alpha [P^e] dA\right] \{\overline{\psi}^e\}$$
$$= \left\{ \begin{array}{l} \displaystyle\int_{v_e} [P^e]^T G_{h_e}\, dv + \\ \displaystyle + \int_{A_\phi} [P^e]^T \alpha \overline{\psi}_{Out}\, dA - \int_{A_h} [P^e]^T \overline{h}_e\, dA \end{array} \right\}, \tag{5.60}$$

and in a matrix form,

$$[G^e] \cdot \{\overline{\psi}^e\} = \{f^e\}. \tag{5.61}$$

We perform the following substitutions in the equation (see also the respective notations):

- $[K^e] = \int_{v_e} [B_e]^T [\Gamma][B_e] dv$ matrix of conductivity,
- $[F^e] = \int_{A_\phi} [P^e]^T \alpha [P^e] dA$ matrix of the surface properties,
- $[G^e] = [K^e] + [F^e]$ generalized matrix of conductivity of the ith finite element,

- $\{f_G^e\} = \int_{v_e} [P^e]^T G_{h_e}\, dv$ load vector of the element due to internal sources (recuperative load).
- $\{f^e\} = \{f_G^e\} + \{f_C^e\} + \{f_{Dr}^e\}$ load vector of the element.
- $\{f_C^e\} = \int_{A_\phi} [P^e]^T \alpha \bar{\psi}_e\, dA$ load vector of convection (Dirichlet load).
- $\{f_C^e\} = \int_{A_\phi} [P^e]^T \bar{h}_e\, dA$ load vector due to direct loading (Neumann's load).

This is the matrix equation sought describing transport in any FE. The solution of Equation 5.61 can be written in the following generalized form:

$$\{\bar{\psi}^e\} = [G^e]^{-1} \cdot \{f^e\}.$$

Note that Equation 5.61 can be used in different coordinate systems, depending on the characteristics of the physical environment in which the ***ETS*** is located. For instance, it gets the following form in a Cartesian coordinate system:

$$\begin{bmatrix} \int_{v_e} \left(\frac{\partial [P^e]^T}{\partial x} \Gamma_x \frac{\partial [P^e]}{\partial x} + \frac{\partial [P^e]^T}{\partial y} \Gamma_y \frac{\partial [P^e]}{\partial y} + \frac{\partial [P^e]^T}{\partial z} \Gamma_z \frac{\partial [P^e]}{\partial z} \right) dx\, dy\, dz \\ + \int_{A_\phi} [P^e]^T \alpha\, dA \end{bmatrix} \cdot \{\bar{\psi}^e\}$$

$$= \left\{ \int_{v_e} [P^e]^T G_{he}\, dx\, dy\, dz + \int_{A_\phi} [P^e]^T \alpha \bar{\psi}_{Out}\, dA - \int_{A_h} [P^e]^T \bar{h}_e\, dA \right\}. \qquad e = 1 \div M \tag{5.62}$$

while in a cylindrical coordinate system, considering axial symmetry, it reads as follows:

$$\left[\int_{v_e} \left(\frac{\partial [P^e]^T}{\partial r} \Gamma_r \frac{\partial [P^e]}{\partial r} + \frac{\partial [P^e]^T}{\partial z} \Gamma_z \frac{\partial [P^e]}{\partial z} \right) dV + \int_{A_\phi} [P^e]^T \alpha\, dA \right] \cdot \{\bar{\psi}^e\}$$

$$= \left\{ \int_{v_e} [P^e]^T G_{he}\, dv + \int_{A_\phi} [P^e]^T \alpha \bar{\psi}_{Out}\, dA - \int_{A_h} [P^e]^T \bar{h}_e\, dA \right\}. \qquad e = 1 \div M$$

The equations thus derived will be employed in Chapter 6, using 1D, 2D, and 3D FEs. The equation formulation will be illustrated via appropriately chosen estimations of transfer in a solid wall, considering boundary condition of real architectural–constructional design (Dimitrov 2013).

5.3.3 Steady Transfer in One-Dimensional Finite Element

One-dimensional FEs are widely used to study transfer, which takes place in bodies with plane, axial, or central symmetry, i.e., when a conducting medium is involved and subsequent boundary conditions are specified. These are the cases when transfer of heat or moisture through external surfaces of buildings or installations is to be assessed (for instance, transfer through the walls of massive reservoirs, chemical reactors, refrigerating chambers, etc.). Then, that specific transfer is known as 1D transfer, and the distribution of the parameters of the ETS is described by means of the ordinary differential equations specified in Table 5.1.

TABLE 5.1

Ordinary Differential Equations for One-Dimensional Transfer

Physical Environment	Differential Equation	Type of Symmetry
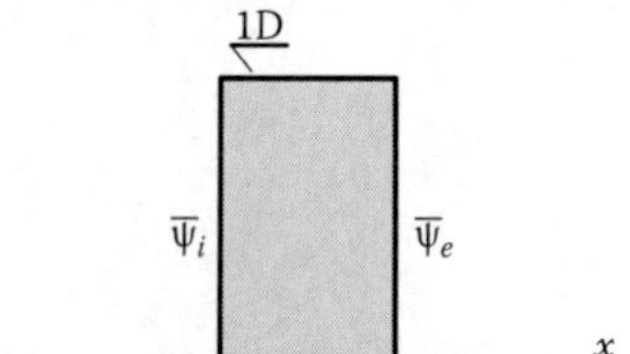	$\frac{d}{dx}\left(\Gamma\frac{d\psi}{dx}\right)=0$	Planar
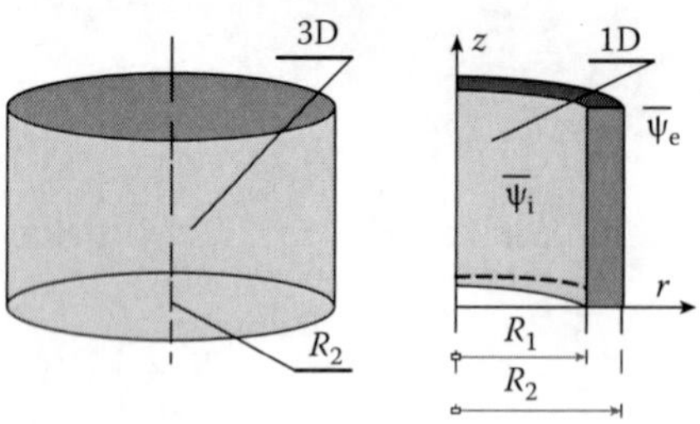	$\frac{1}{r}\frac{d}{dr}\left(r\frac{\Gamma}{\rho C_\psi}\frac{d\psi}{dr}\right)=0$	Axial
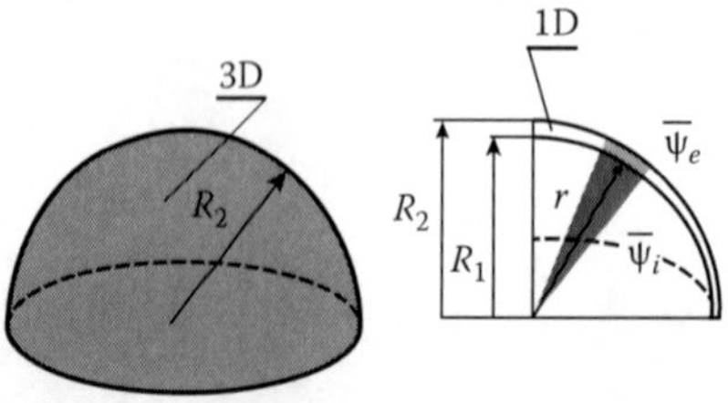	$\frac{1}{r^2}\frac{d}{dr}\left(r^2\frac{\Gamma}{\rho C_\psi}\frac{d\psi}{dr}\right)=0$	Central

If the properties of the physical medium are isotropic, then transfer takes place in one direction only, under the effect of the free energy gradient grad $\psi = d\psi/d\vec{n}_\psi$. Hence, the respective model is symmetric. The common feature of the three cases of symmetry is that transfer takes place along the normal to the isolines of the free energy. Note that the normal coincides with the normal to the external (bounding) surface, that is, it coincides with $\vec{Ox}$ or $\vec{r}$. The integral equations describing the process have identical structure, which follows from Equation 5.61 being valid for the *general 3D case*. Formally, similar integral equations are also used:

- *In a Cartesian coordinate system,*

$$\left(\int_{V_e} \frac{\partial\left[P^e\right]^T}{\partial x}\Gamma_x\frac{\partial\left[P^e\right]}{\partial x}\cdot 1\cdot 1\,dx+\int_{A_\phi}\left[P^e\right]^T\left[P^e\right]\alpha\,dA\right)\cdot\left\{\overline{\psi}^e\right\}$$
$$e=1\div M,\qquad m=i,j \tag{5.63}$$
$$=\int_{V_e}\left[P^e\right]^T G_{h_e}\cdot 1\cdot 1\,dx+\int_{A_\phi}\left[P^e\right]^T\alpha\overline{\psi}_e\,dA-\int_{A_h}\left[P^e\right]^T\overline{h}_e\,dA$$

- *In cylindrical and spherical coordinate systems,*

$$\left(\int_{V_e} \frac{\partial\left[P^e\right]^T}{\partial r}\Gamma_r\frac{\partial\left[P^e\right]}{\partial r}dV+\int_{A_\phi}\left[P^e\right]^T\left[P^e\right]\alpha\,dA\right)\cdot\left\{\overline{\psi}^e\right\}=\int_{V_{ei}}\left[P^e\right]^T G_{h_e}\,dv$$
$$+\int_{A_\phi}\left[P^e\right]^T\alpha\overline{\psi}_e\,dA-\int_{A_h}\left[P^{e(i)}\right]^T\overline{h}_e\,dA.$$
$$e=1\div M,\qquad m=i,j. \tag{5.64}$$

5.3.3.1 Integral Form of the Balance of Energy Transfer through One-Dimensional Finite Element

Consider 1D area (a rod or a pipe) with length $L0$. Performing discretization, it is replaced by N number of two-noded FEs with length $L = L_0/N$ = const; the number of nodes is $n = N + 1$. Apply the classical method of integral forms to the nodes—Section 5.3.

Assume that a constant gradient of the free energy function acts on the 1D area, that is, grad ψ = const. Assume also that it is directed along the normal to the wall surface, from the external to the internal surface, since $\psi_{out} > \psi_{in}$. Flux I of transferred mass, according to Equations 4.4 and 4.5 through 4.7, is directed opposite to the gradient grad ψ as shown in Figure 5.11. The boundary conditions concern the first and the last FEs of the domain shown in

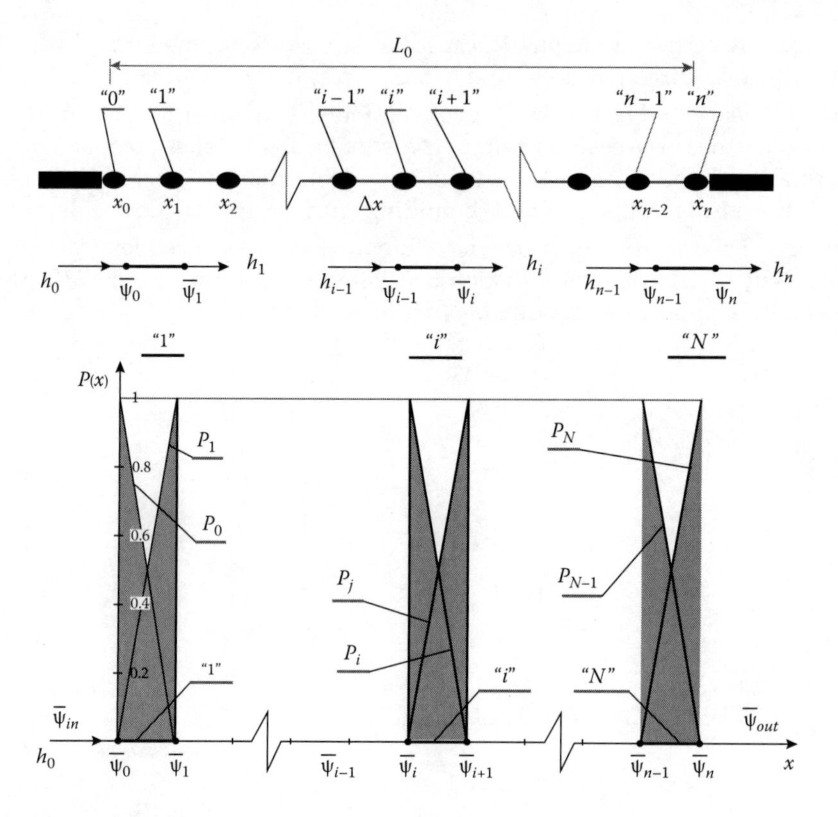

FIGURE 5.11
One-dimensional global area.

Figure 5.11. The balance equation (Equation 5.1) will be valid for each FE $e(i)$ under steady transfer where $[\Xi_g - \Xi_a] = 0$. Then, accounting for the values of density flux of energy, this equation is significantly simplified:

$$I_{in}^{e(i)} + I_{out}^{e(i)} = 0.$$

The balance equation will have the same form for the global 1D area $e(0)$:

$$I_{in}^{e(0)} + I_{out}^{e(0)} = 0. \tag{5.65}$$

Balance equation (Equation 5.65) can be written in the following form for the *facial* FE (denoted by 1 in Figure 5.11):

$$h_{in} - \frac{\Gamma}{L}\left(\overline{\psi}_0 - \overline{\psi}_1\right) = 0.$$

Here, we have the following:

- $I_{in}^{e(1)} = h_{in}A_0$ is a flux flowing into the *facial* FE.
- $I_{out}^{e(1)} = h_{0-1}A_0 = -\frac{\Gamma}{L}(\bar{\psi}_0 - \bar{\psi}_1)A_0$ is a flux flowing out of the *facial* FE ((h_{0-1}) is a specific flux from node "0" to node "1" in compliance with Equation 4.4), and it is proportional to the difference $(\bar{\psi}_0 - \bar{\psi}_1)/L$ being a local estimate of grad ψ in this element.
- h_{in} is a specific flux flowing into the *facial* FE—a boundary condition of element "1" and area $e(0)$.

This can be rewritten as

$$\frac{\Gamma}{L}(\bar{\psi}_0 - \bar{\psi}_1) = h_{in}. \tag{5.66}$$

The balance of the facial element "1" can be expressed alternatively:

$$I_{out}^{e(1)} = -I_1 = -h_1 \cdot A_0$$
$$I_{in}^{e(1)} = h_{0-1}A_0 = -\Gamma(\bar{\psi}_0 - \bar{\psi}_1)A_0,$$

where $I_1(h_1)$ is the flux (specific flux) of matter flowing out of element "**1**" and into element "**2**." Performing the substitution in Equation 5.65, we find an alternative balance equation of element "**1**":

$$-\frac{\Gamma}{L}(\bar{\psi}_0 - \bar{\psi}_1) - h_1 = 0,$$

and hence

$$\frac{\Gamma}{L}(\bar{\psi}_0 - \bar{\psi}_1) = -h_1. \tag{5.67}$$

Equations 5.66 and 5.67 can be united in a matrix form that is more convenient for work:

$$\frac{\Gamma}{L}\begin{bmatrix} 1 & -1 \\ -1 & 1 \end{bmatrix}\begin{Bmatrix} \bar{\psi}_0 \\ \bar{\psi}_1 \end{Bmatrix} = \begin{Bmatrix} h_{in} \\ -h_1 \end{Bmatrix}.$$

A similar relation can be derived for the *i*th FE:

$$\frac{\Gamma}{L}\begin{bmatrix} 1 & -1 \\ -1 & 1 \end{bmatrix}\begin{Bmatrix} \bar{\psi}_j \\ \bar{\psi}_{j+1} \end{Bmatrix} = \begin{Bmatrix} h_{i-1} \\ -h_i \end{Bmatrix}, \tag{5.68}$$

or

$$\left[K^e\right]\left\{\bar{\psi}^e\right\} = \left\{h^e\right\},$$

where h_{i-1} and $\mathbf{h}_i$ are specific fluxes flowing into the ith and $(i + 1)$th FE. Equation 5.68 thus found is a discrete analogue of the integral equation (Equation 5.61) for the "ith" FE. It is found by using the method of the integral forms, and it is valid for 1D steady transfer. Structurally, it is similar to the general matrix equation of the **WRM** (Equation 5.63), but it does not consider in detail the characteristic features of transfer through the 1D element. As seen in the material that follows, it does not account for the area geometry (shape of the cross section of the 1D area, for instance), and thus, it does not reflect the effect of the specific boundary conditions. We shall use Equation 5.68 to illustrate the advantages of the **WRM**.

It will be later seen that the **WRM** enables one to find a much flexible and adequate solutions of various practical problems, which require consideration of various conditions at the boundary of the 1D area. Those solutions can be found after a detailed adaptation of the structure of the general matrix equation (Equation 5.61), which is valid for 1D transfer.

5.3.3.2 Modified Matrix Equation of 1D Transfer

The general structure of Equation 5.61 is kept in the case of 1D transfer, too, but its terms are designed using different approximating functions. Consider that difference on the basis of the general matrix of conductivity $[G^e]$, specified in Equations 5.60 and 5.61.

As assumed, $[G^e]$ consists of two components: the first one $\left[K^{e(i)}\right]$ reflects the physical and geometrical properties of the FE, while the second one $\left[F^{e(i)}\right]$ reflects the physical properties of the FE boundary surface.

As noted, the approximating functions in a linear modification of 1D simplex element in absolute coordinates have the following form (Table 8.1): $P_i^{(e)} = a_i + b_i y = (y_i/L) - (1/L)y$ and $P_j^{(e)} = a_j + b_j y = (y_i/L) - (1/L)y$, and matrices—$[B^e] = 1/L[-1 \quad 1]$ and $[B^e]^T = 1/L\{-1 \quad 1\}$ (Equations 5.50).

The first matrix component of $[G^e]$-conductivity matrix $[K^{(e)}]$ will have the rank, 2 and it will consist of two rows and two columns, reading

$$\left[K^e\right] = \Gamma A_0 \int_0^L \begin{bmatrix} b_i^2 & b_i.b_j \\ b_j.b_i & b_j^2 \end{bmatrix} dx = \Gamma A_0 \int_0^L \begin{bmatrix} \left(-\frac{1}{L}\right)^2 & \left(-\frac{1}{L}\right)\left(\frac{1}{L}\right) \\ \left(\frac{1}{L}\right)\left(-\frac{1}{L}\right) & \left(\frac{1}{L}\right)^2 \end{bmatrix}$$

$$dx = \frac{\Gamma A_0}{L^2} \int_0^L \begin{bmatrix} 1 & -1 \\ -1 & 1 \end{bmatrix} dx = \frac{\Gamma A_0}{L} \begin{bmatrix} 1 & -1 \\ -1 & 1 \end{bmatrix}. \tag{5.69}$$

The second component of $[G^e]$—matrix of the surface properties $[F^e]$—is calculated according to the formula:

$$\left[F^e\right]=\pm\int_{A_{1,\psi}}\left[P^e\right]^T\alpha_{A_{1,\psi}}\left[P^e\right]dA\pm\int_{A_{2,\psi}}\left[P_0^e\right]^T\alpha_{A_{2,\psi}}\left[P_0^e\right]dA\pm\int_{A_{3,\psi}}\left[P_0^e\right]^T\alpha_{A_{3,\psi}}\left[P_0^e\right]dA. \tag{5.70}$$

Sign (+) is valid for convection from the element, while sign (–) is valid for convection to the element.

The matrix of the surface properties $[F^e]$ accounts for the physical conditions of transfer through the boundary, containing the coefficients of transfer α_{A1}, α_{A2}, and α_{A3}, and for the geometry of the boundary surface ($A_\psi = A_{1,\psi} + A_{2,\psi} + A_{3,\psi}$).

Here we have the following:

- A_ψ is the full peripheral surface of the element.
- $A_{1,\psi} = P{\cdot}L$ is the peripheral surface along its perimeter.
- $A_{2,\psi} = A_0$ is the facial surface (at node "i").
- $A_{3,\psi} = A_0$ is the bottom surface (at node "j").
- P is the element perimeter.

If a matrix of the approximating functions is written in absolute coordinates (Dimitrov 2013), then we have

$$\left[P^e\right]=\left[\left(a_i+b_iy\right)\quad\left(a_j+b_jy\right)\right]=\left[\left(\frac{y_i}{L}-\frac{1}{L}y\right)\quad\left(-\frac{y_i}{L}+\frac{1}{L}y\right)\right].$$

- For the facial surface (i.e., for node "i"), $P_i^e=1, \quad a \quad P_j^e=0.$
- For the bottom surface (for node "i + 1"), $P_i^e=0, \quad a \quad P_j^e=1.$

The matrix of the surface conditions $[F^e]$ will get the following form:

$$\left[F^e\right]=\pm\int_{A_{1,\psi}}\begin{Bmatrix}\left(\frac{x_j}{L}-\frac{1}{L}\cdot x\right)\\\left(-\frac{x_i}{L}+\frac{1}{L}\cdot x\right)\end{Bmatrix}\alpha_{A_{1,\psi}}\left[\left(\frac{x_j}{L}-\frac{1}{L}\cdot x\right)\quad\left(-\frac{x_i}{L}+\frac{1}{L}\cdot x\right)\right]dA$$

$$\pm\int_{A_{2,\psi}}\begin{Bmatrix}1\\0\end{Bmatrix}\alpha_{A_{2,\psi}}\begin{bmatrix}1 & 0\end{bmatrix}dA\pm\int_{A_{3,\psi}}\begin{Bmatrix}0\\1\end{Bmatrix}\alpha_{A_{3,\psi}}\begin{bmatrix}0 & 1\end{bmatrix}dA$$

$$=\pm\frac{\alpha_{A_{1,\psi}}\cdot P}{L^2}\int_0^L\begin{bmatrix}\left(x_j-x\right)^2 & \left(x_j-x\right)\left(-x_i+x\right)\\\left(-x_i+x\right)\left(x_j-x\right) & \left(-x_i+x\right)^2\end{bmatrix}dx$$

$$\pm\alpha_{A_{2,\psi}}A_0\begin{bmatrix}1 & 0\\0 & 0\end{bmatrix}\pm\alpha_{A_{3,\psi}}A_0\begin{bmatrix}0 & 0\\0 & 1\end{bmatrix}. \tag{5.71}$$

If however the matrix of the approximating functions is given in natural coordinates $[P^e]=\left[(l_2/L) \quad (l_1/L)\right]$, then we have the following:

- For the facial surface (i.e., for node "i"), $P_i^e=1$ and $P_j^e=0$.
- For the bottom surface (for node "j"), $P_i^e=0$ and $P_j^e=1$.

The matrix of the boundary conditions $[F^e]$ will get the following form:

$$\left[F^e\right]=\pm\alpha_{A_{1,\psi}}P\int_{-1}^{1}\begin{bmatrix}\bar{l}_2^2 & \bar{l}_2\cdot\bar{l}_1\\ \bar{l}_1\cdot\bar{l}_2 & \bar{l}_1^2\end{bmatrix}dl\pm\alpha_{A_{2,\psi}}A_0\begin{bmatrix}1 & 0\\ 0 & 0\end{bmatrix}\pm\alpha_{A_{3,\psi}}A_0\begin{bmatrix}0 & 0\\ 0 & 1\end{bmatrix}. \tag{5.72}$$

Then, the generalized matrix of element conductivity $[G^e]$ will be a sum of $[K^e]$ and $[F^e]$:

$$\left[G^e\right]=\frac{\Gamma\cdot A_0}{L}\begin{bmatrix}1 & -1\\ -1 & 1\end{bmatrix}\pm\frac{\alpha_{A_{1,\psi}}P}{L^2}\int_0^L\begin{bmatrix}(x_j-x)^2 & (x_j-x)(-x_i+x)\\ (-x_i+x)(x_j-x) & (-x_i+x)^2\end{bmatrix}dx$$

$$\pm\alpha_{A_{2,\psi}}A_0\begin{bmatrix}1 & 0\\ 0 & 0\end{bmatrix}\pm\alpha_{A_{3,\psi}}A_0\begin{bmatrix}0 & 0\\ 0 & 1\end{bmatrix}.$$

The load vector $\{f^e\}$, in compliance with definition (Equation 5.61), consists of three components that contain the element boundary conditions:

$$\left\{f^e\right\}=\int_{V_e}\left[P^e\right]^T G_{h_e}\,dV+\int_{A_\psi}\left[P^e\right]^T\alpha\bar{\psi}_e\,dA+\int_{A_{\bar{h}0}}\left[P^e\right]^T\bar{h}_e\,dA. \tag{5.73}$$

The first term in (Equation 5.73) provides information about the recuperation (conversion) of matter within the element. It reads as follows:

$$\left\{f_G^e\right\}=G_{h_e}\cdot A_0\int_0^L\begin{Bmatrix}\left(\dfrac{x_j}{L}-\dfrac{1}{L}\cdot x\right)\\ \left(-\dfrac{x_i}{L}+\dfrac{1}{L}\cdot x\right)\end{Bmatrix}dx=\frac{G_h\cdot A_0}{L}\int_0^L\begin{Bmatrix}x_j-x\\ -x_i+x\end{Bmatrix}dx. \tag{5.74}$$

Considering natural coordinates, matrix $\{f_G^e\}$ takes the following form:

$$\{f_G^e\} = G_{h_e} A_0 \int_{-1}^{1} \begin{Bmatrix} \overline{l_2} \\ \overline{l_1} \end{Bmatrix} dl.$$

The second term in relation (Equation 5.73) contains the boundary conditions of *Dirichlet's type* regarding convection from and to the element through the boundary surfaces ($A_\psi = A_{1,\psi} + A_{2,\psi} + A_{3,\psi}$). It is calculated as

$$\{f_C^e\} = \int_{A_\psi} \left[P^e\right]^T \alpha\overline{\psi}_e \, dA = \pm\alpha_{A_{1,\psi}} \int_{A_{1,\psi}} \left[P^e\right]^T \overline{\psi}_{A_{1,\psi}} \, dA \pm \alpha_{A_{2,\psi}}$$

$$\int_{A_{2,\psi}} \left[P^e\right]^T \overline{\psi}_{A_{2,\psi}} dA \pm \alpha_{A_{3,\psi}} \int_{A_{3,\psi}} \left[P^e\right]^T \overline{\psi}_{A_{3,\psi}} \, dA. \quad (5.75)$$

Vector $\{f_C^{(e)}\}$ gets the following form in absolute coordinates:

$$\{f_C^e\} = \pm\alpha_{A_{1,\psi}} \int_{A_{1,\psi}} \left[P^e\right]^T \alpha_{A_{1,\psi}} \, dA \pm \alpha_{A_{2,\psi}} \int_{A_{2,\psi}} \left[P^e\right]^T \alpha_{A_{2,\psi}} \, dA \pm \alpha_{A_{3,\psi}} \int_{A_{3,\psi}} \left[P^e\right]^T \alpha_{A_{3,\psi}} \, dA.$$

It reads in absolute coordinates as follows:

$$\{f_C^e\} = \pm\alpha_{A_{1,\overline{\psi}}} \overline{\psi}_{A_{1,\overline{\psi}}} P \int_0^L \begin{Bmatrix} \left(\dfrac{x_j}{L} - \dfrac{1}{L}\cdot x\right) \\ \left(-\dfrac{x_i}{L} + \dfrac{1}{L}\cdot x\right) \end{Bmatrix} dx \pm \alpha_{A_{2,\overline{\psi}}} \overline{\psi}_{A_{2,\overline{\psi}}} \int_{A_0} \begin{Bmatrix} 1 \\ 0 \end{Bmatrix} dA \pm \alpha_{A_{3,\overline{\psi}}} \overline{\psi}_{A_{3,\overline{\psi}}} \int_{A_0} \begin{Bmatrix} 0 \\ 1 \end{Bmatrix} dA. \quad (5.76)$$

The sign "+" before the additives refer to the cases of net losses of matter from the element surface. When there is net gain through some of the element surfaces, the sign should be "–."

The third term of the vector equation (Equation 5.73) contains the boundary conditions of *Neumann's type* regarding a direct flux of the controlled matter through the boundary surfaces ($A_h = A_{1,h} + A_{2,h} + A_{3,h}$). The detailed formula of the direct specific flux $\{f_{Dr}^e\}$ has the following form ("–" denotes element net loss):

$$\{f_{Dr}^e\} = \mp \int_{A_{1,h}} \left[P^e\right]^T \overline{h}_{A_{1,h}} \, dA \mp \int_{A_{2,h}} \left[P^e\right]^T \overline{h}_{A_{2,h}} \, dA \pm \int_{A_{3,h}} \left[P^e\right]^T \overline{h}_{A_{3,h}} \, dA. \quad (5.77)$$

To operate with the formula, we rewrite it in absolute coordinates finding

$$\left\{f_{Dr}^{e}\right\}=\mp h_{A_{1,h}}P\int_{0}^{L}\left\{\begin{matrix}\left(\dfrac{x_j}{L}-\dfrac{1}{L}\cdot x\right)\\ \left(-\dfrac{x_i}{L}+\dfrac{1}{L}\cdot x\right)\end{matrix}\right\}dx\mp h_{i-1}\int_{A_0}\begin{Bmatrix}1\\0\end{Bmatrix}dA\mp h_i\int_{A_0}\begin{Bmatrix}0\\1\end{Bmatrix}dA, \quad (5.78)$$

where $\overline{h}_{i-1}$ and $\overline{h}_i$ are the direct fluxes from/to the top and bottom of the "ith" FE. Finally, the load vector $\{f^e\}$ will get the following complex form:

$$\left\{f^{e}\right\}=\frac{G_hA_0}{L}\int_{0}^{L}\begin{Bmatrix}x_j-x\\-x_i+x\end{Bmatrix}dx\pm\alpha_{A_{1,\psi}}\overline{\Psi}_{A_{1,\psi}}P\int_{0}^{L}\left\{\begin{matrix}\left(\dfrac{x_j}{L}-\dfrac{1}{L}.x\right)\\ \left(-\dfrac{x_i}{L}+\dfrac{1}{L}.x\right)\end{matrix}\right\}dx$$

$$\pm\alpha_{A_{2,\psi}}\overline{\Psi}_{A_{2,\psi}}\int_{A_0}\begin{Bmatrix}1\\0\end{Bmatrix}dA\pm\alpha_{A_{3,\psi}}\overline{\Psi}_{A_{3,\psi}}\int_{A_0}\begin{Bmatrix}0\\1\end{Bmatrix}dA$$

$$\mp h_{A_{1,h}}P\int_{0}^{L}\left\{\begin{matrix}\left(\dfrac{x_j}{L}-\dfrac{1}{L}.x\right)\\ \left(-\dfrac{x_i}{L}+\dfrac{1}{L}.x\right)\end{matrix}\right\}dx\mp h_{i-1}\int_{A_0}\begin{Bmatrix}1\\0\end{Bmatrix}dA\mp h_i\int_{A_0}\begin{Bmatrix}0\\1\end{Bmatrix}dA. \quad (5.79)$$

The substitution of Equations 5.69, 5.71, and 5.79 in Equation 5.61 yields the equation of transfer in a simple FE written in Cartesian coordinates—an analogue of Equation 5.68:

$$\left[\begin{matrix}\dfrac{\Gamma\cdot A_0}{L}\begin{bmatrix}1&-1\\-1&1\end{bmatrix}\pm\dfrac{\alpha_{A_{1,\psi}}P}{L^2}\displaystyle\int_{0}^{L}\begin{bmatrix}(x_j-x)^2&(x_j-x)(-x_i+x)\\(-x_i+x)(x_j-x)&(-x_i+x)^2\end{bmatrix}dx\pm\\ \pm\alpha_{A_{2,\psi}}A_0\begin{bmatrix}1&0\\0&0\end{bmatrix}\pm\alpha_{A_{3,\psi}}A_0\begin{bmatrix}0&0\\0&1\end{bmatrix}\end{matrix}\right]\left\{\overline{\Psi}_e\right\}$$

$$=\frac{G_hA_0}{L}\int_{0}^{L}\begin{Bmatrix}x_j-x\\-x_i+x\end{Bmatrix}dx\pm\alpha_{A_{1,\psi}}\overline{\Psi}_{A_{1,\psi}}P\int_{0}^{L}\left\{\begin{matrix}\left(\dfrac{x_j}{L}-\dfrac{1}{L}\cdot x\right)\\ \left(-\dfrac{x_i}{L}+\dfrac{1}{L}\cdot x\right)\end{matrix}\right\}dx\pm\alpha_{A_{2,\psi}}\Psi_{A_{2,\psi}}\int_{A_0}\begin{Bmatrix}1\\0\end{Bmatrix}dA$$

$$\pm\alpha_{A_{3,\psi}}\overline{\Psi}_{A_{3,\psi}}\int_{A_0}\begin{Bmatrix}0\\1\end{Bmatrix}dA\mp h_{A_{1,h}}P\int_{0}^{L}\left\{\begin{matrix}\left(\dfrac{x_j}{L}-\dfrac{1}{L}\cdot x\right)\\ \left(-\dfrac{x_i}{L}+\dfrac{1}{L}\cdot x\right)\end{matrix}\right\}dx\mp h_{i-1}\int_{A_0}\begin{Bmatrix}1\\0\end{Bmatrix}dA\mp h_i\int_{A_0}\begin{Bmatrix}0\\1\end{Bmatrix}dA.$$

(5.80)

We will use relations Equations 5.68 and 5.80 to estimate transfer through 1D simple FE (see Example 8.1 and Dimitrov 2013).

5.3.3.3 Transfer through 1D Simple Finite Element Presented in Cylindrical Coordinates

A cylindrical coordinate system is used when axial symmetry of the physical medium (disc or a low cylinder—see Figure 5.12a through c) is considered. The matrix equation (Equation 5.81), describing transfer in the FE, is similarly written in cylindrical coordinates:

$$\left[G^e\right]_C\left\{\overline{\psi}^e\right\}=\left\{f^e\right\}_C. \tag{5.81}$$

First of all, we shall account for the equation terms, and then we shall give an example involving specific boundary conditions.

As found in Section 5.3.3, the matrix equations in different coordinate systems are formally similar. Such a similarity can be established between Equations 5.80 and 5.81. However, major differences are found in the volume and surface integrals of (Equation 5.61) and in the form of members participating in (Equation 5.81). Initially, we propose a general description of the terms of the matrix equation (Equation 5.81) as an illustration. References (Димитров 2006; Dimitrov 2013) give specific numerical examples.

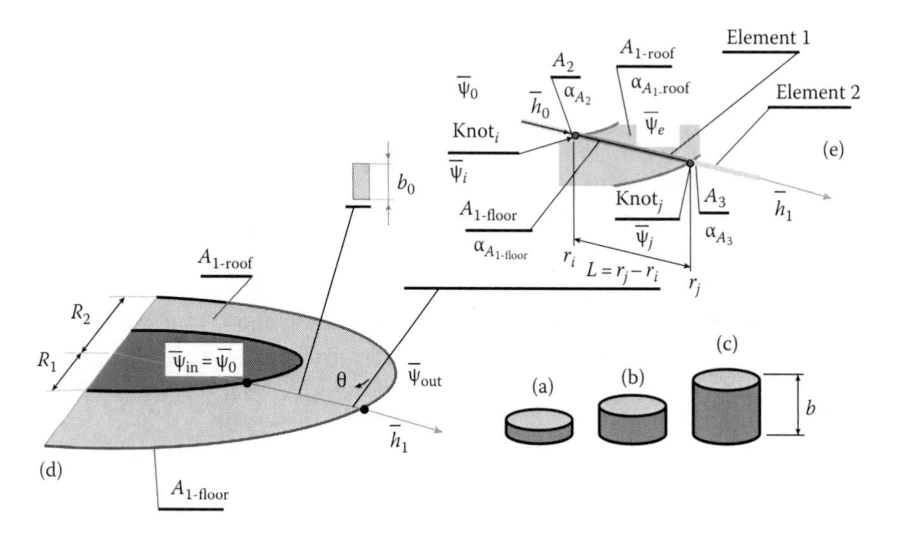

FIGURE 5.12
Boundary conditions in 1D simple finite element in cylindrical coordinates: (a–c) cylindrical elements with different heights; (d) plane of symmetry; (e) two-noded discrete analogue.

Similar to its presentation in Cartesian coordinates, the generalized matrix of conductivity $[G^e]_C$ consists of two components:

- *Matrix of conductivity* $[K^e]_C$, set forth by the equality:

$$\left[K^e\right]_C = \int_{V_{(e)}} \left[B^e\right]_C^T [\Gamma] \left[B^e\right]_C dV.$$

If it is assumed that $r = P_i \cdot r_i + P_j \cdot r_j$, $\left[B^e\right]_C = \frac{1}{(r_j - r_i)}\left[-1 \quad 1\right]$ and $dV = 2\pi \cdot b_0 \cdot dr$ (b_0—disk height), then, $[K^e]_C$ takes the following form:

$$\left[K^e\right]_C = 2\pi \cdot b_0 \left[B^e\right]_C^T [\Gamma] \left[B^e\right]_C \int_{r_i}^{r_j} \left(P_i^e \cdot r_i + P_j^e \cdot r_j\right) dr = \pi \cdot b_0 \cdot [\Gamma] \frac{r_i + r_j}{r_j - r_i} \begin{bmatrix} 1 & -1 \\ -1 & 1 \end{bmatrix}$$

(see Equation 5.69 for comparison).

- *Matrix of the surface properties* $[F^e]_C$, presented as the following:

$$\left[F^e\right]_C = \pm\alpha_{A_{1\text{-}roof}} \int_{A_{1\text{-}roof}} \left[P^e\right]_C^T \left[P^e\right]_C dA \pm \alpha_{A_{1\text{-}floor}} \int_{A_{1\text{-}floor}} \left[P^e\right]_C^T \left[P^e\right]_C dA$$

$$\pm\alpha_{A_2} \int_{A_2} \left[P_{j=0}^e\right]_C^T \left[P_{j=0}^e\right]_C dA \pm \alpha_{A_3} \int_{A_3} \left[P_{i=0}^e\right]_C^T \left[P_{i=0}^e\right]_C dA.$$

Consider the following:

- $dA = 2\pi \cdot r \cdot dr$ in the first and second integrals (referred to FE bottom and top or lower and upper sides).
- $dA = 2\pi \cdot b_0 \cdot dr$ in the third and fourth integrals (referred to FE front and back or fore and rear side).

Then, the matrix of the surface properties is transformed into the following matrix:

$$\left[F^e\right]_C = 2\pi\left(\pm\alpha_{A_{1\text{-}roof}} \pm \alpha_{A_{1\text{-}floor}}\right) \int_0^{r_j - r_i} \begin{bmatrix} r \cdot P_i^2 & r \cdot P_i \cdot P_j \\ r \cdot P_j \cdot P_i & r \cdot P_j^2 \end{bmatrix} dr$$

$$\pm 2\pi \cdot \alpha_{A_2} \cdot b_0 \int_0^{r_i} \begin{bmatrix} 1 & 0 \\ 0 & 0 \end{bmatrix} dr \pm 2\pi \cdot \alpha_{A_3} \cdot b_0 \int_0^{r_j} \begin{bmatrix} 0 & 0 \\ 0 & 1 \end{bmatrix} dr.$$

After the calculation of the first integral (see Example 1 in Appendix 1 in Dimitrov (2013) for details), we find the following final form of that matrix:

$$\left[F^e\right]_C = 2\pi\left(\mp\alpha_{A_{1\text{-}roof}} \mp \alpha_{A_{1\text{-}floor}}\right)\frac{r_j - r_i}{12}\begin{bmatrix}(3r_i + r_j) & (r_i + r_j)\\(r_i + r_j) & (r_i + 3r_j)\end{bmatrix}$$

$$\mp 2\pi\cdot\alpha_{A_2}\cdot b_0\cdot r_i\begin{bmatrix}1 & 0\\0 & 0\end{bmatrix} \mp 2\pi\cdot\alpha_{A_3}\cdot b_0\cdot r_j\begin{bmatrix}0 & 0\\0 & 1\end{bmatrix}$$

(see for comparison equation (Equation 5.71) written in Cartesian coordinates).

- Load vector $\{f^e\}_C$ in Equation 5.81 has three components $\{f_G^e\}_C$, $\{f_C^e\}_C$, and $\{f_{Dr}^e\}_C$, which reflect the three types of boundary conditions valid for the simple FE (see Equation 5.73).

The first $\{f_G^e\}_C$ is found considering recuperated matter in the element (as in Equation 5.74) after substituting dV for volume $dV = 2\pi\cdot r\cdot b_0\cdot dr$:

$$\{f_G^e\}_C = \pm 2\pi\cdot b_0\cdot G_{h_e}\int_{V(e)}\begin{bmatrix}P_i\cdot r\\P_j\cdot r\end{bmatrix}dr = \pm 2\pi\cdot b_0\cdot G_{h_e}\int_{r_j - r_i}\begin{bmatrix}P_i^2\cdot r_i + P_i\cdot P_j\cdot r_j\\P_j\cdot P_i\cdot r_i + P_j^2 r_j\end{bmatrix}dr$$

$$= \pm 2\pi\cdot b_0\cdot G_{h_e}\frac{r_j - r_i}{6}\begin{bmatrix}2r_i + r_j\\r_i + 2r_j\end{bmatrix}$$

(see Example 2 in Appendix 1 in Dimitrov (2013) for details on the integration).

The second component of the load vector—$\{f_C^e\}_C$—reflects the boundary conditions valid for convection through the boundary surfaces of the controlled substance $A_\psi = A_{1\text{-}floor,\psi} + A_{1\text{-}roof,\psi} + A_{2,\psi} + A_{3,\psi}$, and it is calculated using a formula similar to Equation 5.75 but containing four terms:

$$\{f_C^e\}_C = \left(\pm\bar{\psi}_{A_{1\text{-}roof}}\alpha_{A_{1\text{-}roof}} \pm \bar{\psi}_{A_{1\text{-}floor}}\alpha_{A_{1\text{-}floor}}\right)\int_{A_{1,\phi}}\begin{Bmatrix}P_i\\P_j\end{Bmatrix}dA \pm \bar{\psi}_{A_2}\cdot\alpha_{A_2}\int_{A_2}\begin{Bmatrix}1\\0\end{Bmatrix}dA$$

$$\pm\bar{\psi}_{A_3}\cdot\alpha_{A_3}\int_{A_3}\begin{Bmatrix}0\\1\end{Bmatrix}dA.$$

Calculating the second and third integrals, we assume the following:

- $P_i = 1$ and $P_j = 0$ for surface A_2.
- $P_i = 0$ and $P_j = 1$ for surface A_3.

As in the earlier considerations, dA is substituted by corresponding expressions ($dA = 2\pi \cdot r \cdot dr$ for the first surface, and $dA = 2\pi \cdot b_0 \cdot dr$ for the second and third surfaces), and we find the following equality valid for $\{f_C^e\}_C$:

$$\left\{f_C^e\right\}_C = 2\pi\left(\pm\bar{\psi}_{A1\text{-}roof}\alpha_{A1\text{-}roof} \pm \bar{\psi}_{A1\text{-}floor}\alpha_{A1\text{-}floor}\right)\frac{r_j - r_i}{6}\begin{bmatrix} 2r_i + r_j \\ r_i + 2r_j \end{bmatrix}$$

$$\pm 2\pi \cdot \alpha_{A_2} \cdot \bar{\psi}_{A_2} \cdot b_0 \cdot r_i \begin{Bmatrix} 1 \\ 0 \end{Bmatrix} \pm 2\pi \cdot \alpha_{A_3} \cdot \bar{\psi}_{A_3} \cdot b_0 \cdot r_j \begin{Bmatrix} 0 \\ 1 \end{Bmatrix}$$

(in case of net losses at the FE corresponding surface, the sign before the subsequent term in the formula should be "+").

The last component of the load vector $\{f_{Dr}^e\}_C$ expresses the boundary conditions of matter exchange via a direct flux (see the following considerations). We apply a formula similar to formula (Equation 5.77):

$$\left\{f_{Dr}^e\right\}_C = 2\pi\left(\mp h_{A_{1\text{-}roof}} \mp h_{A_{1\text{-}floor}}\right)\int_0^{r_j - r_i}\begin{Bmatrix} P_i\left(P_i \cdot r_i + P_j \cdot r_j\right) \\ P_j\left(P_i \cdot r_i + P_j \cdot .r_j\right) \end{Bmatrix} dr \mp h_{A_2}\int_0^{r_i} 2\pi \cdot b_0 \begin{Bmatrix} 1 \\ 0 \end{Bmatrix} dr$$

$$\mp h_{A_3}\int_0^{r_j} 2\pi \cdot b_0 \begin{Bmatrix} 0 \\ 1 \end{Bmatrix} dr = 2\pi\left(\mp \bar{h}_{A_{1\text{-}roof}} \mp \bar{h}_{A_{1\text{-}floor}}\right)\frac{r_j - r_i}{6}\begin{Bmatrix} 2r_i + r_j \\ r_i + 2r_j \end{Bmatrix}$$

$$\mp 2\pi \cdot r_i \cdot b_0 \cdot \bar{h}_{A_2}\begin{Bmatrix} 1 \\ 0 \end{Bmatrix} \mp 2\pi \cdot r_i \cdot b_0 \cdot \bar{h}_{A_3}\begin{Bmatrix} 0 \\ 1 \end{Bmatrix}.$$

(Compare it to Equation 5.78. Sign "–" is adopted in the case of net loss at a corresponding surface FE.) In a detailed form, the matrix equation of transfer in a 1D simple FE reads as follows in cylindrical coordinates:

$$\begin{bmatrix} \Gamma \cdot b_0 \cdot \dfrac{r_i + r_j}{r_j - r_i}\begin{bmatrix} 1 & -1 \\ -1 & 1 \end{bmatrix} + \left(\pm\alpha_{A_{1\text{-}roof}} \pm \alpha_{A_{1\text{-}roof}}\right)\dfrac{r_j - r_i}{6}\begin{bmatrix} \left(3r_i + r_j\right) & \left(r_i + r_j\right) \\ \left(r_i + r_j\right) & \left(r_i + 3r_j\right) \end{bmatrix} \pm \\ \pm 2 \cdot b_0 \begin{Bmatrix} \mp\alpha_{A2} & 0 \\ 0 & \mp\alpha_{A3} \end{Bmatrix}\begin{bmatrix} \left(r_i + r_j\right) & 0 \\ 0 & \left(r_i + r_j\right) \end{bmatrix} \end{bmatrix}\begin{Bmatrix} \psi_i \\ \psi_j \end{Bmatrix}$$

$$= \frac{r_j - r_i}{3}\left(\mp b_0 \cdot G_h \pm \alpha_{A_{1-roof}} \cdot \bar{\psi}_{A_{1-roof}} \pm \alpha_{A_{1-roof}} \cdot \bar{\psi}_{A_{1-roof}} \mp \bar{h}_{A_{1-roof}} \mp \bar{h}_{A_{1-roof}}\right)\begin{Bmatrix} 2r_i + r_j \\ r_i + 2r_j \end{Bmatrix}$$

$$+ 2 \cdot b_0 \cdot r_i\left(\pm\alpha_{A_2} \cdot \bar{\psi}_{A_2} \mp \bar{h}_{A_2}\right)\begin{Bmatrix} 1 \\ 0 \end{Bmatrix} + 2 \cdot b_0 \cdot r_j\left(\pm\alpha_{A_3} \cdot \bar{\psi}_{A_3} \mp \bar{h}_{A_3}\right)\begin{Bmatrix} 0 \\ 1 \end{Bmatrix}.$$

The FEM gives significantly wider options of setting the boundary conditions in comparison to all other numerical methods (see the examples in Appendix 1 in Dimitrov (2013)).

Although it appears that the design of an FE matrix equation is quite laborious, it becomes a routine procedure bound to automation. Setting forth the solution of the matrix equation in an unfolded form seems appropriate:

- *In Cartesian coordinates,*

$$\{\bar{\Psi}^e\} = \left[\underbrace{\frac{\Gamma\cdot A_0}{L}\begin{bmatrix}1 & -1\\ -1 & 1\end{bmatrix}}_{\text{conductivity matrix}} \pm \underbrace{\begin{matrix}\dfrac{\alpha_{A_{1,\phi}}\cdot P}{L^2}\displaystyle\int_0^L \begin{bmatrix}(x_j-x)^2 & (x_j-x)(-x_i+x)\\ (-x_i+x)(x_j-x) & (-x_i+x)^2\end{bmatrix}dx \pm \\ \pm\alpha_{A_{2,\phi}}A_0\begin{bmatrix}1 & 0\\ 0 & 0\end{bmatrix}\pm\alpha_{A_{3,\phi}}A_0\begin{bmatrix}0 & 0\\ 0 & 1\end{bmatrix}\end{matrix}}_{\text{matrix of surface properties}} \right]^{-1} *$$

$$* \left\{ \begin{matrix} \underbrace{\dfrac{G_{h_e}\cdot A_0}{L}\displaystyle\int_0^L \begin{Bmatrix}x_j-x\\ -x_i+x\end{Bmatrix}dx}_{\text{load due to internal sources}} \pm\alpha_{A_{1,\phi}}\bar{\Psi}_{A_{1,\phi}}P\displaystyle\int_0^L\begin{Bmatrix}\left(\dfrac{x_j}{L}-\dfrac{1}{L}\cdot x\right)\\ \left(-\dfrac{x_i}{L}+\dfrac{1}{L}\cdot x\right)\end{Bmatrix}dx \pm \\ \underbrace{\pm\alpha_{A_0}\cdot\bar{\Psi}_{A_{2,\phi}}\displaystyle\int_{A_0}\begin{Bmatrix}1\\ 0\end{Bmatrix}dA \pm\alpha_{A30}\cdot\bar{\Psi}_{A_{3,\phi}}\displaystyle\int_{A_0}\begin{Bmatrix}0\\ 1\end{Bmatrix}dA \mp}_{\text{load due to convection to/ from the environment}} \\ \underbrace{\pm h_{A_{1,h}}P\displaystyle\int_0^L\begin{Bmatrix}\left(\dfrac{x_j}{L}-\dfrac{1}{L}\cdot x\right)\\ \left(-\dfrac{x_i}{L}+\dfrac{1}{L}\cdot x\right)\end{Bmatrix}dx \mp h_{i-1}\displaystyle\int_{A_0}\begin{Bmatrix}1\\ 0\end{Bmatrix}dA \mp h_i\displaystyle\int_{A_0}\begin{Bmatrix}0\\ 1\end{Bmatrix}dA}_{\text{load due to a direct flux to/ from the environment}} \end{matrix} \right\}.$$

- *In cylindrical coordinates,*

$$\{\bar{\Psi}^e\} = \left(\underbrace{\left(\Gamma\cdot b_0\cdot\frac{r_i+r_j}{r_j-r_i}\right)}_{\text{load due to internal sources}}\cdot\begin{bmatrix}1 & -1\\ -1 & 1\end{bmatrix} + \underbrace{\begin{matrix}\left(\pm\alpha_{A_{1\text{-roof}}}\pm\alpha_{A_{1\text{-floor}}}\right)\dfrac{r_j-r_i}{6}\begin{bmatrix}(3r_i+r_j) & (r_i+r_j)\\ (r_i+r_j) & (r_i+3r_j)\end{bmatrix}\pm \\ \pm 2\cdot b_0\begin{Bmatrix}\pm\alpha_{A2} & 0\\ 0 & \pm\alpha_{A3}\end{Bmatrix}\begin{bmatrix}(r_i+r_j) & 0\\ 0 & (r_i+r_j)\end{bmatrix}\end{matrix}}_{\text{matrix of surface properties}} \right)^{(-1)} *$$

$$* \left(\begin{matrix} \dfrac{r_j-r_i}{3}\left(\mp b_0\cdot G_h \pm\alpha_{A_{1\text{-roof}}}\bar{\Psi}_{A_{1\text{-roof}}}\pm\alpha_{A_{1\text{-roof}}}\bar{\Psi}_{A_{1\text{-roof}}}\mp\bar{h}_{A_{1\text{-roof}}}\mp\bar{h}_{A_{1\text{-roof}}}\right)\begin{Bmatrix}2r_i+r_j\\ r_i+2r_j\end{Bmatrix}+ \\ +2\cdot b_0\cdot r_i\left(\pm\alpha_{A_2}\cdot\bar{\Psi}_{A_2}\mp\bar{h}_{A_2}\right)\begin{Bmatrix}1\\ 0\end{Bmatrix}+2\cdot b_0\cdot r_j\left(\pm\alpha_{A_3}\cdot\bar{\Psi}_{A_3}\mp\bar{h}_{A_3}\right)\begin{Bmatrix}0\\ 1\end{Bmatrix} \end{matrix} \right).$$

Signs "+" or "–" before the particular components depend on *the direction of the transfer fluxes.*

Since 1D FE are used to model transfer in highly idealized conditions, they are scarcely applied or can be used for rather rough engineering estimations (see Example 8.2 as well as Dimitrov 2013).

Note however that 2D FEs are used more often, and they will be an object of our further study.

5.3.4 Steady Transfer in a 2D Finite Element

Two-dimensional (2D) FE are used to study transfer in objects with plane or axial symmetry where the boundary conditions are asymmetric. One can find such objects among the components of building equipment (see for instance the equipment shown in Figure 5.13) or technological facilities such as reservoirs, pools, and silos, etc., operating in regimes, which admit transfer to/from the surroundings.

The common feature of those complex geometrical objects is that the 2D transfer should be discretized, that is,

$$\bar{\psi}(x,y)=\left[\sum_{m=i,j,k}\left[P^e(x,y)\right]^{\bar{\psi}}\left\{\bar{\psi}^e\right\}\right],$$

where

$\left\{\bar{\psi}^e\right\}$ is the vectors containing the estimation of the function of free energy $\bar{\Psi}$ at n nodes

$\left[P^e(x,y)\right]^{\bar{\psi},p,t,V}$ is the matrices containing the 2D approximating functions (of two variables) of the free energy $\bar{\psi}$, pressure p, temperature T, or electrical potential V at the nodes

Studies of transfer in 2D objects aim at

- Design/creation of *patterns/photos* of the fields of pressure, temperature, and function of free energy under specific boundary conditions

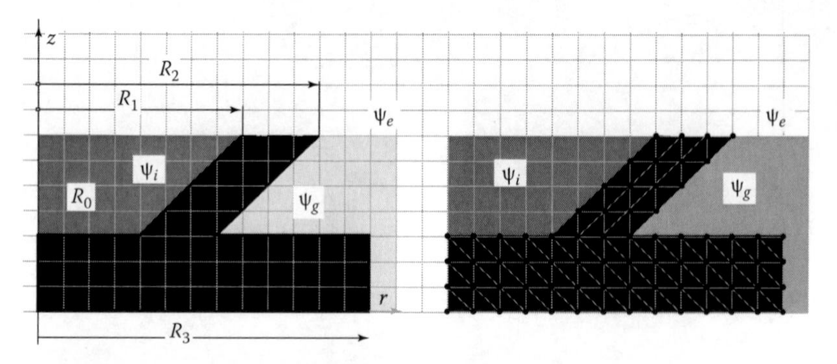

FIGURE 5.13
Boundary conditions in cylindrical coordinates valid for a 2D finite element.

- Improvement of the operational regimes of the studied objects, regarding various boundary conditions and transfer characteristics
- Improvement or optimization of the structure of the physical objects by analyzing various geometries

5.3.4.1 Equation of a 2D Simple Finite Element in Cartesian Coordinates

As disclosed in Section 5.3.1, approximating polynomials of the form (Equation 5.32) is convenient to use when modeling processes that take place in the cases when there is a plane symmetry. In particular, linear 2D polynomial functions have the following form:

$$P_m^e = a_m + b_m x + c_m y \quad m = i, j, k. \tag{5.82}$$

The simple FE is a three-noded triangle (see Figure. 5.8) whose function of free energy $\bar{\psi}$ is interpolated via its values at the three nodes $(\bar{\psi}_i, \bar{\psi}_j, \bar{\psi}_k)$:

$$\bar{\psi}_A(x,y,\tau) \approx \bar{\psi}_A(x,y,\tau) = \sum_{m=i,j,k} P_m^e(x,y)\bar{\psi}_m^e(\tau) = P_i \cdot \bar{\psi}_i + P_j \cdot \bar{\psi}_j + P_k \cdot \bar{\psi}_k$$

as $P_i + P_j + P_k = 1$.

(The approximating functions P_i, P_j, and P_k are given in Equation 5.41.)

The general equation of the FE (Equation 5.61) is actually in the simple 2D case, too. We shall discuss the details of its application in accordance with the structure of the approximating function (Equation 5.82) chosen. At first, focus on the characteristics of the *conductivity matrix* $[K^e]$.

For a 2D simple FE, the matrix rank is 3 (three rows and three columns). The matrix is found after determining its adjugate matrices, that is,

$$\left[K^e\right] = \int_{v^{(e)}} \left[B^e\right]^T [\Gamma]\left[B^e\right] dv = \int_{A_0} \delta\left[B^e\right]^T [\Gamma]\left[B^e\right] dA.$$

Thus, the volume integral is replaced by a surface one where $\left[B^e\right]^T$ and $[B^e]$ are defined in Equation 5.51 by means of coefficients (a_k, b_k, and c_k) of the approximating functions (Equation 5.82). If the element thickness is constant (δ = const), then $[K^e]$ gets the following form:

$$\left[K^e\right] = \delta[\Gamma] A_0 \int_{A_0} \begin{Bmatrix} b_i & c_i \\ b_j & c_j \\ b_k & c_k \end{Bmatrix} \begin{bmatrix} b_i & b_j & b_k \\ c_i & c_j & c_{ik} \end{bmatrix} dA = \delta[\Gamma]$$

$$A_0 \begin{bmatrix} b_i^2 + c_i^2 & b_i b_j + c_i c_j & b_i b_k + c_i c_k \\ b_j b_i + c_j c_i & b_j^2 + c_j^2 & b_j b_k + c_j c_k \\ b_k b_i + c_k c_i & b_k b_j + c_k c_j & b_k^2 + c_k^2 \end{bmatrix}. \tag{5.83}$$

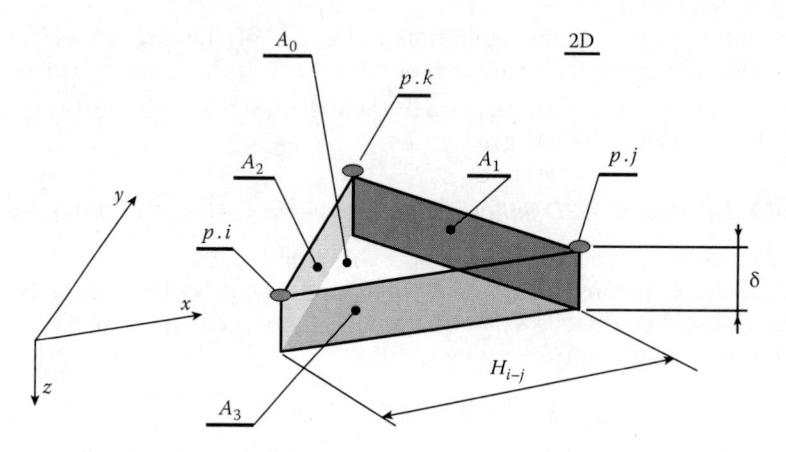

FIGURE 5.14
Boundary conditions in 2D simple finite element in Cartesian coordinates.

The next component in Equation 5.60 is the *matrix of the boundary conditions* $[F^e]$. Compared to 1D FE (Equation 5.70), it is composed of four elements (see Figure 5.14):

$$\left[F^e\right]=\int_{A_\psi}\left[P^e\right]^T\alpha\left[P^e\right]dA=\pm\int_{A_{0,\psi}}\left[P^e\right]^T\alpha_{A_{0,\psi}}\left[P^e\right]dA\pm\int_{A_{1,\psi}}\left[P^e\right]_{i=0}^T\alpha_{A_{1,\psi}}\left[P^e\right]_{i=0}dA$$

$$\pm\int_{A_{2,\psi}}\left[P^e\right]_{j=0}^T\alpha_{A_{2,\psi}}\left[P^e\right]_{j=0}dA\pm\int_{A_{3,\psi}}\left[P^e\right]_{k=0}^T\alpha_{A_{3,\psi}}\left[P^e\right]_{k=0}dA, \tag{5.84}$$

where

A_o is the area of a triangle with apexes i, j, k
$A_1\psi=\delta\cdot H_{j-k}$ is the area of the prism wall (facing node i)
$A_{2,}\psi=\delta\cdot H_{k-i}$ is the area of the wall facing node j
$A_{3,}\psi=\delta\cdot H_{i-j}$ is the area of the wall facing node k
Δ is the prism thickness
H_{i-j}; $H_{i-k}uH_{j-k}$ are distances between nodes

The second specific feature of the 2D case is that zero components emerge when calculating the surface integrals ***I1, I2,*** and ***I3***, in matrices $[P^e]$ and $\left[P^e\right]^T$ due to the properties of the approximating functions. When, for instance, one has to calculate the surface integral along $A_{1,\psi}$ (I_1 facing node i), the elements of matrix $\left[P^e\right]_i$ are zeroed. The same is with the calculation of the surface integrals along $A_{2,\psi}$ and $A_{3,\psi}$, when the respective components of $\left[P^e\right]_j$ and $\left[P^e\right]_k$ are to be eliminated.

Then, the integrals thus designed convert into

$$I_o = \alpha_{A0,\psi} \int_{A0} \begin{bmatrix} P_i^2 & P_i \cdot P_j & P_i \cdot P_k \\ P_j \cdot P_i & P_j^2 & P_j \cdot P_k \\ P_k \cdot P_i & P_k \cdot P_j & P_j^2 \end{bmatrix} dA = \frac{\alpha_{A0,\psi} \cdot A_0}{12} \begin{bmatrix} 2 & 1 & 1 \\ 1 & 2 & 1 \\ 1 & 1 & 2 \end{bmatrix}. \quad (5.85)$$

(Similar solution of the integral is given in Dimitrov (2013), Appendix 1, Example 10.)

Here, A_0 is the area of the triangle formed by nodes i, j, and k.

$$I_1 = \alpha_{A1,\psi} \cdot \delta \int_{H_{j-k}} \begin{bmatrix} 0 & 0 & 0 \\ 0 & P_j^2 & P_j \cdot P_k \\ 0 & P_k \cdot P_j & P_j^2 \end{bmatrix} dH = \frac{\alpha_{A1,\psi} \cdot H_{j-k} \cdot \delta}{6} \begin{bmatrix} 0 & 0 & 0 \\ 0 & 2 & 1 \\ 0 & 1 & 2 \end{bmatrix}. \quad (5.86)$$

$$I_2 = \alpha_{A2,\psi} \int_{H_{k-i}} \begin{bmatrix} P_i^2 & 0 & P_i \cdot P_k \\ 0 & 0 & 0 \\ P_k \cdot P_i & 0 & P_j^2 \end{bmatrix} dH = \frac{\alpha_{A2,\psi} \cdot H_{k-i} \cdot \delta}{6} \begin{bmatrix} 2 & 0 & 1 \\ 0 & 0 & 0 \\ 1 & 0 & 2 \end{bmatrix}. \quad (5.87)$$

$$I_3 = \alpha_{A3,\psi} \int_{H_{i-j}} \begin{bmatrix} P_i^2 & P_i \cdot P_j & 0 \\ P_j \cdot P_i & P_j^2 & 0 \\ 0_i & 0 & 0 \end{bmatrix} dH = \frac{\alpha_{A3,\psi} \cdot H_{i-j} \cdot \delta}{6} \begin{bmatrix} 2 & 1 & 0 \\ 1 & 2 & 0 \\ 0 & 0 & 0 \end{bmatrix}. \quad (5.88)$$

(See Example 8.3 as well as Dimitrov 2013.)

Then, putting relations (Equations 5.85 through 5.88) in relation (Equation 5.84), one gets the following form of the *matrix of the boundary conditions*:

$$\begin{aligned} \left[F^e\right] = &\pm \frac{\alpha_{A0,\psi} \cdot A_0}{12} \begin{bmatrix} 2 & 1 & 1 \\ 1 & 2 & 1 \\ 1 & 1 & 2 \end{bmatrix} \pm \frac{\alpha_{A1,\psi} \cdot H_{j-k} \cdot \delta}{6} \begin{bmatrix} 0 & 0 & 0 \\ 0 & 2 & 1 \\ 0 & 1 & 2 \end{bmatrix} \\ &\pm \frac{\alpha_{A2,\psi} \cdot H_{k-i} \cdot \delta}{6} \begin{bmatrix} 2 & 0 & 1 \\ 0 & 0 & 0 \\ 1 & 0 & 2 \end{bmatrix} \pm \frac{\alpha_{A3,\psi} \cdot H_{i-j} \cdot \delta}{6} \begin{bmatrix} 2 & 1 & 0 \\ 1 & 2 & 0 \\ 0 & 0 & 0 \end{bmatrix}. \end{aligned} \quad (5.89)$$

As a result, one finds the generalized matrix $[G^e] = [K^e] + [F^e]$ of the FE whose rank is also 3. The sign of summation (+) is used in case of net losses via convection from the corresponding boundary. If the element gains mass along some of its coordinate boundaries, the sign should be "–."

Load vector $\{f^e\}$ is composed as a matrix-column of rank 3 applying the traditional matrix form (see Equation 5.74). Note here the specific features similar to those of matrix $[F^e]$. For instance, the load vector due to internal sources gets the following form (see Appendix 1, Example 3 in Dimitrov (2013)):

$$\left\{f_G^e\right\} = \int_{v_e} \left[P^e\right]^T G_{h_e}\, dv = G_{h_e}\delta \int_{A_0} \begin{Bmatrix} P_i \\ P_j \\ P_k \end{Bmatrix} dA = \frac{\delta \cdot G_{h_e} \cdot A_0}{3} \begin{Bmatrix} 1 \\ 1 \\ 1 \end{Bmatrix}.$$

The vector of the *convective load* has four addends corresponding to the four available surfaces, at which *Dirichlet's* convective boundary conditions hold:

$$\left\{f_C^e\right\} = +\int_{A_\psi} \left[P^e\right]^T \alpha\bar{\psi}_e\, dA = \pm\bar{\psi}_{AO,\phi} \int_{A0,\psi} \left[P^e\right]^T \alpha_{A0,\phi}\, dA \pm \bar{\psi}_{A1,\psi} \int_{A1,\psi} \left[P^e\right]^T_{i=0} \alpha_{A1,\psi}\, dA$$

$$\pm \bar{\psi}_{A2,\psi} \int_{A2,\psi} \left[P^e\right]^T_{j=0} \alpha_{A2,\psi}\, dA \pm \bar{\psi}_{A3,\phi} \int_{A3,\psi} \left[P^e\right]^T_{k=0} \alpha_{A3,\phi}\, dA.$$

Similar to the specificities of calculating $[F^e]$, one should foresee the nullification of the approximating functions at the nonadjacent nodes when estimating the second, third, and fourth terms of Equation 5.84. Calculate, for instance, the surface integral over A_1. Then, the approximating function P_i of its nonadjacent node "1" should be nullified. The same condition is valid for integrals along A_2 and A_3. Besides, since $A_1 = \delta H_{j-k}$, $A_2 = \delta H_{k-i}$, and $A_3 = \delta H_{i-j}$, the surface integrals can be transformed into linear ones, so that

$$\left\{f_C^e\right\} = \pm\bar{\psi}_{AO,\phi}\alpha_{A0,\psi} \int_{A0,\psi} \begin{Bmatrix} P_i \\ P_j \\ P_k \end{Bmatrix} dA \pm \bar{\psi}_{A1,\psi}\alpha_{A1,\psi} \int_{A1,\psi} \begin{Bmatrix} 0_i \\ P_j \\ P_k \end{Bmatrix} dA$$

$$\pm \bar{\psi}_{A2,\psi}\alpha_{A2,\phi} \int_{A2,\psi} \begin{Bmatrix} P_i \\ 0 \\ P_k \end{Bmatrix} dA \pm \bar{\psi}_{A3,\psi}\alpha_{A3,\psi} \int_{A3,\psi} \begin{Bmatrix} P_i \\ P_j \\ 0 \end{Bmatrix}$$

$$= \frac{\pm\bar{\psi}_{AO,\psi}\alpha_{A0,\psi}A_0}{3} \begin{Bmatrix} 1 \\ 1 \\ 1 \end{Bmatrix} \pm \frac{\bar{\psi}_{A1,\psi}\alpha_{A1,\psi}\delta H_{j-k}}{2} \begin{Bmatrix} 0 \\ 1 \\ 1 \end{Bmatrix}$$

$$\pm \frac{\bar{\psi}_{A2,\psi}\alpha_{A2,\psi}\delta H_{k-i}}{2} \begin{Bmatrix} 1 \\ 0 \\ 1 \end{Bmatrix} \pm \frac{\bar{\psi}_{A3,\psi}\alpha_{A3,\psi}\delta H_{i-j}}{2} \begin{Bmatrix} 1 \\ 1 \\ 0 \end{Bmatrix}. \tag{5.90}$$

(See Appendix 1, Example 11 in Dimitrov (2013).)

The positive signs before the members of the expression refer to the case of net losses to the surroundings

The *last load vector component* $\left\{f_{Dr}^{e}\right\}$ concerns the implementation of the members of the direct flow through the boundary surfaces of the FE (Neumann's conditions), and it has the same structure as relation (Equation 5.90).

The final vector form of the load of a 2D simple FE $\{f^e\}$ is a sum of the loads along the boundaries between the FE and the surroundings.

$$\left\{f^{e}\right\}=$$

$$=\frac{\delta\cdot G_{h_e}\cdot A_0}{3}\begin{bmatrix}1\\1\\1\end{bmatrix}+\frac{\left(\pm\alpha_{A_0}\cdot\bar{\psi}_{A_0}\mp\bar{h}_{A_0}\right)A_0}{3}\begin{bmatrix}1\\1\\1\end{bmatrix}+\frac{\left(\pm\alpha_{A_{1,\psi}}\cdot\bar{\psi}_{A_{1,\psi}}\mp\bar{h}_{A1}\right)\delta H_{j-k}}{2}\begin{bmatrix}0\\1\\1\end{bmatrix}$$

$$+\frac{\left(\pm\alpha_{A_{2,\psi}}\cdot\bar{\psi}_{A_{2,\psi}}\mp\bar{h}_{A_2}\right)\delta H_{k-i}}{2}\begin{bmatrix}1\\0\\1\end{bmatrix}\pm\frac{\left(\pm\alpha_{A_{3,\psi}}\cdot\bar{\psi}_{A_{3,\psi}}\mp\bar{h}_{A_3}\right)\delta H_{i-j}}{2}\begin{bmatrix}1\\1\\0\end{bmatrix}. \quad (5.91)$$

The final form of the matrix equation of the 2D simple element in Cartesian coordinates is obtained by putting relations Equations 5.83, 5.89, and 5.91 in relation (Equation 5.60).

$$\left(\begin{array}{l}\delta\cdot[\Gamma]\cdot A_0\begin{bmatrix}\left(b_i^2+c_i^2\right) & \left(b_ib_j+c_ic_j\right) & \left(b_ib_k+c_ic_k\right)\\ \left(b_jb_i+c_jc_i\right) & \left(b_j^2+c_j^2\right) & \left(b_jb_k+c_jc_k\right)\\ \left(b_kb_i+c_kc_i\right) & \left(b_kb_j+c_kc_j\right) & \left(b_k^2+c_k^2\right)\end{bmatrix}\pm\\ \pm\frac{\alpha_{A_{0,\psi}}A_0}{12}\begin{bmatrix}2&1&1\\1&2&1\\1&1&2\end{bmatrix}\pm\frac{\alpha_{A_{1,\psi}}H_{j-k}\delta}{6}\begin{bmatrix}0&0&0\\0&2&1\\0&1&2\end{bmatrix}\pm\\ \pm\frac{\alpha_{A_{2,\psi}}H_{k-i}\delta}{6}\begin{bmatrix}2&0&1\\0&0&0\\1&0&2\end{bmatrix}\pm\frac{\alpha_{A_{3,\psi}}H_{i-j}\delta}{6}\begin{bmatrix}2&1&0\\1&2&0\\0&0&0\end{bmatrix}\end{array}\right)\begin{Bmatrix}\bar{\psi}_i\\ \bar{\psi}_j\\ \bar{\psi}_k\end{Bmatrix}.$$

General conductivity matrix

$$=\frac{\delta G_{h_{(i)}}A_0}{3}\begin{Bmatrix}1\\1\\1\end{Bmatrix}+\frac{\left(\pm\alpha_{A_{0,\psi}}\cdot\bar{\psi}_{A_0}\mp\bar{h}_{A_0}\right)A_0}{3}\begin{Bmatrix}1\\1\\1\end{Bmatrix}+\frac{\left(\pm\alpha_{A_{1,\psi}}\cdot\bar{\psi}_{A_{1,\psi}}\mp\bar{h}_{A_1}\right)\delta\cdot H_{j-i}}{2}\begin{Bmatrix}0\\1\\1\end{Bmatrix}$$

$$+\frac{\left(\pm\alpha_{A_{2,\psi}}\cdot\bar{\psi}_{A_{2,\psi}}\mp\bar{h}_{A_2}\right)\cdot\delta\cdot H_{k-i}}{2}\cdot\begin{bmatrix}1\\0\\1\end{bmatrix}\pm\frac{\left(\pm\alpha_{A_{3,\psi}}\cdot\bar{\psi}_{A_{3,\psi}}\mp h_{A_3}\right)\cdot\delta\cdot H_{i-j}}{2}\cdot\begin{bmatrix}1\\1\\0\end{bmatrix}.$$

Load vector

The use of a methodology for the calculation of $\overline{\psi}$ in 2D Cartesian coordinate system is given in Example 8.3.

5.3.4.2 Design of Transfer Equation in Cylindrical Coordinates regarding a Three-Noded 2D Finite Element

As already noted, the matrix equations written in Cartesian and cylindrical coordinates and valid for a 2D FE have similar structure. Matrix of the approximating functions $[P_C^e]$, the gradient matrices $grad\ \psi = \{g\}$, $[B^e]_C$, and $[B^e]_C^T$, is obtained in Section 5.3 (Equations 5.49 and 5.51). Consider the matrix equation of a 2D FE used to discretize a body with axial symmetry in cylindrical coordinates—see Figure 5.15. We will discuss herein the specificity of calculating the surface and volume integrals of the equation members.

The geometric parameters of the elements in a cylindrical coordinate system obey relationships specified by the approximation functions:

$$\begin{Bmatrix} r \\ z \end{Bmatrix} = \begin{bmatrix} P_i & 0 & P_j & 0 & P_k & 0 \\ 0 & P_i & 0 & P_j & 0 & P_k \end{bmatrix} \cdot \begin{Bmatrix} r_i \\ z_i \\ r_j \\ z_j \\ r_k \\ z_k \end{Bmatrix}, \tag{5.92}$$

where r, z (i, j, k) are the absolute coordinates of nodes i, j, and k.

It follows from Equation 5.92 that

$$r = P_i \cdot r_i + P_j \cdot r_j + P_k \cdot r_k, \tag{5.93}$$

which can be used in the calculation of the members of the matrix equation in cylindrical coordinates.

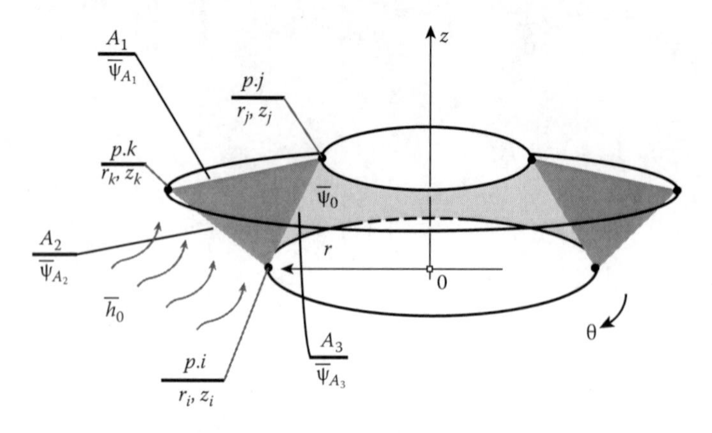

FIGURE 5.15
Boundary conditions in a 2D simplex finite element in cylindrical coordinates.

Which are the specific features of the calculation of FE conductivity matrix $[K^e]_C$? The volume integral in the definition expression (Equation 5.60) transforms into a surface one $\int_{A_0} r\left[B^e\right]_C^T\left[B^e\right]_C dA$, after substituting the elementary volume dv for volume $dv = 2\pi \cdot r \cdot d$A. Note that equality (Equation 5.93) is valid for r, and equality (Equation 5.51) is valid for $[B^e]_C$ and $[B^e]_C^T$. Then, we find

$$\left[K^e\right]_C = 2\pi \cdot \Gamma A_0 \frac{\left(r_i + r_j + r_k\right)}{3}\begin{bmatrix} b_i^2 + c_i^2 & b_i b_j + c_i c_j & b_i b_k + c_i c_k \\ b_j b_i + c_j c_i & b_j^2 + c_j^2 & b_j b_k + c_j c_k \\ b_k b_i + c_k c_i & b_k b_j + c_k c_j & b_k^2 + c_k^2 \end{bmatrix}. \quad (5.94)$$

The earlier presentation differs from that in relation (Equation 5.83) by coefficient $(r_i + r_j + r_k)/3$ only (see Example 12 in Appendix 1 in Dimitrov (2013)).

The surface integral participating in the matrix of the surface properties $[F^e]_C$ converts into a linear one by means of the substitution $dA = 2\pi \cdot r \cdot dH$:

$$\left[F^e\right] = \pm 2\pi \cdot \alpha_{A_{j-ik}} \int_{H_{j-k}} \begin{bmatrix} 0 & 0 & 0 \\ 0 & \left(P_j^3 \cdot r_j + P_j^2 \cdot P_k \cdot r_k\right) & \left(P_j^2 \cdot P_k \cdot r_j + P_j \cdot P_k^2 \cdot r_k\right) \\ 0 & \left(P_k P_j^2 \cdot r_j + P_k^2 \cdot P_j \cdot r_k\right) & \left(P_j \cdot P_k^2 \cdot r_j + P_k^3 \cdot r_k\right) \end{bmatrix} dH$$

$$\pm 2\pi \cdot \alpha_{A_{i-k}} \int_{H_{i-k}} \begin{bmatrix} \left(P_i^3 \cdot r_i + P_i^2 \cdot P_k \cdot r_k\right) & 0 & \left(P_i^2 P_k \cdot r_i + P_i \cdot P_k^2 \cdot r_k\right) \\ 0 & 0 & 0 \\ \left(P_k P_i^2 \cdot r_j + P_k^2 \cdot P_i \cdot r_k\right) & 0 & \left(P_i \cdot P_k^2 \cdot r_j + P_k^3 \cdot r_k\right) \end{bmatrix} dH$$

$$\pm 2\pi \cdot \alpha_{A_{i-j}} \int_{H_{i-j}} \begin{bmatrix} \left(P_i^3 \cdot r_i + P_i^2 \cdot P_j \cdot r_j\right) & \left(P_i^2 P_j r_j + P_i \cdot P_j^2 \cdot r_j\right) & 0 \\ \left(P_i^2 P_j \cdot r_j + P_j^2 \cdot P_i \cdot r_j\right) & \left(P_i P_j^2 \cdot r_i + P_j^3 \cdot r_j\right) & 0 \\ 0 & 0 & 0 \end{bmatrix} dH$$

$$= \pm \frac{2\pi \cdot \alpha_{j-k} \cdot H_{j-k}}{12} \begin{bmatrix} 0 & 0 & 0 \\ 0 & \left(3r_j + r_k\right) & \left(r_j + r_k\right) \\ 0 & \left(r_j + r_k\right) & \left(r_j + 3r_k\right) \end{bmatrix}$$

$$\pm \frac{2\pi \cdot \alpha_{k-i} \cdot H_{k-i}}{12} \begin{bmatrix} \left(3r_i + r_k\right) & 0 & \left(r_i + r_k\right) \\ 0 & 0 & 0 \\ \left(r_i + r_k\right) & 0 & \left(r_i + 3r_k\right) \end{bmatrix}$$

$$\pm \frac{2\pi \cdot \alpha_{i-j} \cdot H_{i-j}}{12} \begin{bmatrix} \left(3r_i + r_j\right) & \left(r_i + r_j\right) & 0 \\ \left(r_i + r_j\right) & \left(r_i + 3r_j\right) & 0 \\ 0 & 0 & 0 \end{bmatrix}. \quad (5.95)$$

The expression (Equation 5.95) is found considering that the approximating functions $[P^e]$ of nodes not belonging to the surface are zeroed over the integration areas—surfaces A_i, A_j, and A_k (see Example 8.4 as well as Dimitrov 2013). Sign (+) refers to element net losses at the corresponding surface.

The element of *the generalized matrix of conductivity* $[G^e]$ is calculated using methods similar to those used in the previous cases $\left(\left[G^e\right]=\left[K^e\right]+\left[F^e\right]_{j-k}+\left[F^e\right]_{i-k}+\left[F^e\right]_{i-j}\right)$. The calculation of load vector $\{f^e\}$ in cylindrical coordinates has the same characteristic features as the calculation of the matrix of the surface properties matrix (Equation 5.91), since the general form of vector $\{f^e\}$ has the same structure as that written in Cartesian coordinates. Those features are outlined when specifying the detailed form of the vector components.

Consider, for instance, the vector of matter recuperation $\{f_G^e\}_C$. Note that the transformation of the volume integral into a surface one yields an additional term that changes the structure of the sub-integral function:

$$\left\{f_G^e\right\}=2\pi G_{h(e)}\int_{A_0} r\left[P^e\right]^T dA=2\pi G_{h_e}\int_{A_0}\left[P_i\cdot r_i+P_j\cdot r_j+P_k\cdot r_k\right]\begin{Bmatrix}P_i\\P_j\\P_k\end{Bmatrix}dA$$

$$=2\pi\cdot G_{h_e}\int_{A_0}\begin{Bmatrix}P_i^2\cdot r_i+P_i\cdot P_j\cdot r_j+P_i\cdot P_k\cdot r_k\\P_jP_i\cdot r_i+P_j^2\cdot r_j+P_j\cdot P_k\cdot r_k\\P_kP_i\cdot r_i+P_k\cdot P_j\cdot r_j+P_k^2\cdot r_k\end{Bmatrix}dA=\frac{\pi\cdot G_{h_e}\cdot A_0}{6}\begin{bmatrix}2\cdot r_i+r_j+r_k\\r_i+2\cdot r_j+r_k\\r_i+r_j+2\cdot r_k\end{bmatrix}.$$

(Details concerning the integration are given in Example 13, Appendix 1 in Dimitrov (2013).)

We get the form of the vector of convective load $\left\{f_C^e\right\}_C$ in cylindrical coordinates after transforming the surface integral into a linear one, using again equality $dA = 2\pi\cdot r\cdot dH$:

$$\left\{f_C^e\right\}=\bar{\psi}_{A_\psi}\int_H\left[P^e\right]^T dA=\pm2\pi\cdot\alpha_{A_1}\cdot\bar{\psi}_{A_1}\int_{H_{j\text{-}k}}\left[0+P_j\cdot r_j+P_k\cdot r_k\right]\begin{Bmatrix}0\\P_j\\P_k\end{Bmatrix}dH$$

$$\pm2\pi\cdot\alpha_{A_2}\cdot\bar{\psi}_{A_2}\int_{H_{k\text{-}i}}\left[P_i\cdot r_i+0+P_k\cdot r_k\right]\begin{Bmatrix}P_i\\0\\P_k\end{Bmatrix}dH$$

$$\pm2\pi\cdot\alpha_{A_3}\cdot\bar{\psi}_{A_3}\int_{H_{ji\text{-}ji}}\left[P_i\cdot r_i+P_j\cdot r_j+0\right]\begin{Bmatrix}P_i\\P_j\\0\end{Bmatrix}dH$$

$$=\frac{\pm\pi\cdot\alpha_{A_1}\cdot\bar{\psi}_{A_1}\cdot H_{j-k}}{3}\begin{bmatrix}0\\2r_j+r_k\\r_j+2\cdot r_k\end{bmatrix}\pm\frac{\pi\cdot\alpha_{A_2}\cdot\bar{\psi}_{A_2}\cdot H_{i-k}}{3}\begin{bmatrix}2r_i+r_k\\0\\r_i+2r_k\end{bmatrix}$$

$$\pm\frac{\pi\cdot\alpha_{A_3}\cdot\bar{\psi}_{A_3}\cdot H_{i-j}}{3}\begin{bmatrix}2r_i+r_j\\r_i+2r_j\\0\end{bmatrix}. \tag{5.96}$$

(See the integration details in Example 6, Appendix 1 in Dimitrov (2013).)

Finally, the *vector of load* due to the direct flux $\left\{f_{Dr}^{e}\right\}_C$ has the same structure in cylindrical coordinates as that specified by equality (5.96), and it has the following final form:

$$\left\{f_{Dr}^{e}\right\}_C = \mp\frac{\pi\cdot\bar{h}_{A_1}\cdot H_{j-k}}{3}\begin{bmatrix}0_j\\2r_j+r_k\\r_j+2r_k\end{bmatrix}\mp\frac{\pi\cdot\bar{h}_{A_2}\cdot H_{k-i}}{3}\begin{bmatrix}2r_i+r_k\\0\\r_i+2r_k\end{bmatrix}\mp\frac{\pi\cdot\bar{h}_{A_3}\cdot H_{i-j}}{3}\begin{bmatrix}2r_i+r_j\\r_i+2r_k\\0\end{bmatrix}.$$

Consider area V_e. If there is net gain of matter at some of the surfaces of the rotating element, change the sign before the corresponding term.

The final form of the *load vector* of a 2D FE in cylindrical coordinates reads as follows:

$$\left\{f^{e}\right\}_C = \pm\frac{2\pi\cdot G_h\cdot A_0}{6}\begin{bmatrix}2r_i+r_j+r_k\\r_i+2r_j+r_k\\r_i+r_j+2r_k\end{bmatrix}+\frac{\pi\left(\pm\alpha_{A_1}\cdot\bar{\psi}_{A_1}\mp h_{A_1}\right)H_{j-k}}{3}\begin{bmatrix}0\\2r_j+r_k\\r_j+2r_k\end{bmatrix}$$
$$+\frac{\pi\left(\pm\alpha_{A_2}\cdot\bar{\psi}_{A_2}\mp h_{A_2}\right)H_{i-k}}{3}\begin{bmatrix}2r_i+r_k\\0\\r_i+2r_k\end{bmatrix}+\frac{\pi\left(\pm\alpha_{A_3}\cdot\bar{\psi}_{A_3}\mp h_{A_3}\right)H_{i-j}}{3}\begin{bmatrix}2r_i+r_j\\r_i+2r_j\\0\end{bmatrix}. \tag{5.97}$$

The matrix equation of the 2D simple FE is obtained after putting relations (Equations 5.94, 5.95, and 5.97) in relation (Equation 5.60):

$$\left[\begin{array}{l}2\pi\cdot\Gamma\cdot A_0\cdot\frac{r_i+r_j+r_k}{3}\begin{bmatrix}\left(b_i^2+c_i^2\right) & \left(b_ib_j+c_ic_j\right) & \left(b_ib_k+c_ic_k\right)\\\left(b_jb_i+c_jc_i\right) & \left(b_j^2+c_j^2\right) & \left(b_jb_k+c_jc_k\right)\\\left(b_kb_i+c_kc_i\right) & \left(b_kb_j+c_kc_j\right) & \left(b_k^2+c_k^2\right)\end{bmatrix}\pm\\ \pm\frac{2\pi\cdot\alpha_{j-k}\cdot H_{j-k}}{12}\begin{bmatrix}0 & 0 & 0\\0 & \left(3r_j+r_k\right) & \left(r_j+r_k\right)\\0 & \left(r_j+r_k\right) & \left(r_j+3r_k\right)\end{bmatrix}\pm\frac{2\pi\cdot\alpha_{i-k}\cdot H_{i-i}}{12}\begin{bmatrix}\left(3r_i+r_k\right) & 0 & \left(r_i+r_k\right)\\0 & 0 & 0\\\left(r_i+r_k\right) & 0 & \left(r_i+3r_k\right)\end{bmatrix}\pm\\ \pm\frac{2\pi\cdot\alpha_{i-j}\cdot H_{i-j}}{12}\begin{bmatrix}\left(3r_i+r_j\right) & \left(r_i+r_j\right) & 0\\\left(r_i+r_j\right) & \left(r_i+3r_j\right) & 0\\0 & 0 & 0\end{bmatrix}\end{array}\right]\begin{Bmatrix}\bar{\psi}_i\\\bar{\psi}_j\\\bar{\psi}_k\end{Bmatrix}$$
$$=\pm\frac{2\pi\cdot G_h\cdot A_0}{6}\begin{bmatrix}2r_i+r_j+r_k\\r_i+2r_j+r_k\\r_i+r_j+2r_k\end{bmatrix}+\frac{\pi\left(\pm\alpha_{A_1}\cdot\bar{\psi}_{A_1}\mp h_{A_1}\right)H_{j-k}}{3}\begin{bmatrix}0\\2r_j+r_k\\r_j+2r_k\end{bmatrix}+$$
$$+\frac{\pi\left(\pm\alpha_{A_2}\cdot\bar{\psi}_{A_2}\mp h_{A_2}\right)H_{i-k}}{3}\begin{bmatrix}2r_i+r_k\\0\\r_i+2r_k\end{bmatrix}+\frac{\pi\left(\pm\alpha_{A_3}\cdot\bar{\psi}_{A_3}\mp h_{A_3}\right)H_{i-j}}{3}\begin{bmatrix}2r_i+r_j\\r_i+2r_j\\0\end{bmatrix}. \tag{5.98}$$

Calculations involving specific values of the node coordinates and specific boundary conditions will be performed in a typical example given in (see Example 8.4 as well as Dimitrov 2013).

5.3.5 Transfer through a 3D Simple Finite Element

5.3.5.1 Design of the Matrix Equation of Transfer in Cartesian Coordinates

As discussed in Section 5.3, the simple 3D FE consists of four points—see Figures 5.5 and 5.16. The interpolated values of the characteristics of the ETS $\bar{\psi}(x,y,z), \bar{p}(x,y,z), \bar{V}(x,y,z)$, and $u\bar{T}(x,y,z)$ are estimated by means of their values $\bar{\psi}_m, \bar{p}_m, \bar{V}_m\ u\bar{T}_m$, $m = i, j, k, l$ at the nodes of the FE via Equation 5.28, namely, the following:

$$\bar{\psi}(x,y,z) = \bar{\psi}_i \cdot P_i^{\bar{\psi}} + \bar{\psi}_j \cdot P_j^{\bar{\psi}} + \bar{\psi}_k \cdot P_k^{\bar{\psi}} + \bar{\psi}_l \cdot P_l^{\bar{\psi}} = \left[P^e\right]_{\bar{\psi}} \left\{\bar{\psi}^e\right\},$$
$$\bar{p}(x,y,z) = \bar{p}_i \cdot P_i^p + \bar{p}_j \cdot P_j^p + \bar{p}_k \cdot P_k^p + \bar{p}_l \cdot P_l^p = \left[P^e\right]_{\bar{p}} \left\{\bar{p}^e\right\},$$
$$\bar{V}(x,y,z) = \bar{V}_i \cdot P_i^{\bar{V}} + \bar{V}_j \cdot P_j^{\bar{V}} + \bar{V}_k \cdot P_k^{\bar{V}} + \bar{V}_l \cdot P_k^{\bar{V}} = \left[P^e\right]_{\bar{V}} \left\{\bar{V}^e\right\},$$
$$\bar{T}(x,y,z) = \bar{T}_i \cdot P_i^T + \bar{T}_j \cdot P_j^T + \bar{T}_k \cdot P_k^T + \bar{T}_k \cdot P_k^T = \left[P^e\right]_{\bar{T}} \left\{\bar{T}^e\right\}.$$

Matrix $[P^e]_{\bar{\psi},\bar{p},\bar{V},\bar{T}}$ contains the linear approximating functions of $\bar{\psi}_m, \bar{p}_m, \bar{V}_m$, and $\bar{T}_m$, defined for the eth FE by means of Equation 5.52:

$$\begin{bmatrix} P_i & P_j & P_k & P_l \end{bmatrix} = \frac{1}{6v_0} \begin{bmatrix} a_i & b_i & c_i & d_i \\ a_j & b_j & c_j & d_j \\ a_k & b_k & c_k & d_k \\ a_l & b_l & c_l & d_l \end{bmatrix}^T \cdot \begin{Bmatrix} 1 \\ x \\ y \\ z \end{Bmatrix}.$$

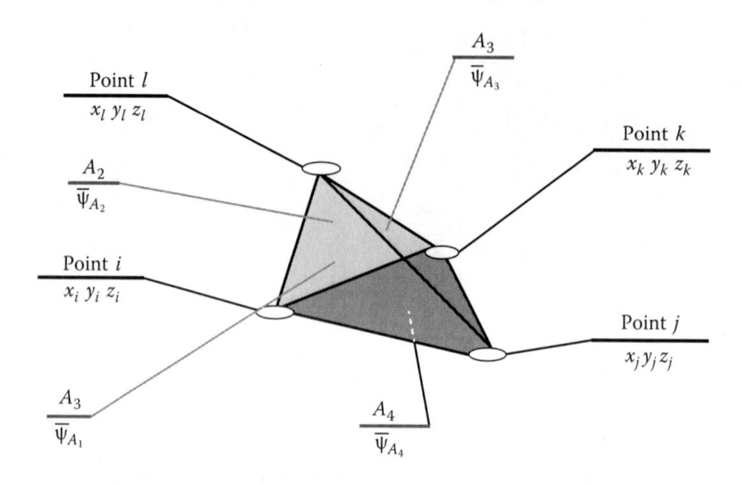

FIGURE 5.16
Three-dimensional simple finite element in Cartesian coordinates.

Coefficients a_m, b_m, c_m, d_m ($m = i, k, k, l$) are calculated using the node coordinates (see Equation 5.47).

Considering a 3D simple FE, the conductivity matrix $[K^e]$ is calculated using the definition equality (see Equation 5.60):

$$[K^e] = \int_{V^e} [B^e]^T [\Gamma][B^e] dV,$$

where matrices $[B^e]^T$ and $[B^e]$ are found using Equation 5.53, and the expression for $[K^e]$ gets the following form:

$$[K^e] = [\Gamma] \cdot v_0 \begin{bmatrix} b_i^2 + c_i^2 + d_i^2 & b_i \cdot b_j + c_i \cdot c_j + d_i \cdot d_j & b_i \cdot b_k + c_i \cdot c_k + d_i \cdot d_k & b_i \cdot b_l + c_i \cdot c_l + d_i \cdot d_l \\ b_j \cdot b_i + c_j \cdot c_i + d_j \cdot d & b_j^2 + c_j^2 + d_j^2 & b_j \cdot b_k + c_j \cdot c_k + d_j \cdot d_k & b_j \cdot b_l + c_j \cdot c_l + d_j \cdot d_l \\ b_k \cdot b_i + c_k \cdot c_i + d_k \cdot d_i & b_k \cdot b_j + c_k \cdot c_j + d_k \cdot d_j & b_k^2 + c_k^2 + d_k^2 & b_k \cdot b_l + c_k \cdot c_l + d_k \cdot d_l \\ b_l \cdot b_i + c_l \cdot c_i + d_l \cdot d_i & b_l \cdot b_j + c_l \cdot c_j + d_l \cdot d_j & b_l \cdot b_k + c_l \cdot c_k + d_l \cdot d_k & b_l^2 + c_l^2 + d_l^2 \end{bmatrix}. \tag{5.99}$$

The matrix of the surface properties is found on the basis of the respective definition specified by Equation 5.60, considering that the approximating functions are zeroed at nodes not belonging to the integration area:

$$\left[F^e\right] = \mp \alpha_{A_1} \int_{A_1} \begin{bmatrix} 0 & 0 & 0 & 0 \\ 0 & P_j^2 & P_j \cdot P_k & P_j \cdot P_l \\ 0 & P_k \cdot P_j & P_k^2 & P_k \cdot P_l \\ 0 & P_l \cdot P_j & P_l \cdot P_k & P_l^2 \end{bmatrix} dA \mp \alpha_{A_2} \int_{A_2} \begin{bmatrix} P_i^2 & 0 & P_i \cdot P_k & P_i \cdot P_l \\ 0 & 0 & 0 & 0 \\ P_k \cdot P_i & 0 & P_k^2 & P_k \cdot P_l \\ P_l \cdot P_i & 0 & P_l \cdot P_k & P_l^2 \end{bmatrix} dA$$

$$\mp \alpha_{A_3} \int_{A_3} \begin{bmatrix} P_i^2 & P_i \cdot P_j & 0 & P_i \cdot P_l \\ P_j \cdot P_i & P_j^2 & 0 & P_j \cdot P_l \\ 0 & 0 & 0 & 0 \\ P_l \cdot P_i & P_l \cdot P_j & 0 & P_l^2 \end{bmatrix} dA \mp \alpha_{A_4} \int_{A_4} \begin{bmatrix} P_i^2 & P_i \cdot P_j & P_i \cdot P_k & 0 \\ P_j . P_i & P_j^2 & P_j \cdot P_k & 0 \\ P_k \cdot P_i & P_k \cdot P_j & P_k^2 & 0 \\ 0 & 0 & 0 & 0 \end{bmatrix} dA$$

$$= \mp \frac{\alpha_{A_1} \cdot A_1}{12} \begin{bmatrix} 0 & 0 & 0 & 0 \\ 0 & 2 & 1 & 1 \\ 0 & 1 & 2 & 1 \\ 0 & 1 & 1 & 2 \end{bmatrix} \mp \frac{\alpha_{A_2} \cdot A_2}{12} \begin{bmatrix} 2 & 0 & 1 & 1 \\ 0 & 0 & 0 & 0 \\ 1 & 0 & 2 & 1 \\ 1 & 0 & 1 & 2 \end{bmatrix}$$

$$\mp \frac{\alpha_{A_3} \cdot A_3}{12} \begin{bmatrix} 2 & 1 & 0 & 1 \\ 1 & 2 & 0 & 1 \\ 0 & 0 & 0 & 0 \\ 1 & 1 & 0 & 2 \end{bmatrix} \mp \frac{\alpha_{A_4} \cdot A_4}{12} \begin{bmatrix} 2 & 1 & 1 & 0 \\ 1 & 2 & 1 & 0 \\ 1 & 1 & 2 & 0 \\ 0 & 0 & 0 & 0 \end{bmatrix}.$$

Note that they can be composed using the following 3D matrix:

$$[M] = \begin{bmatrix} 2 & 1 & 1 \\ 1 & 2 & 1 \\ 1 & 1 & 2 \end{bmatrix}$$

and inserting a zero row and zero column, which contain numbers corresponding to the lower index of the respective matrix $\left[M_k^0\right]$, $k = 1, 2, 3, 4$ (see Example 3 in Appendix 1 in Dimitrov (2013)).

Similarly, one can derive formulas of the components of the FE load vector. We introduce the following notations to provide a compact form of the formulas:

$$\vec{t} = \begin{Bmatrix} 1 \\ 1 \\ 1 \\ 1 \end{Bmatrix}, \quad \vec{t}_1 = \begin{Bmatrix} 0 \\ 1 \\ 1 \\ 1 \end{Bmatrix}, \quad \vec{t}_2 = \begin{Bmatrix} 1 \\ 0 \\ 1 \\ 1 \end{Bmatrix}, \quad \vec{t}_3 = \begin{Bmatrix} 1 \\ 1 \\ 0 \\ 1 \end{Bmatrix}, \quad \vec{t}_4 = \begin{Bmatrix} 1 \\ 1 \\ 1 \\ 0 \end{Bmatrix}$$

- *FE load due to internal matter recuperation*

$$\left\{f_G^e\right\} = G_{h_e} \int_{V_e} \begin{Bmatrix} P_i \\ P_j \\ P_k \\ P_l \end{Bmatrix} dV = \frac{G_{h_e} \cdot V_0}{4} \vec{t}. \tag{5.100}$$

- *Load due to convection at the boundary surfaces*

$$\left\{f_C^e\right\} = \pm \frac{\alpha_{A_1} \overline{\varphi}_{A_1} A_1}{3} \vec{t}_1 \pm \frac{\alpha_{A_2} \overline{\varphi}_{A_2} A_2}{3} \vec{t}_2 \pm \frac{\alpha_{A_3} \overline{\varphi}_{A_3} A_3}{3} \vec{t}_3 \pm \frac{\alpha_{A_4} \overline{\varphi}_{A_4} A_4}{3} \vec{t}_4$$

$$= \pm \frac{1}{3} \sum_{k=1}^{4} \left(\alpha_{A_k} \overline{\varphi}_{A_k} A_k\right) \vec{t}_k. \tag{5.101}$$

- *Load due to a direct flux along the boundary surfaces*

$$\left\{f_C^e\right\} = \mp \frac{\overline{h}_{A_1} A_1}{3} \vec{t}_1 \mp \frac{\overline{h}_{A_2} A_2}{3} \vec{t}_2 \mp \frac{h_{A_3} A_3}{3} \vec{t}_3 \mp \frac{\overline{h}_{A_4} A_4}{3} \vec{t}_4 = \pm \frac{1}{3} \sum_{k=1}^{4} \left(\overline{h}_{A_k} A_k\right) \vec{t}_k. \tag{5.102}$$

The generalized formula of the *load vector* is found by summing the last three formulas (Equations 5.100 through 5.102):

$$\{f^e\} = \frac{G_h V_0}{4}\vec{t} + \frac{\left(\pm\alpha_{A_1}\bar{\varphi}_{A_1} \mp \bar{h}_{A_1}\right)A_1}{3}\vec{t}_1 + \frac{\pm\alpha_{A_2}\bar{\varphi}_{A_2} \mp \bar{h}_{A_2}A_2}{3}\vec{t}_2$$
$$+ \frac{\left(\pm\alpha_{A_3}\bar{\varphi}_{A_3} \mp \bar{h}_{A_3}\right)A_3}{3}\vec{t}_3 + \frac{\left(\pm\alpha_{A_4}\bar{\varphi}_{A_4} \mp \bar{h}_{A_4}\right)A_4}{3}\vec{t}_4$$
$$= \frac{G_h V_0}{4}\vec{t} + \frac{1}{3}\sum_{k=1}^{4}\left(\pm\alpha_{A_k}\bar{\varphi}_{A_k} \mp \bar{h}_{A_k}\right)A_k\vec{t}_k.$$

The matrix equation of a *3D simple* FE in Cartesian coordinates has the following form:

$$\left(\frac{\Gamma}{36v_0}\begin{bmatrix} b_i^2+c_i^2+d_i^2 & b_i\cdot b_j+c_i\cdot c_j+d_i\cdot d_j & b_i\cdot b_k+c_i\cdot c_k+d_i\cdot d_k & b_i\cdot b_l+c_i\cdot c_l+d_i\cdot d_l \\ b_j\cdot b_i+c_j\cdot c_i+d_j\cdot d & b_j^2+c_j^2+d_j^2 & b_j\cdot b_k+c_j\cdot c_k+d_j\cdot d_k & b_j\cdot b_l+c_j\cdot c_l+d_j\cdot d_l \\ b_k\cdot b_i+c_k\cdot c_i+d_k\cdot d_i & b_k\cdot b_j+c_k\cdot c_j+d_k\cdot d_j & b_k^2+c_k^2+d_k^2 & b_k\cdot b_l+c_k\cdot c_l+d_k\cdot d_l \\ b_l\cdot b_i+c_l\cdot c_i+d_l\cdot d_i & b_l\cdot b_j+c_l\cdot c_j+d_l\cdot d_j & b_l\cdot b_k+c_l\cdot c_k+d_l\cdot d_k & b_l^2+c_l^2+d_l^2 \end{bmatrix} \mp \right.$$
$$\left. \pm\frac{\alpha_{A_1}\cdot A_1}{12}\begin{bmatrix}0&0&0&0\\0&2&1&1\\0&1&2&1\\0&1&1&2\end{bmatrix} \pm\frac{\alpha_{A_2}\cdot A_2}{12}\begin{bmatrix}2&0&1&1\\0&0&0&0\\1&0&2&1\\1&0&1&2\end{bmatrix} \pm\frac{\alpha_{A_3}\cdot A_3}{12}\begin{bmatrix}2&1&0&1\\1&2&0&1\\0&0&0&0\\1&1&0&2\end{bmatrix} \pm\frac{\alpha_{A_4}\cdot A_4}{12}\begin{bmatrix}2&1&1&0\\1&2&1&0\\1&1&2&0\\0&0&0&0\end{bmatrix}\right)\begin{Bmatrix}\bar{\psi}_i\\\bar{\psi}_j\\\bar{\psi}_k\\\bar{\psi}_l\end{Bmatrix}$$

General conductivity matrix a

$$= \frac{G_h\cdot v_0}{4}\begin{Bmatrix}1\\1\\1\\1\end{Bmatrix} + \frac{\left(\pm\alpha_{A_1}\cdot\bar{\psi}_{A_1} \mp \bar{h}_{A_1}\right)\cdot A_1}{3}\begin{Bmatrix}0\\1\\1\\1\end{Bmatrix} + \frac{\left(\pm\alpha_{A_2}\cdot\bar{\psi}_{A_2} \mp \bar{h}_{A_2}\right)\cdot A_2}{3}\begin{Bmatrix}1\\0\\1\\1\end{Bmatrix}$$

load vectors

$$+ \frac{\left(\pm\alpha_{A_3}\cdot\bar{\psi}_{A_3} \mp h_{A_3}\right)\cdot A_3}{3}\begin{Bmatrix}1\\1\\0\\1\end{Bmatrix} + \frac{\left(\pm\alpha_{A_4}\cdot\bar{\psi}_{A_4} \mp \bar{h}_{A_4}\right)\cdot A_4}{3}\begin{Bmatrix}1\\1\\1\\0\end{Bmatrix}. \tag{5.103}$$

See the numerical example in Example 8.5 as well as Dimitrov (2013), which illustrates the design of the earlier equation.

6

Initial and Boundary Conditions of a Solid Wall Element

The physical processes running at the boundary of an electrothermodynamic system (*ETS*) and determining the free energy function were discussed in Chapters 2 and 3 (Section 3.3). Here, however, we will discuss an additional form of interaction between the surroundings and the building envelope specifying respective boundary conditions. In particular, this is the process of matter (energy and mass) transfer. It employs the free energy *resources* of the *ETS* to eject matter (energy and mass) through the envelope, and it can be assessed by employing popular analytical or graphical engineering techniques, including the use of psychometric diagrams of the state of air (Тихомиров 1981; Богословский 1982). Note also that energy transfer opposes the free energy function gradient.

6.1 Effects of the Environmental Air on the Building Envelope

Consider Point *A* of the facial surface of a facade wall. Consider also a fluid flow running through an infinitesimally small area around Point *A* with dimensions *dx*, *dy*, and *dz*. Assume flow and surface temperatures $T_A = F(x,z)_{y=\delta+\Delta}$ and T_w, respectively. Using the state diagram of vapor–air mixture (see Figures 6.1 and 6.2), we can foresee the following effects:

- *Wall dehumidification*: Moisture transfer from the wall to the surrounding air (see Figure 6.1)
- *Wall humidification*: Moisture extraction from the surrounding air and *supply to the wall* by a secondary vertical air convection (see Figure 6.2)

Mass (moisture) transfer to/from the building envelope takes place mainly in winter and in transitional seasons when there is negative temperature

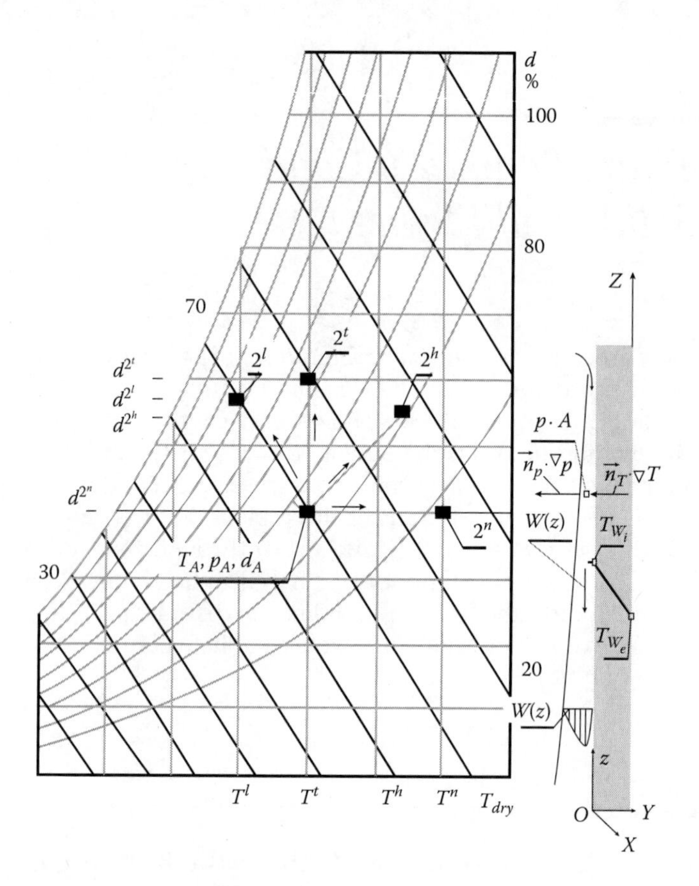

FIGURE 6.1
Wall dehumidification via change of the state of a vapor–air mixture along a cold vertical wall.

gradient along axis Oy at Point A (grad $T_A < 0$). It results in surface and structural damage of the facade elements and is of essential practical interest.

6.1.1 Mass Transfer from the Building Envelope (*Wall Dehumidification, Drying*)

Theoretically, this phenomenon may occur at the wall internal surface, assuming a *negative pressure gradient* $(\text{grad}\, p)_A$ along axis Oy at Point A, that is, $p_A < p_w$ (see Figure 6.1). Wall dehumidification takes place when the working fluid—vapor–air mixture—meets the vertical wall as a result of convection running along the wall height. Inversing the physical conditions, such a phenomenon may occur at the wall external surface. Except for transfer of vapor and heat from the wall, latent heat transfer owing to vapor energy also takes place. As seen in Figure 6.1, these processes determine the boundary conditions valid for the wall element and described in the next section.

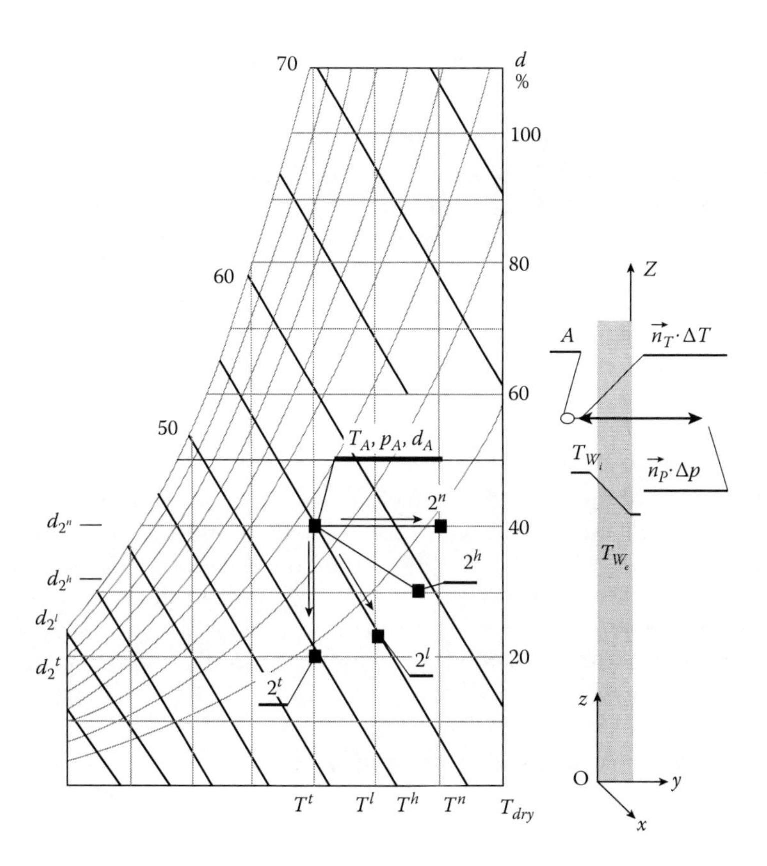

FIGURE 6.2
Wall humidification.

6.1.1.1 Processes Running at a Cold Wall ($T_A \geq T_{w_i}$)

6.1.1.1.1 *Wall Heating under Moisture Equilibrium within the Working Fluid (Direction A-2^h)*

This is a *border* process developing within a fluid flowing past a cold wall (see Figure 6.1). Although the fluid temperature $T_A = T_A(z)$ decreases to T^n, fluid moisture contents d_A remain constant ($d_A = d_{2^n} = \text{const}\%$). In this case, *dry heating* in the vicinity of Point A takes place. Heat gained is equal to the difference between enthalpies $h = \left(h_A - h_{w_{2^n}}\right)$, setting forth a second-order boundary condition (see Equation 6.5).

The process can also run in the opposite direction, when pressure gradient changes its sign ($(\text{grad}\, p)_A > 0$).

6.1.1.1.2 *Wall Cooling under Moisture Extraction by Air (Direction A-2^h)*

Under the adopted conditions ($\text{grad}\, p_A < 0$) and $\text{grad}\, T_A < 0$, the fluid flowing past the bordering wall heats up and takes moisture from the wall under the effect of $(\text{grad}\,p)_A$. Moisture extracted in the vicinity of Point A is estimated as $d_{2^h} - d_A$, and it presents the flux of mass taken from the wall. The specific *wall–fluid* heat flux is equal to the difference between enthalpies $\left(h_A - h_{w_{2^h}}\right)$, and it cools down the wall. Moisture and heat transfer to/from the wall determines the boundary conditions of operation of the wall element.

Consider the limit case when the process develops along line A-2^t, where $T_A = T_w$. Then, moisture transfer from the wall and into the surrounding air takes place under negative gradient of pressure $(\text{grad}\,p)_A$ within the boundary layer only. Here, moisture gained by the air in the vicinity of Point A is $\left(d_{2^t} - d_A\right)$. It is in fact moisture lost by the wall, and wall's energy loss amounts to $\left(h_A - h_{2^t}\right)$ owing to evaporation.

6.1.1.1.3 *Adiabatic Processes (Direction (A – 2^l))*

Adiabatic loss of moisture (wall drying) and its absorption by a working fluid (air) flowing past the wall (h_A = const) at Point A is only possible under positive pressure gradient $(\text{grad}\,p)_A$ in the boundary layer. As seen in the state diagram (see Figure 6.1), the direction of moisture release to the air is $(A - 2^l)$. Then, air will cool down absorbing released moisture under positive pressure gradient $(\text{grad}\,p)_A$, and the amount of moisture taken from the wall without energy loss $\left(h_{2^l} - h_A = 0\right)$ will be $d_A - d_{2^l}$.

6.1.1.2 Processes Running at a Cold Wall ($T_w < T_A$)

6.1.1.2.1 *Wall Heating and Humidification (under Air Cooling and Moisture Release) (Direction A-2^h)*

This particular case of interaction between the building envelope and the surroundings (air) considers positive pressure gradient in the vicinity of Point A, that is, $(\text{grad}\,p_A > 0)$. Air heats up and dries, and the wall cools down and absorbs released moisture under positive pressure gradient $(\text{grad}\,p)_A$. Wall humidification takes place when the working fluid flows past a vertical wall due to free convection. Moisture extracted from the air is absorbed by the envelope in the vicinity of Point A. It amounts to $\left(d_{2^h} - d_A\right)$, while the specific heat flux directed outward can be measured by the difference between the enthalpies $\left(h_A - h_{w_{2^h}}\right)$. Considering the limit case (when the process develops along line A-2^t, where $T_A = T_w$), moisture is absorbed by the wall owing to the negative pressure gradient $(\text{grad}\,p)_A$ in the wall boundary layer only. Here, moisture gained by the air in the vicinity of Point A is $\left(d_{2^t} - d_A\right)$, and the specific heat flux (energy gain) is presented by the enthalpy difference $\left(h_A - h_{2^t}\right)$. The energy gain is due to the latent heat of phase transition released via the water vapor in the air and absorbed by the wall.

6.1.1.2.2 *Transfer under Adiabatic Conditions (Direction A-2^l)*

Mass transfer from the ambient air to the building envelope is due to negative temperature gradient and positive pressure gradient in the vicinity of

Point A belonging to the boundary layer. Basically, mass transfer to the wall runs along line (A–2^l) (see Figure 6.2). In that case, the warming air $\left(T_A \rightarrow T_{2^i}\right)$ descends along the wall releasing its moisture under the effect of the positive pressure gradient $(\text{grad}\,p)_A$, and moisture is absorbed by the wall. The specific flux of absorbed mass is defined by the difference $d_A - d_{2^l}$.

Hence, energy balance is established (condition $h_A - h_{2^l} = 0$ is satisfied, the energy of air heating is equal to the heat of phase transition extracted by the water vapors, and the heat flux is directed to the envelope. These energy interactions (energy and mass transfer, in particular) between the envelope and the ambient air take place within the boundary layer. They are caused by change of the values of the *ETS* generalized potential resulting from change of the ambient conditions (change of the solar radiation, change in the ambient temperature, or emergence of convective currents). Thus, the phenomena occurring at the envelope/surrounding interface specify the boundary conditions of subsequent problems of envelope design and analysis of *ETS* behavior.

6.2 Various Initial and Boundary Conditions of Solid Structural Elements

As is known, to solve the differential equation of transfer, one needs to specify subsequent boundary conditions. Three types of boundary conditions are known, first-, second-, and third-order boundary conditions, while the last ones are also known as *mixed boundary conditions*. A first-order boundary condition specifies temperature/pressure distribution or the function of free energy at the wall boundary. It is called the Dirichlet boundary condition having the form

$$F_1(\vec{x}) = T(\vec{x}), \quad F_1'(\vec{x}) = p(\vec{x}), \quad F_1''(\vec{x}) = V(\vec{x})u, \quad F_1'''(\vec{x}) = \psi(\vec{x}).$$

Consider bodies with planar symmetry. Note that there are different distributions of the generalized forces (surface temperature, pressure, electric potential, or free energy), depending on different impacts acting on both sides of the wall elements.

Figure 6.3a through c illustrate different temperature distributions at the boundary:

- Uniform distribution (see Figure 6.3a):

$$T_w = \text{const}, \quad p_w = \text{const}, \quad V_w = \text{const}\, u, \quad \psi_w = \text{const};$$

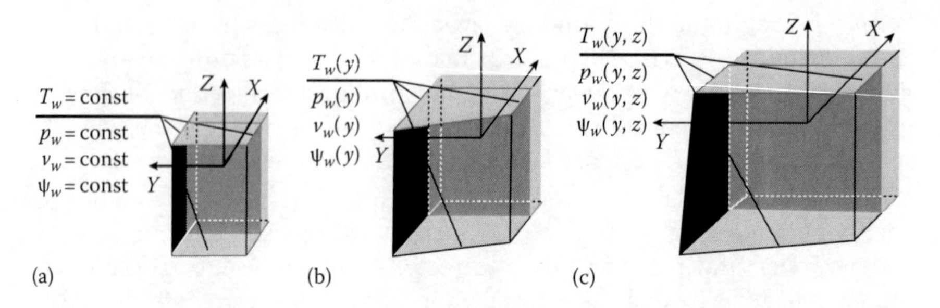

FIGURE 6.3
Boundary conditions at a vertical wall: (a) uniform distribution of the independent variables, (b) linear distribution of the independent variables, (c) two-dimensional distribution of the independent variables.

- Linear distribution of the independent variables (see Figure 6.3b):

$$T_w = f(y), \quad p_w = f_1(y), \quad V_w = f_2(y)u, \quad \psi_w = f_3(\psi);$$

- Two-dimensional distribution of the independent variables (see Figure 6.3c):

$$T_w = f(y,z), \quad p_w = f_1(y,z), \quad V_w = f_2(y,z)u, \quad \psi_w = f_3(y,z).$$

For elements with axial and central symmetry, the boundary conditions are specified in appropriately chosen coordinate systems, for instance, in a cylindrical coordinate system ($T_{WO} = f(\theta)$, and $T_{WO} = f(r, \theta)$) or in a spherical coordinate system ($T_{WO} = f(\theta, \lambda)_\rho$ and $T_{WO} = f(r, \theta, \lambda)$). Quite often, the state variable distribution along the boundary of a wall element with area A_0 is nonlinear, and it requires full balance of mass transfer to the surroundings.

The second-order boundary condition is specified by a distribution of the flow density vector, and it is also known as *Neumann boundary* condition:

$$F_2(\vec{x}) = \vec{h}(\vec{x}) = \left(-[\Gamma]\frac{\partial \psi(\vec{x})}{\partial \vec{n}_\psi} \right). \tag{6.1}$$

It requires knowing the specific heat flux $\vec{q}$, density of the electric current, rate of matter delivery $\vec{g}$, or density $\vec{h}$ of the total matter transferred through the entire wall surface ($\vec{x}$ is the radius vector of Point *A* of the boundary surface; see Figure 5.2). These boundary conditions cannot be specified arbitrarily, and they require total balance of mater (energy and mass) in the *ETS*.

The third-order boundary condition is a combination between boundary conditions of the first and second order, that is, a combination of Dirichlet and Neumann conditions:

$$F_3(\vec{x}) = a(\vec{x}) \cdot \psi(\vec{x}) + b(\vec{x}) \cdot \frac{\partial}{\partial \tau} \psi(\vec{x}). \tag{6.2}$$

It is specified in areas subjected to combined impacts (for instance, solar radiation and forced and natural convection).

As noted, the differential and integral equations of transfer require exact formulation of the initial and boundary conditions of a specific element. To account for the *initial conditions* of the wall element, the divergent form of the equation of transfer concerning a homogeneous area should be written in the form

$$\frac{\partial\psi}{\partial\tau}-\frac{[\Gamma]}{C_\psi\cdot\rho}\operatorname{div}(\operatorname{grad}\psi)=\frac{G_h}{C_\psi\rho}=0. \tag{6.3}$$

Integrating along the boundary of area R and applying Gauss theorem (Equation 3.41), we find the following equality:

$$\frac{\partial}{\partial\tau}\int_{\nu}\psi d\nu-\frac{[\Gamma]}{C_\psi\cdot\rho}\int_{A_0}\vec{n}_\psi\cdot(\operatorname{grad}\psi)dA-\int_{\nu}\frac{G_h}{C_\psi\rho}d\nu=0. \tag{6.4}$$

Its integration within the interval $\tau_b\le\tau\le\tau_{end}$ yields an equation that gives a balance form of the general law of conservation of transferred matter (energy and mass) within area R:

$$\left(\frac{\partial}{\partial\tau}\int_{\nu}\psi d\nu\right)_{End}-\left(\frac{\partial}{\partial\tau}\int_{\nu}\psi d\nu\right)_{St}-\frac{[\Gamma]}{C_\psi\rho}\int_{\tau_H}^{\tau_{kp}}\int_{A_0}\frac{\partial\psi}{\partial\vec{n}_\psi}dAd\tau-\frac{1}{C_\psi\rho}\int_{\tau_H}^{\tau_{kp}}\int_{v}G_hd\nu d\tau=0,\ \rightarrow$$

$$\left(\frac{\partial}{\partial\tau}\int_{\nu}\psi d\nu\right)_{end}-\left(\frac{\partial}{\partial\tau}\int_{\nu}\psi d\nu\right)_{b}-\frac{[\Gamma]}{C_\psi\rho}\int_{\tau_b}^{\tau_{end}}\int_{A_0}\frac{\partial\psi}{\partial\vec{n}_\psi}dAd\tau-\frac{1}{C_\psi\rho}\int_{\tau_b}^{\tau_{end}}\int_{\nu}G_hd\nu d\tau=0, \tag{6.5}$$

where

$$\left(\frac{\partial}{\partial\tau}\int_{\nu}\psi d\nu\right)_{St}\rightarrow\left(\frac{\partial}{\partial\tau}\int_{\nu}\psi d\nu\right)_{b}$$

$$\left(\frac{\partial}{\partial\tau}\int_{\nu}\psi d\nu\right)_{End}\rightarrow\left(\frac{\partial}{\partial\tau}\int_{\nu}\psi d\nu\right)_{end}$$

$$\frac{[\Gamma]}{C_\psi\rho}\int_{\tau_H}^{\tau_{kp}}\int_{A_0}\frac{\partial\psi}{\partial\vec{n}_\psi}dAd\tau\rightarrow\frac{[\Gamma]}{C_\psi\rho}\int_{\tau_b}^{\tau_{endp}}\int_{A_0}\frac{\partial\psi}{\partial\vec{n}_\psi}dAd\tau$$

$$\frac{1}{C_\psi\rho}\int_{\tau_H}^{\tau_{kp}}\int_{\nu}G_hd\nu d\tau\rightarrow\frac{1}{C_\psi\rho}\int_{\tau_b}^{\tau_{end\,p}}\int_{\nu}G_hd\nu d\tau$$

The aforementioned equation expresses the sum of the three quantities of matter: the first one is mass/energy accumulated in area R, the second one is matter (mass/energy) transferred through the boundary surface A_0, and the third one is recuperated matter (energy and mass) generated by the internal source. Note that the sum should be zero in the time interval $\tau_b \le \tau \le \tau_{end}$. The equation is usually used to calculate the amount of the recuperated matter (energy and mass) within area R at the final moment τ_{end}. Quantity $\left(\frac{\partial}{\partial \tau}\int_{v} \psi dv\right)_{end}$ is involved in various applied problems, and it can be found from Equation (6.2):

$$\left(\frac{\partial}{\partial \tau}\int_{v} \psi dv\right)_{end} = \left(\frac{\partial}{\partial \tau}\int_{v} \psi dv\right)_{b} + \frac{[\Gamma]}{C_{\psi}\rho}\int_{\tau_b}^{\tau_{end}}\int_{A_0} \frac{\partial \psi}{\partial \vec{n}_{\psi}} dA d\tau + \frac{1}{C_{\psi}\rho}\int_{\tau_b}^{\tau_{end}}\int_{v} G_h dv d\tau \qquad (6.6)$$

The first two terms of the equation right-hand side account for the initial (the first term) and the boundary (the second term) conditions of transfer. However, as architects' practice proves, solid external elements of building walls undergo various impacts depending on their location, orientation, surroundings topology, and, last but not least, astronomical time and meteorological conditions. For instance, components of the eastern, southern, and western facades of buildings, as well as the roof structure, undergo heat exchange via radiation and convection, while exchange intensity varies within a 24 h period and during the whole season. In particular, building northern facade undergoes convective heat exchange and not exchange due to direct radiation.

On the other hand, floor solid elements exchange matter (energy and mass) via transfer. If the element performs radiative heat exchange, it seems useful to employ a modified form of the third type of boundary conditions:

$$F_3^M(\vec{x}) = a(\vec{x})\cdot(\psi(\vec{x}) + \psi_r(\vec{x})) + b(\vec{x})\cdot\frac{\partial}{\partial \tau}\psi(\vec{x}), \qquad (6.7)$$

where ψ_r is a correction of ψ, due to radiation heat exchange taking place at the wall surface.

6.3 Design of Boundary Conditions of Solid Structural Elements

Boundary and initial conditions should be chosen such as to satisfy the balance equations of matter (energy and mass) conservation within a control volume of the surrounding environment, and the element overall dimensions should be finite. Boundary and initial conditions (Equations 6.1 and 6.2) of the differential equations of transfer provide values of the terms

$$\frac{[\Gamma]}{C_{\psi}\rho}\int_{\tau_{\mathcal{H}}}^{\tau_{kp}} \frac{\partial\psi}{\partial n}d\tau \quad \text{and} \quad \left(\frac{\partial}{\partial\tau}\psi\right)_{\mathcal{H}}.$$

The first term expresses matter (energy and mass) transfer to the surroundings in the vicinity of a point, while the second one is the initial rate of variation of the generalized state parameter ψ. The boundary conditions of the integral equation (Equation 6.5) should enable one to find the RHS members of Equation 6.6.

6.3.1 Boundary Conditions of Convective Transfer Directed to the Wall Internal Surface

Consider an air layer enveloping a solid wall element. The respective boundary conditions involve *ETS* state variables in planes $y = 0$ and $y = \delta_{Fl}$ (see Figures 6.1 and 6.2), and they are specified by means of functional relations of the forms

$$T_0 = F_1(x,z)_{y=\delta} \quad \text{and} \quad T_0'' = F_1(x,z)_{y=0}, \tag{6.8}$$

$$p_0 = F_2(x,z)_{y=\delta} \quad \text{and} \quad p_0'' = F_2(x,z)_{y=0}, \tag{6.9}$$

$$V_0 = F_3(x,z)_{y=\delta} \quad \text{and} \quad V_0'' = F_3(x,z)_{y=0}, \tag{6.10}$$

$$\psi_0 = F_4(T_0,p_0,V_0)_{y=\delta} \quad \text{and} \quad \psi_0'' = F_4(T_0,p_0,V_0)_{y=0}, \tag{6.11}$$

$$w_0 = F_5(x,z)_{y=\delta} \quad \text{and} \quad w_0'' = F_5(x,z)_{y=0}. \tag{6.12}$$

Some parameters of the *ETS* in area $R((0 \le x \le L), (0 \le y \le B), (0 \ge z \ge H))$, such as

$$T = T(x,y,z), \quad p = p(x,y,z), \quad V = V(x,y,z), \quad \psi = \psi(x,y,z),$$
$$\text{and} \quad \vec{w} = \vec{w}(x,y,z), \tag{6.13}$$

can be valid for the whole premise, with facade element that is the wall under consideration. Note that they can be numerically found.

The differential equation describing convective transfer within the boundary layer is derived on the basis of the divergence transfer equation (Equation 6.3). Note that an equation of continuity of an incompressible fluid should be added, multiplied by the *ETS* generalized state variable for an arbitrary point, ψ:

$$\operatorname{div}(\psi\cdot\vec{W}) = 0 \rightarrow \frac{\partial\psi\cdot w_x}{\partial x} + \frac{\partial\psi\cdot w_y}{\partial y} + \frac{\partial\psi\cdot w_z}{\partial z} = 0. \tag{6.14}$$

The new divergence transfer equation including convective transfer components takes the form

$$\frac{\partial\psi}{\partial\tau}+\mathrm{div}(\psi\cdot\vec{W})-\frac{[\Gamma]}{C_\psi\cdot\rho}\mathrm{div}(\mathrm{grad}\,\psi)-\frac{G}{C_\psi\cdot\rho}=0,$$

or

$$\frac{\partial\psi}{\partial\tau}+\frac{\nabla}{C_\psi\rho}\left(\rho\cdot\vec{W}-[\Gamma]\nabla\psi\right)-\frac{G_h}{C_\psi\rho}=0. \tag{6.15}$$

Integrating Equation 6.15 within volume ν, we get an integral form of transfer with convective members:

$$\frac{\partial}{\partial\tau}\int_\nu\psi d\nu+\frac{1}{C_\psi\cdot\rho}\int_\nu\nabla\left(\psi\cdot\vec{W}-[\Gamma]\nabla\psi\right)d\nu-\frac{1}{C_\psi\cdot\rho}\int_\nu G_h d\nu=0.$$

The application of Gauss theorem yields other forms:

- *Vector form*:

$$\frac{\partial}{\partial\tau}\int_\nu\psi d\nu+\frac{1}{C_\psi\cdot\rho}\int_{A_0}\vec{n}\left(\psi\cdot\vec{V}-[\Gamma]\nabla\psi\right)dA-\frac{1}{C_\psi\cdot\rho}\int_\nu G_h d\nu=0. \tag{6.16}$$

- *Gradient form*:

$$\frac{\partial}{\partial\tau}\int_\nu\psi d\nu+\frac{1}{C_\psi\cdot\rho}\int_{A_0}\frac{\partial}{\partial n}\left(\psi\cdot\vec{W}-[\Gamma]\frac{\partial\psi}{\partial n}\right)dA-\frac{1}{C_\psi\cdot\rho}\int_\nu G_h d\nu=0. \tag{6.17}$$

- *In a Cartesian coordinate system*:

$$\frac{\partial}{\partial\tau}\int_\nu\psi dxdtdz+\int_{A_0}\left[w_x\cdot\psi+w_y\cdot\psi+w_z\cdot\psi-[\Gamma]\left(\vec{i}\,\frac{\partial\psi}{\partial x}+\vec{j}\cdot\frac{\partial\psi}{\partial y}+\vec{k}\cdot\frac{\partial\psi}{\partial z}\right)\right]dA-;$$

$$-\int_\nu G_h dxdydz=0. \tag{6.18}$$

- *In a cylindrical coordinate system*:

$$\frac{\partial}{\partial\tau}\int_\nu\psi\cdot rdrd\theta dz+\int_{A_0}\left[w_r\cdot\varphi+w_\theta\cdot\varphi+w_z\cdot\varphi-[\Gamma]\left(n_r\cdot\frac{\partial\psi}{\partial r}+n_\theta\cdot\frac{1}{r}\frac{\partial\psi}{\partial\theta}+n_z\cdot\frac{\partial\psi}{\partial z}\right)\right]dA-;$$

$$-\int_\nu G_h\cdot rdrd\theta dz=0. \tag{6.19}$$

- *In a spherical coordinate system:*

$$\frac{\partial}{\partial\tau}\int_{v}\psi\cdot r^{2}\cdot\cos\theta drd\theta d\lambda +$$

$$-\int_{A_0}\left[w_r\cdot\psi+w_\theta\cdot\psi+w_\lambda\cdot\psi-[\Gamma]\left(n_r\cdot\frac{\partial\psi}{\partial r}+n_\theta\cdot\frac{1}{r}\frac{\partial\psi}{\partial\theta}+n_\lambda\cdot\frac{1}{r\cdot\sin\theta}\cdot\frac{\partial\psi}{\partial\lambda}\right)\right]dA.$$

$$-\int_{v}G_h\cdot r^{2}\cdot\cos\phi drd\theta d\lambda=0 \tag{6.20}$$

Quantity $\vec{W}$ in these relations is the flow speed, and its presentation depends on the coordinate system used:

$$\vec{w}=w_x\cdot\vec{i}+w_y\cdot\vec{j}+w_z\cdot\vec{k}; \tag{6.21}$$

$$\vec{w}=w_r\cdot\vec{n}_r+w_\theta\cdot\vec{n}_\theta+w_z\cdot\vec{n}_z; \tag{6.22}$$

$$\vec{w}=w_r\cdot\vec{n}_r+w_\theta\cdot\vec{n}_\theta+w_\lambda\cdot\vec{n}_\lambda. \tag{6.23}$$

Integrate relations Equations 6.16 through 6.20 within the time interval $\tau_b\le\tau\le\tau_{end}$. Solve the equation found with respect to $\left(\int\psi dV\right)_{end}$. Then we get the presentation

$$\left(\int\psi dv\right)_{kp}=\left(\int\psi dv\right)_{H}-\int_{\tau_H}^{\tau_{kp}}\left[\frac{1}{C_\psi\cdot\rho}\int_{A_0}\vec{n}_\psi\cdot\left(\psi\cdot\vec{w}-[\Gamma]\nabla\psi\right)dA\right]d\tau$$

$$+\int_{\tau_H}^{\tau_{kp}}\left[\frac{1}{C_\psi\cdot\rho}\int_{v}G_h dv\right]d\tau. \tag{6.24}$$

The first integral at the RHS of Equation 6.24 expresses *initial mass concentration*: the second one is the estimation of matter (energy and mass) transfer through the area boundary via convection, that is, this is the *boundary condition* of transfer; while the third integral gives an estimation of matter (energy and mass) generated by the internal source within the wall spatial area R. Equality (Equation 6.24) shows that the integral equations can be used to define the initial and boundary conditions at the boundary of the wall element shown in Figure 6.3.

6.3.2 Boundary Conditions at the Wall External Surface

Consider a wall element. Note that the direction of the pressure gradient grad$p$ is of essential importance when matter (energy and matter) transfer runs within the element, while the element location on the building facade specifies the gradient direction. This is so since the distribution of pressure applied on the building circumferential surface has a rather complex character. It varies in time and depends on a number of factors:

- General building architectonics
- Dynamic pressure applied on the facade walls owing to wind load $p_d = 0,5\rho w_e^2$
- Temperature of the working fluid inside the building T_i
- Temperature of the working fluid in the area past the building T_e

Figure 6.4 shows a configuration of the field of nondimensional difference between pressure acting on the external wall surface p_e and pressure acting on the internal wall surface p_i: $\Delta\overline{p} = 2(p_e - p_i)/\rho w_e^2$ of a cylinder (see Figure 6.4a) and a parallelepiped (see Figure 6.4b). These bodies are considered as models of buildings outflowed by a fluid ($T_i = T_e$).

As seen in Figure 6.5, the building's walls underwent pressure whose magnitude and sign vary. These change the pressure gradient $(\text{grad}\, p)_{A_i}$ and energy flux direction and intensity, accordingly.

Two types of wall elements are worth noting:

1. Elements operating under positive pressure difference $\Delta\overline{p}$, that is, under positive gradient $(\text{grad}\, p)_{A_i}$ (see Figures 6.4 and 6.5)
2. Elements operating under negative pressure difference $\Delta\overline{p}$, that is, under negative gradient $(\text{grad}\, p)_{A_i}$

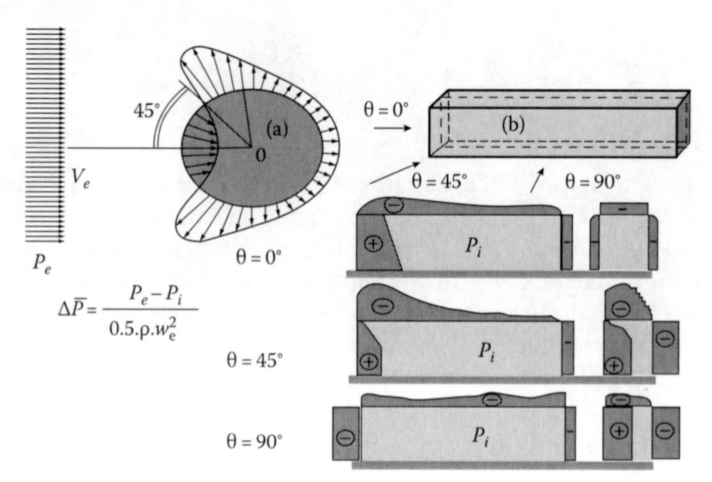

FIGURE 6.4
Change of the boundary conditions due to building aerodynamics: (a) Axes symmetrical building and (b) cuboid building.

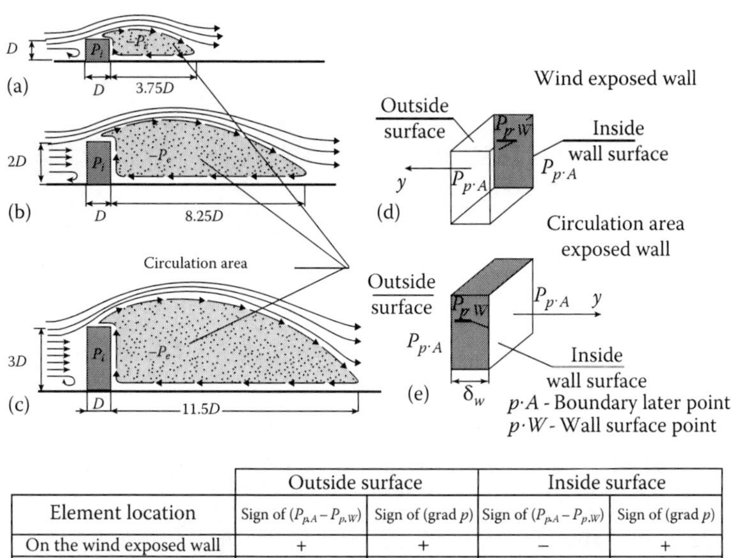

Element location	Outside surface		Inside surface	
	Sign of $(P_{p.A} - P_{p.W})$	Sign of (grad p)	Sign of $(P_{p.A} - P_{p.W})$	Sign of (grad p)
On the wind exposed wall	+	+	−	+
On the circulation area exposed wall	−	−	+	−

(f)

FIGURE 6.5
Boundary conditions determined by building overall dimensions and location of a structural element on the building facade: (a–c) Change in the length of *circulation zone*, depends on height of the building; (d, e) scheme components; and (f) table with the signs of pressure differences according to facade's element placement.

A facade wall may undergo different impacts depending on its orientation and the angle θ of wind attack (see Figure 6.5.). The possible scenarios are illustrated in Figure 6.5f. The distribution of $\Delta\bar{p}$ on the roof plate is also affected by the angle of attack θ. Pressure head $\Delta\bar{p}$ is always negative since $\Delta\bar{p} < 0$, and hence, the roof plate undergoes a negative pressure gradient and $p_i > p_e$. Therefore, transfer from the building inside to the surroundings is less likely to operate through the roof. Yet the inverse transfer is quite possible.

The change of the magnitude of $(\text{grad}\, p)_{A_i}$ for different facade elements is specified by the location and dimensions of the area of air circulation formed at the building leeward side. Figure 6.5 shows that the dimensions of the circulation area depend on building size and aerodynamic shadow of the neighboring structures. Figure 6.5a illustrates how the storey height and wall element location affect the magnitude and direction of the dynamic pressure gradient.

A facade element can be installed on an arbitrary building side, which can be windward or leeward depending on the landscape character, local micro- and macroclimate including the so-called wind rose, proximity to large 3D structures, etc.

Anyhow, when analyzing a wall element, one should consider its location height, orientation, exposure to large wind loads, and pressure difference

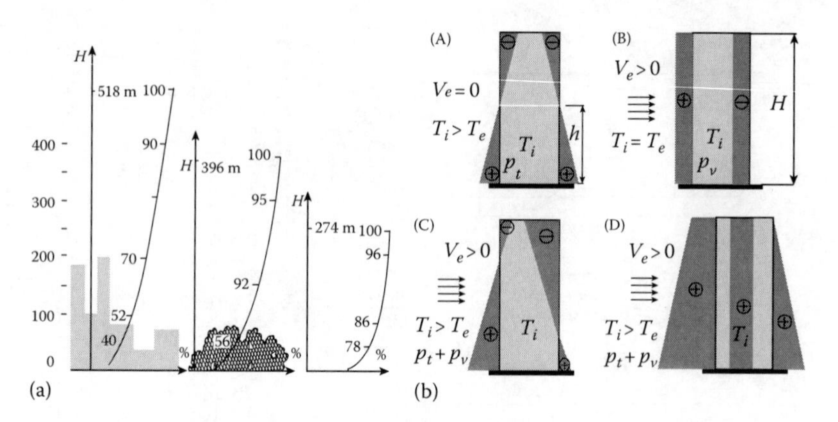

FIGURE 6.6
Boundary conditions of the building cover: (a) Due to the wind and (b) b) due to temperature difference $T_e - T_i$.

between the building inside and outside. In addition, the physical picture is significantly complicated by the so-called stack building effect due to the difference between the inside and outside temperatures. This is shown in Figure 6.6b, where two main cases are compared—nonisothermal and isothermal distribution of pressure (see Figure 6.6b(A) and b(B)). Nonisothermal distribution of $\Delta\overline{p}$ arises as a result of the different density of the inside and outside air. Wall openings at height h equalize pressure $p_e = p_i$, thus controlling the distribution of $\Delta\overline{p}$ on the building walls (see Figure 6.6b(A) and b(C)). The complex effect of air density difference and wind-generated dynamic pressure is shown in Figure 6.6b(A) and b(D).

As stated in Chapter 5, one of our tasks is to illustrate the application of the method of integral forms to various finite elements using Equation 5.60. Hence, Appendix A (Dimitrov 2013) presents generation of matrix equations of 1D, 2D, and 3D discrete models, considering complex boundary conditions. Initially, the boundary conditions of an individual finite element are set forth in a dimensionless form accounting for the three types of FE discussed in Chapter 6. Finally, the application (Chapter 8) illustrates the capabilities of the method to find the distribution of the *ETS* parameters in global 1D, 2D, and 3D domains (see Examples 8.6 and 8.7, and Dimitrov 2013, in particular, the free energy generalized potential and the vector quantities of transfer).

Chapter 7 sets forth some engineering problems that have emerged during the study of the interaction between the building envelope and the surroundings. It also outlines some practical solutions. We have presented the material in the form of a number of *engineering methodologies* intended to assist the user. These are, for instance,

- Calculation of economically optimal thickness of insulation of the building envelope, using indexes and classes of energy efficiency

- Window sizing
- Calculation of the dimensions of the solar shading devices
- Data processing needed to establish the occurrence of infiltration/exfiltration through the envelope and air ducts of the air conditioners using the Delta_Q method
- Assessment of building ecological sustainability and energy system using the class and index of environmental sustainability
- Selection of the building structure accounting for energy efficiency

Although the methodologies do not directly involve transfer models, they are in relation with the material discussed so far, since they have induced the model design. Moreover, modern calculation technique attributes to the increase of the efficiency of the engineering approach to transfer phenomena.

7

Engineering Methods of Estimating the Effect of the Surroundings on the Building Envelope: Control of the Heat Transfer through the Building Envelope (Arrangement of the Thermal Resistances within a Structure Consisting of Solid Wall Elements)

Since its emergence as an independent science at the end of the nineteenth century when the first energy projects were in progress, building thermotechnics and power engineering passed through boyhood and later matured together with electrotechnical engineering and material science. Various modern energy technologies were designed at that time and various problems of the type *how it is made* were solved. New materials (including foaming urethanes, wadding, reflectors, selective and electrochromatic coatings, aero gel plates, phase-exchanging materials) used in civil engineering were successfully incorporated into the structure of building envelopes, while tests of newly produced materials are underway nowadays. Modern level of scientific studies and technologies provides the necessary prerequisites so that *building heat techniques and power engineering* would enter a new development stage—the efficient use of primary energy sources. Despite the widening scope of modern technologies, new intelligent materials should be used in order to conform to the building growing needs of energy, that is, *energy transfer through the envelope should be appropriately controlled,* and the use of primary energy sources should be efficient. Those arguments have already been discussed in Chapter 1, in relation with the illustration in Figure 1.6.

If not appropriately made use of, the energy flux inflowing from the surroundings could present a significant energy load applied on the building system in the course of regulation of the thermal and visual comfort. During the time of low-cost energy carriers (the whole twentieth century), the energy of the surrounding spontaneously supplied to the building was considered as an *unwanted external impact.* Design aimed at its neutralization or significant reduction by the envelope structure, whereas models of the envelope operation as an *energy barrier* or an *adiabatic insulation* were designed.

To operate efficiently as an *energy barrier,* a solid element of the envelope should have a special structure to reduce the expected energy fluxes (with specific energy spectra and orientation). This could be done by a specific arrangement of thermal resistances within a unilateral *energy barrier* and along the free energy gradient (as shown in Figure 7.1a for instance). Position at first a reflective resistance (position 3) and then a volume one (position 2) and a capacitive one (position 1). Within a bilateral *energy barrier* (see Figure 7.1b), however, the capacitive resistance 1 should be positioned between two pairs of reflective and volume resistances. In order to behave as an efficient *energy barrier,* a thick component of the building envelope should be specifically composed to reduce the expected energy flows (with defined energy spectra and orientation) via tracking the resistance in the direction of the thermal capacity.

An idea of passive regulation of *inflowing and outflowing* energy fluxes emerged during the next stage of architectural interpretation of the building energy functions (the 1980s). It treated the building envelope as an

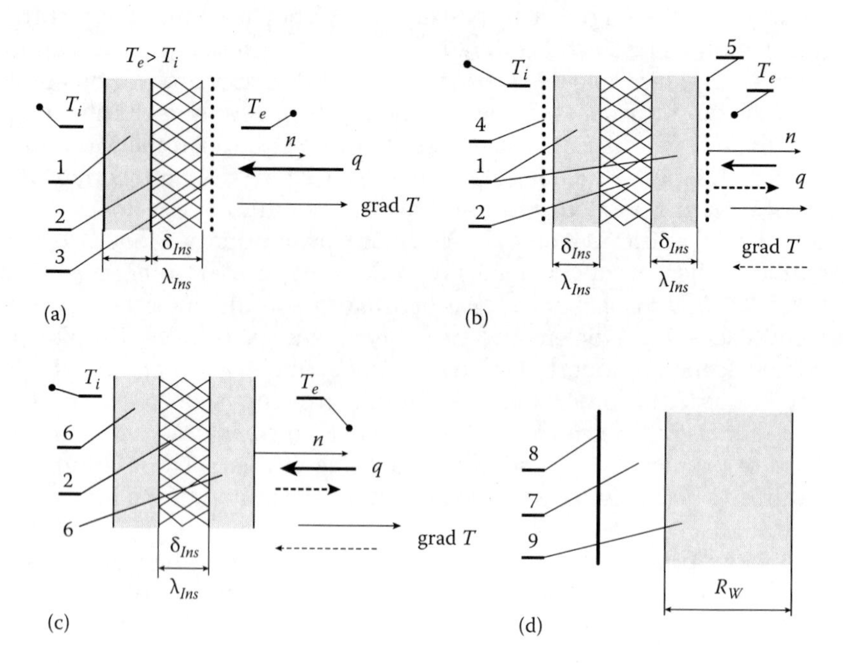

FIGURE 7.1
Arrangement of adiabatic and filter envelopes: (a) unilateral heat barrier; (b) bilateral heat barrier; (c) bilateral filter; (d) Trombe/Michelle wall: (1) capacity; (2) Fourier insulation; (3) reflective resistance; (4) wide-spectrum reflective resistance; (5) IR long-wave filter; (6) capacitive resistances; (7) air crevice (insulation); (8) window pane (UV/IR filter); (9) heat capacity.

energy filter, shifting in time the inflowing energy fluxes or reducing their intensity, if necessary.

Calculate, fabricate, and assemble an element of the building envelope. Then, it is expected that the element would *operate* without significant modification of its functional characteristics during the whole life of the envelope system, following a fixed operational program (i.e., a passive element). A building designed with an envelope operating as a filter of natural energy is called *passive*—the energy can be absorbed, if needed, or reflected. Most often, the components of the building envelope are assembled such as to *shift* the energy fluxes in the daytime (during the interval 10–16 h)—reduce the energy inflow in the daytime or accumulate energy for night consumption. This can be achieved by a selection and use of elements fabricated from materials with *spectrally determined* characteristics. Facade *spectral determination* is attained by specific selection of the *capacitive* and *volume* resistance, $R_C = 1/(C_v \cdot \omega_0)$ and $R_{Wall} = \delta_{Ins}/\lambda_{Ins}$, arranged along the photon flux (see Figure 7.1c and d) or by the use of special materials to filter the radiation energy flux. Classical materials such as reinforced concrete, ceramic plates, rock, or a combination of them with large *thermal mass* are used to fabricate envelope solid elements and change the heat capacity C_V, kJ/K m^3. Materials with varying phase can also be used for that purpose at temperatures, typical for the envelope operation (290 ÷ 300 K, for instance).

We may assume that structural material science and technology are close to the design of materials with *flexible* energy conduction (heat of electric conductivity, for instance), with controllable energy capacity (latent and enthalpy), with *spectral-sensitive* characteristics (not only transparency) and with a capacity to entirely change their properties under external programmed impacts (under varying electromagnetic fields, in particular). Then, a larger base of envelope elements should be available in order to model the operation of the envelope as an *intelligent membrane*, which would adapt to specific conditions depending on the energy needs at a certain moment. The design of structural elements with intelligent physical properties is a serious challenge to the structural material science requiring the introduction of nontraditional nanotechnologies.

The following two approaches are assumed to be essential from a methodological point of view:

1. Account for the integrity of the building envelope (the function of its elements should be complex-protective, *energetic*, illuminating)
2. Introduction of *intelligence* of the envelope components (their characteristics should vary upon variation of the input conditions)

This chapter proposes a survey of *our methods* applicable in the calculation of building envelope and based on estimates of the running energy transformations.

7.1 Calculation of the Thermal Resistance of Solid Structural Elements

Pursuing maximal useful effect, specialists face the problem of maximal efficient use of the available energy resources during the execution of a specific energy project. However, this is the main task of constructional projects, too, where the inevitable problem of building envelope design arises. Note also that the envelope design faces esthetic and ecological issues, together with pure energy-related ones.

The oldest and traditional strategy of regulation of the surroundings—ETS energy exchange is to fabricate ETS adiabatic envelope as an absolute barrier to the energy flux. To operate as a barrier, an element belonging to an adiabatic envelope should particularly conform to the temperature gradient (see Figure 7.1a and b). When the temperature gradient does not change its direction (unilateral transfer), the arrangement of the wall layers along grad T is capacitive layer, volume (Fourier) layer and reflective layer (see Figure 7.1a). When the direction of the temperature gradient varies cyclically, the arrangement of the thermal resistances along the normal is as follows (see Figure 7.1b):

- Reflective one resisting the infrared waves (position 4)
- Volume (Fourier) one with δ_{Ins} (position 1)
- Thermocapacitive one ($\tau_{Charge} \cong 12$ h) (position 2)
- Reflective one resisting the direct and diffusive solar radiation (position 5)

One may face a specific problem consisting in how to calculate the layer thickness and especially the thickness of the volume resistance layer. The elementary logic of preparing an *entire insulation* or the argument that the thermal resistance should tend to ∞ are incorrect, nonprofitable, and contra-productive in some cases. There are two sets of contra-argument and restrictions imposed on the *excessive insulation* of facilities and components:

1. Economical restrictions or *economically profitable thermal and optimal resistance* —R_{ec}
2. *Physical* restriction or *critical thickness of the thermal resistance*

The economical restriction imposed on the excessive insulation is based on the social and private *investment interest*. It is the sum of the cost of the deposited insulating layer with thickness δ_{Ins} together with the initial

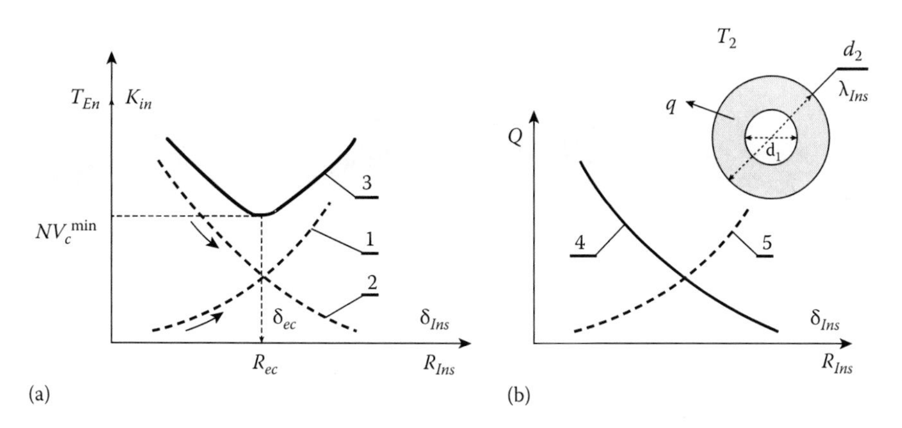

FIGURE 7.2
Restrictions imposed on the envelope dimensions: (a) optimization of the mean expenses NV_c; (b) specification of envelope critical thickness. (1) initial investment $K_{In} = f(\delta_{Ins})$; (2) annual energy expenses $T_{En} = f(\delta_{Ins})$; (3) NV_c—investment present value; (4) $q = f(\delta_{Ins})$; (5) $Q = f(\delta_{Ins})$.

investment (K_{In}), on the one hand, and annual operating payments for energy supply (T_{En}), on the other hand. That sum should be minimal. It reads

$$NV_c = K_{In} + e^{-1} * T_{En}, \tag{7.1}$$

where

$K_{In} = f(\delta_{Ins})$ is the initial investment K_{In} depending on the thickness of the insulating layer (see curve 1 in Figure 7.2a)

$T_{En} = f(\delta_{Ins})$ is the cost of energy supply depending on the thickness of the insulating layer, ***EU***/year (see curve 2 in Figure 7.2a)

e is the discount factor of the current costs at the moment of the initial investment*

NV_c is the net value, ***EU/levs*** (see curve 3 in Figure 7.2a)

Optimizing the functional $NV_c = f(\delta)$, we find the cost of the thickness of the thermal insulating layer ($\delta = \delta_{ec}$), where NV_c is minimum (see Figure 7.2a). It is called *economically profitable thickness* of the insulating layer. If $\delta \geq \delta_{ec}$, the object is excessively insulated. Insulation thickness higher than δ_{ec} is economically *unjustified*, resulting in unduly *high initial investments* and making pointless the strive for absolute adiabatic insulation.

To provide investor-friendly conditions, we will employ a specific methodology to find δ_{ec}, indexes (IEE_{Env}) and classes of energy efficiency (CEE_{Env})

* The discount factor *transforms* future payments into current ones.

(see Chapter 4) (Dimitrov 2013). It facilitates the communication between professionals and investors, while thickness δ_{ec} has the form

$$\delta_{ec} = \frac{\lambda_{Ins}}{U_{Ref}} * IEE_{Env}^{-0,5} = \lambda_{Ins} * R_{Ref} * IEE_{Env}^{-0,5}. \tag{7.2}$$

Here,

- $U_{Ref} = R_{Ref}^{-1}$, W/m² K is the recommended value of the coefficient of thermal conductance/resistance conforming to the energy standard* and the National Building Code, issued by the Department of Regional Regulations
- λ_{Ins}, W/m K is the coefficient of conductivity of the used insulating material
- IEE_{Env} is the index of energy efficiency, required by the investor in accordance with the adopted class of energy efficiency (CEE_{Env}), for example, for CEE_{Env} "A"—$IEE_{Env} = 0.5$, for *CEE* "B"—$IEE_{Env} = 0.75$, for *CEE* "C"—$IEE_{Env} = 1.0$, etc.

To design the insulation, the investor should declare CEE of the building envelope, only (e.g., *CEE* "A," "B," or "C"), considering the envelope index IEE_{Env} (as specified in a normative correlation table) (Димитров 2008a). Example 8.9 is a numerical illustration of the use of Equation 7.2.

Figure 7.3 shows the values of δ_{ec} depending on IEE_{Env} (or *CEE*) for reference values of the thermal resistance pursuant to the energy standard R_{Ref} and within the interval $1.66 \le R_{Ref} \le 5.0$ m²K/W for $\lambda_{Ins} = 0.03$ W/m K. Data prove that the values of δ_{ec} for the high class of energy efficiency (*A*) grow significantly (0.07 ÷ 0.12 m). For classes lower than *B*, the thickness of the economically profitable thermal insulation is within limits $0.03 \le \delta_{ec} \le 0.08$ m. For mass construction, however, thickness larger than $\delta_{ec} = 0.08$ m is not expedient. Buildings of energy efficiency classes *D* ÷ *G*, whose insulation is thinner than 0.05 m, are bound to repair (to architectural structural recovery).

Additionally, Figure 7.3 illustrates the effect of the envelope index of energy efficiency on the thickness of the insulating layer δ_{ec} regarding different values of the thermal resistance of the spherical standard $R_{SpSt_{ec}}$, which varies within limits $1.66 \le R_{SpSt_{ec}} \le 5.0$ m²K/W. Hence, the thickness of the insulating layer grows significantly at high-efficiency classes and values of the spherical standard. Classes lower than *B* (i.e., $IEE_{Env} > 0.77$ and $R_{SpSt_{ec}} \le 2.5$ m²K/W) have economically advantageous thickness of the insulation layer within limits $0.03 \le \delta_{ec} \le 0.08$ m. Hence, we may assume that the

* An imaginary building without glazed elements, with volume equal to the designed one and spherical shape (with minimal surface area) is taken as a standard—the so-called *Spherical standard*.

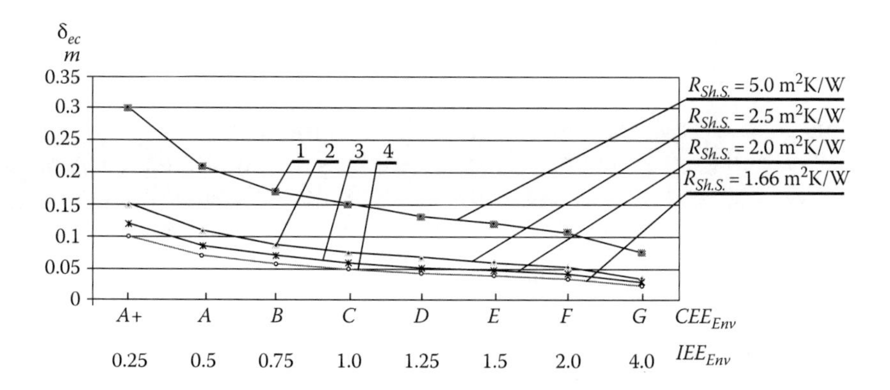

FIGURE 7.3
Maximal admissible thickness of the thermal insulation. (Values of δ_{ec}, calculated for different thermal resistance of the energy standard R_{Ref}, depending on the class CEE_{Env} and index IEE_{Env} of the energy efficiency of an envelope manufactured from an insulating material with λ_{Ins} = 0.03 W/mK).

economically advantageous thickness of the insulation layer *should not exceed* $\delta_{Ins} \cong 0.08$ m. Note that the insulation layer should be fabricated from a material with heat conductivity coefficient $\lambda_{Ins} \cong 0.03$ mK/W. Figure 7.3 shows that buildings of efficiency classes *F* and *G*, having insulation layers with thickness less than 0.05 m, are bound to *envelope repair.*

The second type of restrictions holds for axisymmetric objects (pipes), where a heat carrier with temperature T_1 flows (see Figure 7.2b). Consider a pipe, insulated by a layer with thickness δ_{Ins} releasing heat to the surroundings (with temperature T_2). Then, two opposite tendencies are observed:

1. *The intensity of the specific heat flux* $q(\delta_{Ins})$ ⇓ decreases* with the increase of δ_{Ins} due to the growing resistance R_{Wall}.
2. On the other hand, the intensity of the *heat flux* directed to the surroundings $Q(\delta_{Ins})$ ⇑ increases with the increase of δ_{Ins} due to the increase of the efficient outside surface of the pipe envelope.†

Obviously, a limit (*critical*) value of δ_{Ins} exists, and its exceeding is senseless and not physically justified. When $0.5U_{Ins}(d_{mp} + 2\delta_{Ins})/\lambda_{Ins} \leq 1$, heat exchange becomes less intensive. Increase of the intensity of the heat flux is observed in the opposite case, which seems to be a paradox beyond the aforementioned considerations. The limit thickness δ_{Lim} of the pipe insulation can be assessed via

$$\delta_{Lim} \leq \lambda_{Ins}/U_{Out} - 0.5d_{Pipe}.$$

* $q = (T_1 - T_2) \Big/ \left[\ln\left(1 + \frac{\delta_{Ins}}{d_{pipe}}\right) \right] \Big/ 2\lambda_{Ins}$.

† $A_{eff} = 2\pi L_{eff} \cdot (d_{pipe} + 2\delta_{Ins})$.

When $\delta \gg \delta_{Lim}$, heat exchange intensifies and the deposition of an insulating layer seems senseless. In the opposite case, the thermal insulation operates accordingly.

Until recently, the normative requirements to the thermotechnical properties of building structures were reduced to the fulfillment of condition $T_{Wall} \geq T_{Sat}$, which prevented the condensation of water vapors of the ambient air (Димитров 2006). At that time, the cost of the thermal energy was quite low and there was no need of introducing normative requirements to the heat exchange of the whole building, although some authors (Димитров 1986; Димитров and Александров 1988) introduced building general characteristics.

An idea about the introduction of regulations of the heat exchange of the whole building and not that of its individual components emerged in the 1980s of the past century. Simultaneously, the normative requirements to the components of the envelope structure were revised. Table 7.1 presents modifications made in the leading EU countries (the numerators were valid till 1978, while denominators are coefficients of heat transfer of the respective building components, newly introduced at the end of the past century—U_{Wall}, U_{Roofl}, and U_{Floor}). Smaller local values of the heat transfer coefficients are valid since 1998 (Димитров 2003). As seen, the norms of the EU members, together with the Bulgarian ones, are reduced by 300%–400%. This is probably done with an idea that those normative documents will be usable in the next 15–20 years.

The new modern methods of calculating the thickness of the insulating layer in a projected envelope are discussed in (Димитров 2011a) and require an optimization of

$$NV_C = \sum_{1}^{3} K_i + \sum_{1}^{4} T_j * e_j^{-1}. \tag{7.3}$$

TABLE 7.1

Normative Values of the Heat Transfer Coefficient in EC

Country	Walls (W/Km²)	Windows (W/Km²)	Roof Structure (W/Km²)	Floor Structure (W/Km²)
Denmark	0.42/0.3	3.0/2.5	0.37/0.2	0.55/0.3
Germany	1.57/0.47	5.23/3.02	0.81/0.38	1.01/0.47
Italy	1.39/0.36	6.05/3.33	2.03/0.32	1.47/0.7
Norway	0.58/0.27	3.14/2.33	0.47/0.2	0.7/0.24
Great Britain	1.7/0.55	5.68/2.68	1.42/0.32	1.0/0.5
Bulgaria	$(R_n)^{-1}$/0.5	3.2/2.65	—/0.3	—/0.5

As shown in (Димитров 2011), modifying the functional equation (Equation 7.3), and finding its first derivative with respect to the thermal resistivity of the building envelope, one obtains the following formula for the evaluation of the *cost-effective coefficient of the building envelope resistance*:

$$R_{ec}^{Env} = \left(4.84 * IEE_{Env} * v_{Buil}^{0.666}\right)^{-0.5} \sqrt{\left(\sum_{i=1}^{3} \left(F_{K_i}\right) + F_{T_1} + \sum_{j=2}^{4} \left(\left(\frac{1-(1+i)^{n_j}}{i}\right) F_{T_j}\right)\right)}, \tag{7.4}$$

where functions F_{K_i} and F_{T_j} are given in Table 7.9 in (Димитров 2011). The formula of R_{ec}^{Env} can be simplified if it is interpreted as a product of two variables—the energy efficiency index IEE_{Env} and the cost-effective coefficient of thermal resistance of the spherical standard $R_{SpSt_{ec}}$ (see Equation 7.2).

By definition (Dimitrov 2006), the spherical standard is an imaginary building with volume identical to that of the building under design, but with standard cost-effective coefficient of thermal resistance, where the index of energy efficiency (IEE_{Env}) is $IEE_{Env} = 1$. Hence, Equation 7.1 is used to evaluate the cost-effective resistance of the energy standard R_{ec} and it will be modified as $R_{ec} \Rightarrow R_{SpSt_{ec}}$

$$R_{SpSt_{ec}} = \left(4.84 * v_{Buil}^{0.666}\right)^{-0.5} \sqrt{\left(\sum_{i=1}^{3} \left(F_{K_i}\right) + F_{T_1} + \sum_{j=2}^{4} \left(\left(\frac{1-(1+i)^{n_j}}{i}\right) F_{T_j}\right)\right)}. \tag{7.5}$$

So, to estimate the value of the cost-effective thermal resistance of the designed envelope, one should use the relationship

$$R_{ec}^{Env} = IEE_{Env}^{-0.5} * R_{SpSt_{ec}}, \tag{7.6}$$

indicating that the economically advantageous value of the total thermal resistance of the building envelope depends on both the index of energy efficiency IEE_{Env} (ordered by the investor) and the cost-effective resistance of the energy standard R_{SpSt} (normatively defined).

Replace R_{ec}^{Env} in Equation 7.6 by $\delta_{ec}/\lambda_{Ins}$ and $R_{SpSt_{ec}}$ by R_{Ref}. Then, Equation 7.3 is obtained. It is used to calculate the insulation thickness (see the examples in Table 7.8—row no. 2).

Following the strategy of creating a cost-effective insulation of the building envelope, we assume that to be sufficiently insulated a building should satisfy the inequality

$$U_{Buil} \leq U_{Env}^{Rec}, \tag{7.7}$$

where

U_{Buil} is the *assessed value of* the total thermal conductivity coefficient of the designed building envelope (current value)—see (Dimitrov 2009)

U_{Env}^{Rec} is the *recommended value* of the total thermal transitivity coefficient of the building envelope

Using the definition expression* (Dimitrov 2013), $U_{Env}^{Rec} = IEE_{Env} * U_{ShSt}$ is designed as a function of the index of energy efficiency ($U_{Env}^{Rec} = IEE_{Env} * U_{ShSt}$) and $U_{Env}^{Rec} = IEE_{Env} * U_{ShSt}$, respectively. It takes the following form:

$$U_{Env}^{Rec} = IEE_{Env} * U_{ShSt}. \tag{7.8}$$

After putting the expression for the *spherical standard* (Equation 226; Dimitrov 2013) in Equation 7.8, we find

$$U_{Env}^{Rec} = 4.84 \frac{V_{Buil}^{0.666}}{A_0 R_{Ref}} * IEE_{Env}, \quad \text{or} \tag{7.9}$$

$$U_{Env}^{Rec} = 4.84 U_{Ref} \frac{V_{Buil}^{0.666}}{A_0} * IEE_{Env}, \tag{7.10}$$

where

U_{SpSt} is the building envelope index of energy efficiency†

U_{SpSt} is the thermal conductivity coefficient of the energy standard called *spherical standard* "U_{SpSt}–value"‡ (see Equation 226 Dimitrov [2013])

A_0 and V_{Buil} are the areas of the surrounding surface and volume of the designed building

$U_{Ref} = (R_{Ref})^{-1}$ is the *benchmark* (reference value) of the thermal conductance/resistance coefficient of the energy standard (subject to periodic normalization in DBA; yet, the currently used value is $R_{Ref} = 2.5 U_{Env}^{Rec}$)

---

* The index of energy efficiency— $IEE_{Env} = U_{Env}/U_{ShSt}$.

† See IEE_{Env} in the correlation table—Table 8, Appendix (Dimitrov 2013).

‡ It expresses the maximum value of the total thermal conductivity coefficient of the audited/designed building, keeping its volume but minimizing its surrounding surface.

In the proposed methodology, the main definitive relationships (Equations 7.9 and 7.10) are used to calculate the recommended value of *the total thermal conductivity coefficient* of the building envelope U_{Env}^{Rec}, which is a function of the class U_{Env}^{Rec} and index (U_{Env}^{Rec}) of energy efficiency.

The use of Equation 7.10 to calculate U_{Env}^{Rec} is illustrated in the first row of Table 7.2. The original data used in Table 7.2 and concerning our estimations are building volume 567 m^3, area of the surrounding surface 450 m^2, and energy efficiency class of envelope *B*.

As seen in Table 7.2, the recommended value of the total thermal conductivity coefficient of the building envelope is $U_{Env}^{Rec} = 0.26\ \text{W}/(\text{m}^2\text{K})$. In the course of the architectural design, the envelope characteristic U_m should be varied in order to attain the calculated value of the envelope total coefficient of thermal conductivity.

TABLE 7.2

Examples for the Application of the Suggested Methodology

Variable:	Used Formulas or Expressions (Equation)	Examples
Recommended value U_{Env}^{Ref}. of the total thermal conductivity coefficient	(7.10)	$U_{Env}^{Rec} = 4.84 \dfrac{567.0^{0.67}}{450.0 * 2.5} * 0.875 = 0.26\ \text{W}/(m^2\text{K})$ (for class B – $IEE_{Env} = 0.875$)
Estimated amount of energy for building heating according to CEE_{Env}	(7.11)	$E_{Buil}^{Heating} = 0.0424 * 10^6 \dfrac{567.0^{0.666}}{2.5} * 0.875 * 2400 = 2.424$ TJ year (if HDD = 2400 DD)
Estimated amount of energy for building cooling (annual) according to CEE_{Env}	(7.12)	$E_{Buil}^{Heating} = 0.0424 * 10^6 \dfrac{567.0^{0.666}}{2.5} * 0.875 * 1200 = 1.212$ TJ year (if CDD = 1200 DD)
Estimated energy capacity of the heat generator	(7.13)	$P_{Buil}^{Heating} = 4.84 * \dfrac{567.0^{0.666}}{2.5} * 0.875 * 38 = 0.26 * 450 * 38 = 4446\ \text{W}$
Estimated power needed by the building heating pumps	(7.14)	$P_{El}^{Heating\ Pump} = 9.68 * \dfrac{P_{Buil}^{0.666}}{R_{Ref}} * IEE_{Env} * \Delta T_{Max}^{Witer}\left(1 - \dfrac{T_1}{T_2}\right) = 4446 * 2 * \left(1 - \dfrac{278}{318}\right) = 1118.5\ \text{W}$
Estimated energy capacity of the building chiller	(7.15)	$P_{Buil}^{Ciller} = 4.84 * \left(\dfrac{567.0^{0.666}}{2.5}\right) * 0.875 * 12 = 0.26 * 450 * 12 = 1067\ \text{W}$

Obeying condition (Equation 7.9), the other energy-related characteristics of the building are found via the following formulas:

- Estimation of the *energy spared* in building heating (annually*), according to CEE_{Env}:

$$E_{Buil}^{Heating} = U_{Env}^{Ref} * A_0 * HDD = 0.0424 * 10^6 * \frac{V_{Buil}^{0.666}}{R_{Ref}} * IEE_{Env} * HDD; \quad (7.11)$$

- Estimation of the *energy spared* in building cooling (annually†), according to CEE_{Env}:

$$E_{Buil}^{Cooling} = U_{Env}^{Ref} * A_0 * CDD = 0.0424 * 10^6 * \frac{V_{Buil}^{0.666}}{R_{Ref}} * IEE_{Env} * CDD, \ J; \quad (7.12)$$

- Estimation of the *energy capacity* of the building thermal generator(boiler)‡:

$$P_{Buil}^{Heat\ Generator} = U_{Env}^{CEE} * A_0 * \Delta T_{Max}^{Winter} = 4{,}84 * \frac{V_{Buil}^{0{,}666}}{R_{ref}} * IEE_{Env} * \Delta T_{Max}^{Winter} \quad (7.13)$$

- Estimation of the power needed by the heat pumps:

$$P_{El}^{HeatPump} = P_{Heating}^{HeatPump} \eta_{HP}^{-1} = P_H^{HP} * \left(a * \eta_{Carnot}\right)^{-1}; \quad (7.14)$$

- Estimation of the *energy capacity* of the building chiller§:

$$P_{Buil}^{Chiller} = U_{Env}^{CEE} * A_0 * \Delta T_{Max}^{Summer} = 4{,}84 * \frac{V_{Buil}^{0{,}666}}{R_{ref}} * IEE_{Env} * \Delta T_{Max}^{Summer} \quad (7.15)$$

The specified formulas are useful in the analysis of the variants of the building preliminary design. Such data should be provided by designers or investors so that clients can make an appropriate choice. To illustrate the capabilities of the proposed methodology, we will discuss investments in the town of Varna, Eu.

The initial data are given in Table 7.3. The information indicates that the class of energy efficiency of the building envelope is a powerful tool to control the building initial investments. The initial expectation is that an increase in CEE_{Env} from E to A would raise the initial investments (effectively by about

* HDD—the annual heating day-degrees of the building location area.

† CDD—the annual cooling day-degrees of the building location area.

‡ $\Delta T_{Max}^{Winter} = \left(T_{Indoor} - T_{Environment}^{Winter}\right)$, °C.

§ $\Delta T_{Max}^{Summer} = \left(T_{Environment}^{Summer} - T_{Indoor}\right)$, °C.

TABLE 7.3

An Analysis of the Variants of the Building Preliminary Design

	Class of Envelope EE $CEE_{Env} \Rightarrow E$ ($IEE_{Env} = 1.75$)	**Class of Envelope** EE $CEE_{Env} \Rightarrow A$ ($IEE_{Env} = 0.625$)	**Difference**
Insulating layer δ_{Ins}	0.06 m (Equation 7.2)	0.1 m (Equation 7.2)	+0.04 m or (+40%)
Value of II	19.5 Eu/m² Neopor BASF 20	23.5 Eu/m²	+9212 Eu or (+20%)
Heat generator	37.84 kW (Equation 7.13)	13.56 kW (Equation 7.13)	−24.3 kW or (−64.2%)
Value of the H.G.	1725 Eu 41 kW-Wolf FNG	1060 Eu 17 kW-Wolf FNG	−665 Eu or (−38.5%)
Estimated amount of heating energy	12.33 TJ (Equation 7.11)	4.4 TJ (Equation 7.11)	−7.93 TJ or (−64.3%)
Energy expenses	143.74 * 10³ Eu/year (0.0417 Eu/kWh- Rilagas)	51.11 * 10³ Eu/year (0.0417 Eu/kWh- Rilagas)	−92.63 * 10³ Eu/year or (−64.4%)
Estimated amount of cooling energy	6.16 TJ (Equation 7.12)	2.2 TJ (Equation 7.12)	−3.96 TJ or (−65.4%)
Energy expenses	85.5 * 10³ Eu/year (0.05 Eu/kWh)	26.61 * 10³ Eu/year (0.05 Eu/kWh)	−58.89 * 10³ Eu/year or (−68.9%)

Initial parameters of the investment in the town of Varna-Bulgaria, Eu (2400 HDD; 1200 CDD; $\Delta T^{Max}_{Heating} = 31°C$; $\Delta T^{Max}_{Cooling} = 12°C$): $V_{Buil} = 6912\ m^3$, $A_0 = 2303\ m^2$, $R_{Ref} = 2.5\ m^2K/W$ and $\lambda_{Ins} = 0.03\ W/mK$ (A duck system of AC with length of 100 m is considered); ("+" stands for cost, and "−" for saving).

≈ +20% for thermal insulation). Data in Table 7.3, however, show that the initial investment would drop significantly by reducing the demand for power of the components of the energy system (including supply of heat generator (≈ −39%), supply of heaters (≈ −60%), ducts (≈ −20%), cables (≈ −35%), etc.). That reduction is of the same order, and the raise of the insulation investments can be fully compensated in some cases.

Naturally, however, most savings during the building exploitation are due to the reduction of the operating costs of energy supply (≈ −64% per year).

(See Example 8.8).

7.2 Solar Shading Devices (Shield) Calculation

The proposed method is based on the arsenal of the computational mathematics. From a mathematical point of view, the classical method of *shadow masks* consists in a projection of the contour of the solar shading devices on the facade plane of the light-transmitting aperture (plane yOz; see Figure 7.4a) (Dimitrov and Kolev 1990). Performing successive approximations (iterations)

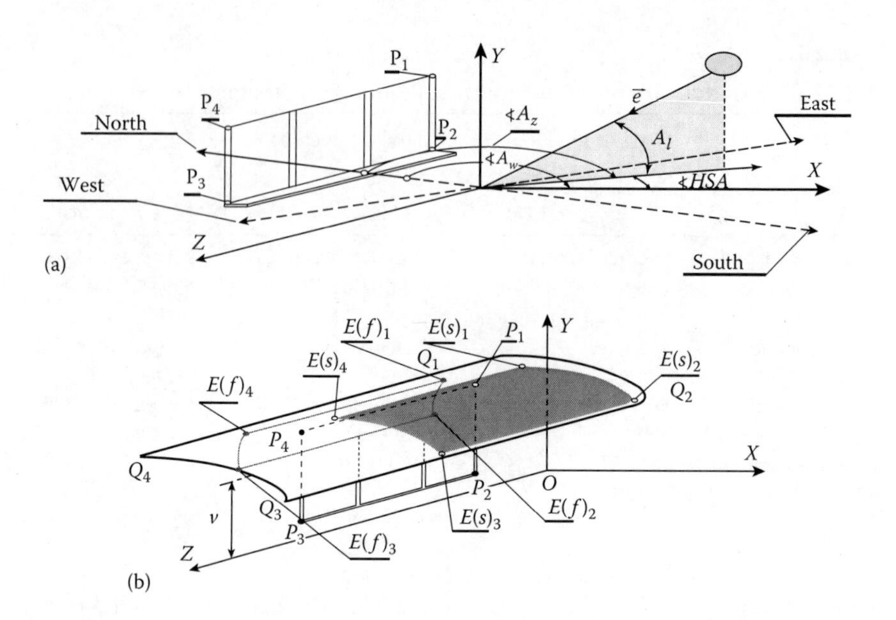

FIGURE 7.4
Inversion of the designed object (the window contours are projected on the solar shading devices): (a) *solar geometry* on the facade; (b) determination of the window projections on the first and last day of solar protection.

the aim is to cover the whole area $P_1P_2P_3P_4$ by the fallen shadow. Yet, manual calculations are time and labor consuming and boring. The originality of the proposed method is based on the *mathematical inversion of the projection plane*. Thus, calculations are made easier and the numerical algorithm, discussed later, can be automated (Dimitrov and Kolev 1990).

We propose here to project the contour of the light transmitting aperture on the surface of the solar shield. That surface is specified as an arbitrary second-degree one. In fact, the proposed method is an inversion of the known classical methods. The equation of the shading devices surface will have the form

$$a_0 + a_1 \cdot x + a_2 \cdot y + a_3 \cdot z + a_4 \cdot x^2 + a_5 \cdot y^2 + a_6 \cdot z^2 = 0. \tag{7.16}$$

Coefficients $a_{i(i\,=\,1\div 6)}$ are calculated pursuant to the formulas specified in Table 7.12, Appendix and following the respective solar shading devices stereometry selected by the designer: a sphere, a paraboloid, an ellipsoid, or a hyperboloid.

Practically, the shading devices stereometry are selected via menus. For instance, if the designer intends to shape the solar shading devices as an elliptic cylinder with an axis parallel to Oz (case 3 in Table 7.12, Appendix), coefficient k should assume values $k > 1$ (e.g., $k = 2$), while height V and radius l (see Figure 7.4) are specified in a dialog box on the screen. Then, coefficients in Equation 7.16 read $a_1 = a_3 = a_6 = 0$; $a_0 = 4(V^2-l^2)$; $a_2 = -8V$; $a_4 = a_5 = 4$, that is, the shading devices surface will be expressed by the following equation:

$$-8 \cdot V \cdot y + 4 \cdot x^2 + 4 \cdot y^2 = 4 \cdot \left(l^2 - V^2\right). \tag{7.17}$$

The design of the contour of the light-transmitting aperture $P_1P_2P_3P_4$ projected on the surface described by Equation 7.17 (e.g., see Figure 7.4b) can be performed using vector $\vec{e}\ (\lambda, \mu, \gamma)$, which is collinear with the sun's rays (see Figure 7.4).

Due to the enormous distance to the sun, the solar flux is considered to be a bundle of parallel lines (a light shaft) with direction cosines λ, μ, and γ:

$$\lambda(\tau) = -\cos\left(A_l(\tau)\right) \cdot \cos\left(A_z(\tau) - A_w\right)$$

$$\mu(\tau) = -\sin\left(A_l(\tau)\right)$$

$$\gamma(\tau) = \cos\left(A_l(\tau)\right) \cdot \sin\left(A_z(\tau) - A_w\right),$$

where $A_l(\tau)$ and $A_z(\tau)$ are sun's height and azimuth angle found at the start ($\tau = \tau_S$) and end ($\tau = \tau_f$) of shading (τ is current time) [hour, day].

The used coordinate system (see Figure 7.4a) *Oxyz* is with right-handed orientation, and it is fixed to the facade of the shading devices light-transmitting aperture (the facade plane coincides with the coordinate plane *yOz*). Facade position is arbitrarily chosen, and it is defined by angle ($<A_w$), concluded between the normal to the façade plane (axis *Ox*, respectively) and the north direction, *N*. The *shaded* light-transmitting aperture $P_1P_2P_3P_4$ should be projected on the shading devices surface specified by Equation 7.16 as follows:

- Draw a bundle of parallel lines (4 lines in this case) collinear with the directory of the sun's rays ($\vec{e}$) passing through characteristic points of the contour of the light-transmitting aperture points $P_{i(i = 1 \div 4)}$.
- Find the coordinates of the ray *pierce* points on the surface described by Equation 7.21.
  Find the shielded area $E_1E_2E_3E_4(s)$.

Figure 7.4b shows points $E_{1(s)}$; $E_{2(s)}$; $E_{3(s)}$; $E_{4(s)}$ at which four straight lines passing through knots P_1, P_2, P_3, and P_4 of the light-transmitting aperture *punch* the shading elliptic cylinder, and the lines are collinear with the sun's rays $\vec{e}$ at moment $\tau = \tau_s$ (shading initial moment). Connect those points and find the boundary of the shielded area $E_1E_2E_3E_4(s)$, which will cast a total shadow on the light-transmitting aperture at $\tau = \tau_s$. If the solar shading covers all punch points in the time interval $\tau_s \leq \tau \leq \tau_f$, it will overshadow the light-transmitting aperture at all intermediate positions, that is, it will do the job. In other words, the shadow of the solar device will cover line P_2P_3 or will fall below it during the period $\Delta\tau = \tau_f - \tau_s$ [hour, day].

Practically, the screen design consists in cutting an area bounded by lines Q_2Q_3 and Q_1Q_4 in Figure 7.4 and part of the surface described by Equation 7.16. Note that Q_2Q_3 and Q_1Q_4, parallel to the axis of the elliptic cylinder, while the newly found contour is plotted by a solid line.

We shall use in what follows a specific algorithm to find the coordinates of the cross points $E_{i(\tau)(\tau_S \le \tau \le \tau)}$ belonging to the shading area. The equation of a family of lines parallel to vector $\vec{e}\left(\lambda_{(\tau)}, \mu_{(\tau)}, \gamma_{(\tau)}\right)$ reads

$$y = Y_{E_i} + \mu_{(\tau)}\left(x - X_{E_i}\right)/\lambda_{(\tau)} \tag{7.18}$$

$$z = Z_{E_i} + \gamma_{(\tau)}\left(x - X_{E_i}\right)/\lambda_{(\tau)},$$

where

$X_{E_i}, Y_{E_i}, Z_{E_i}$ are coordinates of points belonging to the shading devices area and satisfying Equation 7.16

x, y, z are current coordinates of points of the straight lines

If the aforementioned lines pass through the points of the contour of the light-transmitting aperture $P_i\left(X_{p_i}, Y_{p_i}, Z_{p_i}\right)$, the current coordinates will *assume values* $x = X_{p_i} = 0$, $y = Y_{p_i}$, and $z = Z_{p_i}$. Equations 7.18 will read

$$Y_{p_i} = Y_{E_i} + \mu_{(\tau)} \cdot X_{E_i}/\lambda_{(\tau)} \tag{7.19}$$

$$Z_{p_i} = Z_{E_i} + \gamma_{(\tau)} X_{E_i}/\lambda_{(\tau)}. \tag{7.20}$$

Solving the system (Equations 7.19 and 7.20), we find the coordinates of the cross points $X_{E(i)}^{(\tau)}$, $Y_{E(i)}^{(\tau)}$, and $Z_{E(i)}^{(\tau)}$:

$$X_{E_{1,2}(i)}^{(\tau)} = \left(-b \pm \sqrt{b^2 - 4a.b}\right)/2a; \tag{7.21}$$

$$Y_{E_{1,2}(i)}^{(\tau)} = Y_{P(i)} - \mu_{(\tau)} \cdot X_{E_{1,2}(i)}^{(\tau)}/\lambda_{(\tau)};$$

$$Z_{E_{1,2}(i)}^{(\tau)} = Z_{P(i)} - \gamma_{(\tau)} X_{E_{1,2}(i)}^{(\tau)}/\lambda_{(\tau)};$$

$$a = a_4 + a_5\left(\mu_{(\tau)}/\lambda_{(\tau)}\right)^2 + a_6\left(\gamma_{(\tau)}/\lambda_{(\tau)}\right)^2;$$

$$b = a_1 - a_2 \cdot \mu_{(\tau)}/\lambda_{(\tau)} - a_3 \cdot \gamma_{(\tau)}/\lambda_{(\tau)} - 2a_5 \cdot Y_{p_i} \cdot \mu_{(\tau)}/\lambda_{(\tau)} - 2a_6 \cdot Z_{p_i} \cdot \gamma_{(\tau)}/\lambda_{(\tau)};$$

$$c = a_0 - a_2 \cdot Y_{P(i)} + a_3 \cdot Z_{P(i)} + a_5 \cdot Y_{P(i)}^2 + a_6 \cdot Z_{P(i)}^2$$

$$\text{for } X_{E_i} > 0 \quad \text{and} \quad Y_{E_i} > 0.$$

Considering condition $\left(D = \sqrt{b^2 - 4ac} = 0\right)$ of the existence of a real solution of the system, we find the minimal dimensions of the solar shading devices—parameters V and l—while the rest of the input parameters remain constant. If the building esthetics is not satisfactory, the designer may vary other input parameters, thus changing the total appearance of the facade (see Table 8.13).

The computerized design of the solar shading devices is carried out in four steps:

1. For given $\lambda = \lambda_{(s)}$, $\mu = \mu_{(s)}$, and $\gamma = \gamma_{(s)}$, calculate the coordinates of points $E_{1(s)}$, $E_{2(s)}$, $E_{3(s)}$, and $E_{4(s)}$ (see Figure 7.4).
2. Find the coordinates of the sun protection line and those at the final moment of sun protection $E_{1(f)}$, $E_{2(f)}$, $E_{3(f)}$, and $E_{4(f)}$ (see Figure 7.4).
3. Cut a section from the sun-protecting surface including both areas $E_1E_2E_3E_4$ (s) and $E_1E_2E_3E_4$ (f).
4. Visualize the structure thus compounded on the monitor screen.

When dealing with solar shading devices surfaces of type 5 or 7 (Table 8.13), one should find the coordinates of points of the shading devices line $E_i^{(2)}$, only, using the condition $E_i^{(2)} \rightarrow \min$ (see Figure 7.5a through c).

When the used configurations are inappropriate from a structural or esthetic point of view, the computer design would enable one to modify the configurations by intersecting them with an additional plane similar to that shown in Figure 7.6 and analytically described by the equation

$$x.tg(\alpha) + y - V_1 = 0,$$

where

V_1 is height of the intersecting line
α is the slope of the intersecting plane with respect to the horizon

The surfaces thus found are shown in Figure 7.6a and b. The proposed new method makes possible the design of original facades with efficient functionality.

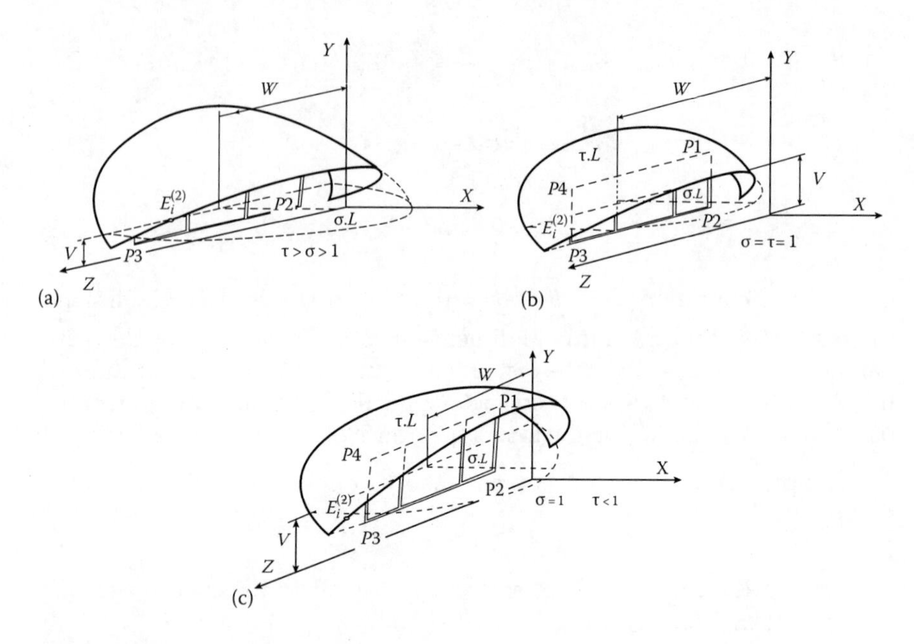

FIGURE 7.5
Solar shading devices: (a) parabolic cylinder with an axis parallel to Oy; (b) hemisphere with a plane parallel to Oz; (c) ellipsoid with a main axis parallel to Oz.

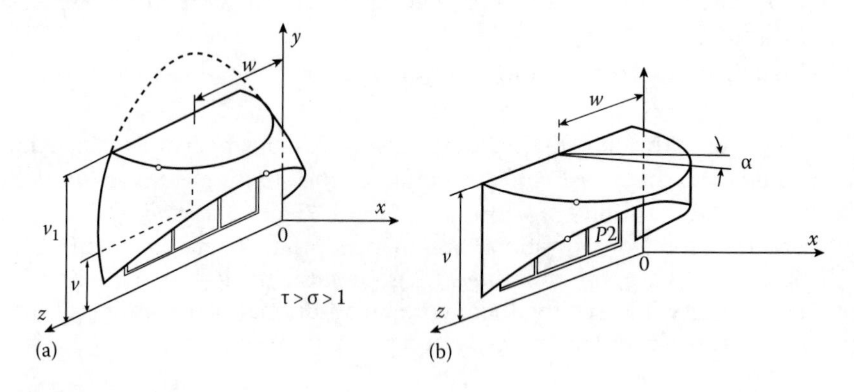

FIGURE 7.6
Solar shading devices: (a) parabolic cylinder with an axis parallel to Oy; (b) cylinder with an axis parallel to Oy.

7.3 Modeling of Heat Exchange between a Solar Shading Device, a Window, and the Surroundings

Traditional architectural solutions pursuing shading devices of light-transmitting apertures (windows, glazed doors, etc.) from direct solar radiation consider assembly of concrete peaks (see Figure 7.7) fixed to the facade by

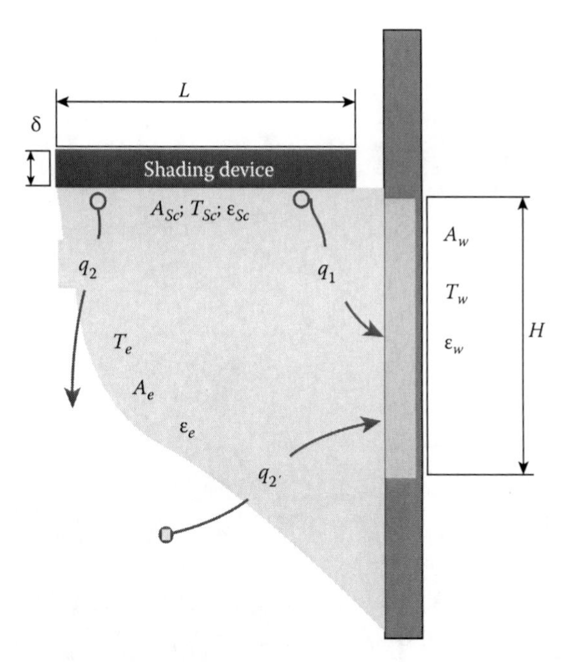

FIGURE 7.7
Radiant heat exchange between a solar shading device, a window, and the surroundings: (1) solar shading device; (2) surroundings; (3) window.

means of a cantilever support. Those shading devices operate accordingly overshadowing the light-transmitting aperture and barring the sunlight. Note that having undergone heating the shading devices start releasing heat to the surroundings and to the facade and the light-transmitting aperture. Hence, a secondary radiant flux occurs. The intensity of radiation increases with the increase of shading devices current temperature (temperature exceeding 120°C has been measured). The secondary radiant heat flux, having flown into the building, additionally loads the building systems, and the additional energy load should be eliminated (conditioned or ventilated) consuming additional primary energy.

The problem of modeling radiant heat exchange in a closed volume bounded by a solar shading device, a window, and the surroundings arose from the idea to incorporate a material with exchanging phase and melting temperature of $T_{Mell} \cong 22 \div 24$ °C into the shield bulk (see Figure 7.7). The aim was to avoid increase of the shading surface temperature over T_{Mell}, and thus reduce the effect of secondary radiation and prevent its penetration into the building.

Since a cavern is formed within the control volume (closed between the shield, the window and the surroundings), it seems convenient to model the real physical processes using a three-link electric circuit shown in Figure 7.8a.

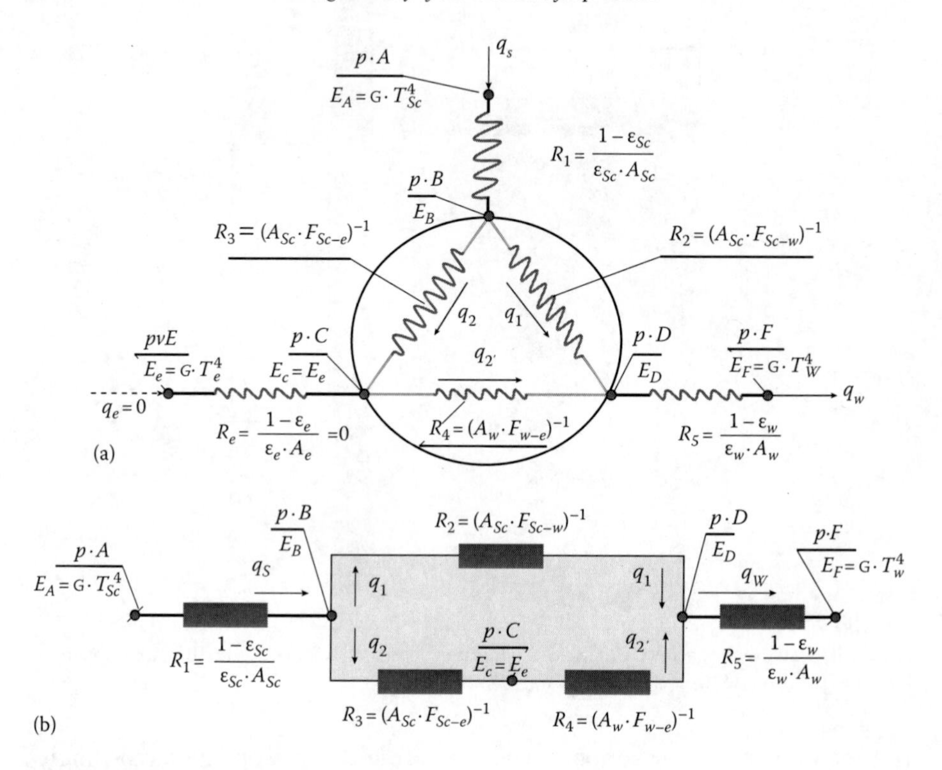

FIGURE 7.8
Equivalent electrical analog system: (a) Three-link analog scheme of modeling the heat exchange in a 3D closed volume (a cavern); (b) two-pole electrical equivalent.

The values of the electric resistance are shown in the two-pole scheme Figure 7.8b, keeping the following restrictions:

1. The radiation coefficient of a window pane is varied within the interval $0.9 \leq \varepsilon_W \leq 0.95$ (Incropera and DeWitt 1985). Resistance R_5 is expressed as $R_5 = (1 - \varepsilon_W)/\varepsilon_W A_W$. Other resistances are specified in Figure 7.8b.
2. The angle coefficients of radiation are found using the following relations: $F_{Sc-W}A_{Sc} = F_{W-Sc}A_W$; $F_{Sc-W} + F_{Sc-e} = 1$; and $F_{W-Sc} + F_{W-e} = 1$ (Incropera and DeWitt 1985).
3. Resistance R_e is zero since $A_e \Rightarrow \infty \left(R_e = (1-\varepsilon_e)/\varepsilon_e A_e\right)$ as seen in Figure 7.8a.

TABLE 7.4

Quantities of Heat-Electrical Analogy

	Thermal Value	Electrical Value
1	Specific heat flux running from the solar shield to the window $\vec{q}_1$	Current $\vec{\delta}_1$
2	Specific heat flux running from the solar shield to the surroundings $\vec{q}_2$	Current $\vec{\delta}_2$
3	Specific heat flux running from the window to the surroundings $\vec{q}_2$	Current $\vec{\delta}_2$
4	Temperature of the solar shield T_{Sc}	Electric potential at *p*. *A* -E_A
5	Temperature of the widow T_W	Electric potential at *p*. *F* -E_F
6	Temperature of the surroundings T_e	Electric potential at *p*. *E* -E_E

To find the solar fluxes running from the shield to the window (q_1) and to the surroundings (q_2), as well as the flux running between the window and the surroundings (q_2), we should calculate the currents in the equivalent electrical scheme.

For facility's sake, Table 7.4 specifies similar links between the thermal and electric quantities used in Figure 7.8. The next steps are (1) design the electrothermal analogue model, employing the balance equations pursuant to the two laws of Kirchhoff, and (2) compose an algebraic system containing the electrical analogues of the thermal characteristics of the radiant heat exchange in a 3*D* cavern (*D* is the number of the heat exchanging surfaces forming the ETS control boundaries).

Applying the first law of Kirchhoff along the three-link contour, we find the following three algebraic equations:

$$\delta_S = \delta_1 + \delta_2 - \text{electric currents at point } B; \tag{7.22}$$

$$\delta_W = \delta_1 + \delta_3 - \text{electric currents at point } D; \tag{7.23}$$

$$\delta_e = \delta_3 - \delta_2 - \text{electric currents at point } C. \tag{7.24}$$

Applying the second law of Kirchhoff, we find

$$\delta_S * R_1 + \delta_1 * R_2 + \delta_W * R_6 = E_A - E_F; \tag{7.25}$$

$$\delta_e * R_5 + \delta_3 * R_4 + \delta_W * R_6 = E_E - E_F; \tag{7.26}$$

$$\delta_e * R_5 + \delta_2 * R_3 + \delta_S * R_1 = E_A - E_E. \tag{7.27}$$

7.3.1 Mathematical Model

Equations 7.22 ÷ 7.27 present a linear system whose matrix form reads

$$\{\delta\}*[R]=\{\Delta E\}, \tag{7.28}$$

where

{δ} is the vector of electric current density
[R] is the matrix of electric resistance
{ΔE} is the matrix-vector of the electric potential

The thermal analogue of Equation 7.28 can be written as

$$\{q\}*[R]=\{\Delta T\}, \tag{7.29}$$

where

{q} is the vector column of the specific heat fluxes (unknown quantity)
[R] is the matrix of the thermal resistances
{ΔT} is the matrix vector of the generalized potential (temperature at the control boundary)

Note that quantities {q}, [R], and {ΔE} read as follows in our case:

$$q=\{q\}=\begin{Bmatrix} q_1 \\ q_2 \\ q_3 \\ q_e \\ q_S \\ q_W \end{Bmatrix},\quad R=[R]=\begin{bmatrix} 0 & 1 & -1 & 1 & 0 & 0 \\ 1 & 1 & 0 & 0 & -1 & 0 \\ 1 & 0 & 1 & 0 & 0 & 1 \\ R_2 & 0 & 0 & 0 & R_1 & R_6 \\ 0 & 0 & R_4 & R_5 & 0 & R_6 \\ 0 & R_3 & 0 & R_5 & R_1 & 0 \end{bmatrix},$$

$$\text{and}\quad \Delta T=\{\Delta T\}=\begin{Bmatrix} 0 \\ 0 \\ 0 \\ T_A-T_F \\ T_E-T_F \\ T_A-T_E \end{Bmatrix}.$$

The solution of the linear system is found in the form

$$\begin{Bmatrix} q_1 \\ q_2 \\ q_3 \\ q_e \\ q_S \\ q_W \end{Bmatrix} = \begin{bmatrix} 0 & 1 & -1 & 1 & 0 & 0 \\ 1 & 1 & 0 & 0 & -1 & 0 \\ 1 & 0 & 1 & 0 & 0 & 1 \\ (A_{Sc} * F_{Sc-W})^{-1} & 0 & 0 & 0 & \frac{1-\varepsilon_{Sc}}{\varepsilon_{Sc} * A_{Sc}} & \frac{1-\varepsilon_W}{\varepsilon_W * A_W} \\ 0 & 0 & (A_W * F_{W-e})^{-1} & \frac{1-\varepsilon_e}{\varepsilon_e * A_e} & 0 & \frac{1-\varepsilon_W}{\varepsilon_W * A_W} \\ 0 & (A_{Sc} * F_{Sc-e})^{-1} & 0 & \frac{1-\varepsilon_{ec}}{\varepsilon_e * A_e} & \frac{1-\varepsilon_{Sc}}{\varepsilon_{Sc} * A_{Sc}} & 0 \end{bmatrix}^{-1} * \begin{Bmatrix} 0 \\ 0 \\ 0 \\ T_A - T_F \\ T_E - T_F \\ T_A - T_E \end{Bmatrix}. \tag{7.30}$$

The performed simulations using Equation 7.30 on the double standard glass by varying the temperature of the solar shield T_A in the range $20°C \le T_A \le 130°C$ indicate that the use of sunscreens, made of a phase-changing material, increases the efficiency of their operation, reducing the radiation heat transfer through the windows with thirty percent (30%), compared to those made of traditional building materials (concrete, stone, or metal).

7.4 Design of Minimal-Admissible Light-Transmitting Envelope Apertures Using the Coefficient of Daylight (CDL)

7.4.1 Energy and Visual Comfort

Quite often, the ideas of the designers' community about interior comfort are focused only on the visual (illumination) comfort of the building interior. Buildings, which are open to the exterior and to the beautiful nature seen through the window, provoke visitors' applause. Yet, visitors hardly guess at the *reverse side of the medal—the cost of energy* needed to heat the premises to provide thermal comfort, that is, the cost of the esthetic pleasure.

Invoices for supplied and consumed energy implicitly include not only the cost of daytime artificial lighting but also that of illumination during the dark hours, both supplied by the wiring for electricity. There are also two *dark sides* of the visual comfort provided by large windows: (1) they deteriorate the thermal comfort within a large perimeter since they decrease the radiation temperature of the inhabited area T_R; depending on the bioclimatic schedule, *cold currents* emerge creating a sense of cold.

Another *secret* of the large windows is that they add *much more kWh* to the final balance of the energy consumption as compared to *solid* walls—the latter could be transformed into *adiabatic barriers* to the thermal changes. This is due to the fact that glazed building components, such as windows, shop windows, glazed doors, transom windows, etc., are classical examples of thermal bridges whose thermal resistance is several times lower than that of solid building components. Moreover, heat propagation a priori *finds* components with weaker resistance. Prior to the discussion of *energy hunger* of the envelope-glazed components, we shall formulate the problems of visual comfort from an ergonomic point of view.

The available illumination, brightness, illuminance direction, and interior components of different color are among the basic factors that affect the sense of visual comfort. Illumination discomfort exists when:

- The external contrast is strong owing to the brightness of various surfaces within the range of vision
- The inhabitant is under stress or with tired eyes
- The inhabitant is preoccupied with reading or focusing on small objects

Habitation of illumination-discomfortable premises yields *loss of ability to work, headache, total weakness,* and eventually *contrast sensitivity loss* and *eye injury.*

The basic requirement to an inhabited area for visual comfort is that sufficient illumination of the work surfaces should be available. Hence, there are different illumination levels regulations in different states. They reflect the respective *living standard* and the general geographic conditions (sky brightness and lighting, local vegetation, and landscape tonality). A national standard on *illuminance and brightness* has been adopted in our county. For instance, luminance level of premises of residential buildings should be 100 lx, while that of libraries –300 lx, etc. That illuminance guarantees visual comfort to 90% of the inhabitants (only 10% of them feel discomfort).

Hence, providing visual comfort is an important architectural engineering task, which should be part of the preliminary building design. It should be the basic focus of three specialists: an architect (in charge of the building esthetics), an electrical engineer (in charge of the artificial lighting), and a thermal comfort designer (in charge of the window dimensions, window thermal characteristics, and natural lighting of premises with windows).

As the architectural engineering practice shows, the use of *hybrid lighting systems* is *advisable* in Europe.* Their energizing is double: once due to the sunshine and next due to the available energy (electrical or chemical carried by gaseous carriers such as singas, *urban* gas, propane–butane, or natural gas). Such lighting systems comprise

- A window lighting system (facade and upper illumination) assembled within the building perimeter or on the roof.
- An artificial lighting (the most often used illuminators are assembled within the building core). Illuminators are switched on during the dark hours.

The basic problem consists in that the modification of the illumination systems (increase or decrease of their efficiency) ambiguously affects the power needed for their energizing.

The problem is illustrated in Figure 7.9. Consider solar energy $T_v = I_e/I_i$ (including the visible light) transmitted through a window. If it increases, the energy consumed by the electrical lighting will decrease (curve 2), while energy consumed by the air-conditioning system will increase (curve 1). The use of glazed elements in mass constructions and increase of their total

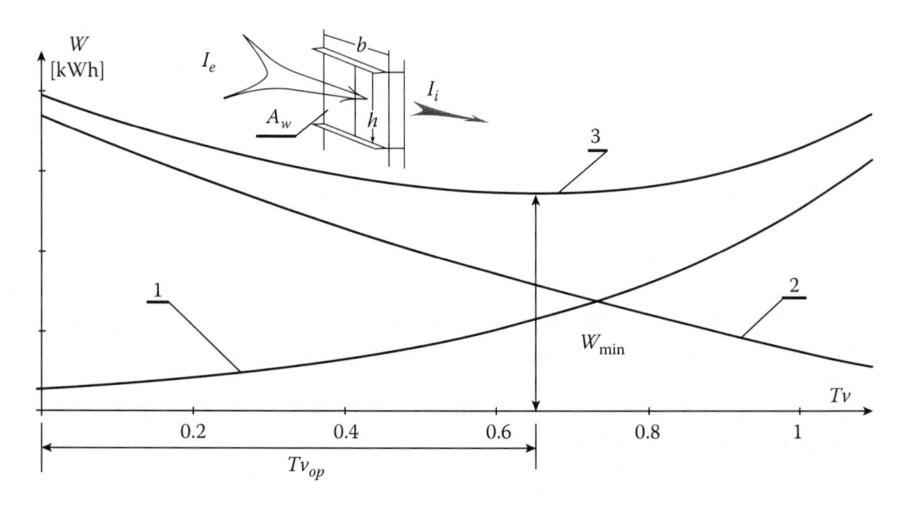

FIGURE 7.9
Effect of the coefficient of transparency of glazed structural elements on the amount of used energy: (1) air-conditioning; (2) illumination; (3) total energy consumption.

* Artificial illuminators are mostly used in public buildings in countries with equatorial, tropical, subtropical, and polar climate.

dimensions (surface areas) and transparency yields decrease of power, consumed for energizing the artificial illumination system. Yet, the required power and the energy consumed by the thermal comfort system increase (thermal gains due to sunshine and energy losses due to thermal conductance in cold days and at night increase).

The synthesis of such *hybrid lighting systems* (combining natural and artificial illumination) with a task of keeping minimal initial investments and minimal current overhead expenses for consumed energy and service proves to be *quite a complex problem.* This is best illustrated by the operation of the *system for automatic regulation* of the building visual comfort (see Figure 7.10). We use here our simulation estimates ordered by ZEH with headquarters in Las Vegas and performed using the Energy10 software. Figure 7.10 shows that there is a controllable difference between the moment of switch off of the *artificial lighting system* and that of *switch on* of the natural illumination through the windows. It is called nullband, deadband, zero energy band, or zero power demand zone (ZPDZ) and used to avoid incident switch on under random fluctuations of sensor signals. Such instability may occur as a result of an inappropriately selected spot of sensor assembly, unexpected energy impacts due to open doors and windows, direct radiation, etc. *The width of the nulband can be manipulated* (narrowed or widened). The larger the width B_{ZPDZ}, the lesser the energy consumed by the artificial lighting installation.

The basic factors affecting B_{ZPDZ} and the illuminance of the building area are window surface areas A_w (with overall dimensions h and b) and transparency T_v. The larger the window height H (surface area A_w, respectively), the larger the depth of sunlight penetration into the building. Thus, one can better meet the natural illuminance requirements and the nulband will grow wider. Moreover, there will be no need of prolonged use of *artificial* illumination energy in that building area, and the illumination overhead expenses will be lower.

On the other hand, the use of larger windows instead of solid walls will raise the building cost since a glazed envelope is two to three times more expensive than the solid one.* Besides, the coefficient of heat transfer of a larger window exceeds that of a solid wall, yielding larger thermal losses/gains. The thermal installation of buildings supplied with large windows is more powerful, more expensive, and consumes more energy needed for the maintenance of thermal comfort. Hence, another factor should be accounted for, that is, the thermal energy cost or the ratio between costs of illumination and heating (air-conditioning) energy.† Consider an unjustified decrease of the window overall dimensions. Then, although the improvement of the thermal comfort and drop of its cost, *more* additional energy will be needed

* We do not consider here luxury walls covered with polished stone.

† To solve the optimization problem of calculating the window overall dimensions, one should model the energy exchange between the building and the surroundings. Software of the type EnergyPlus and ESP-r can be used for that purpose.

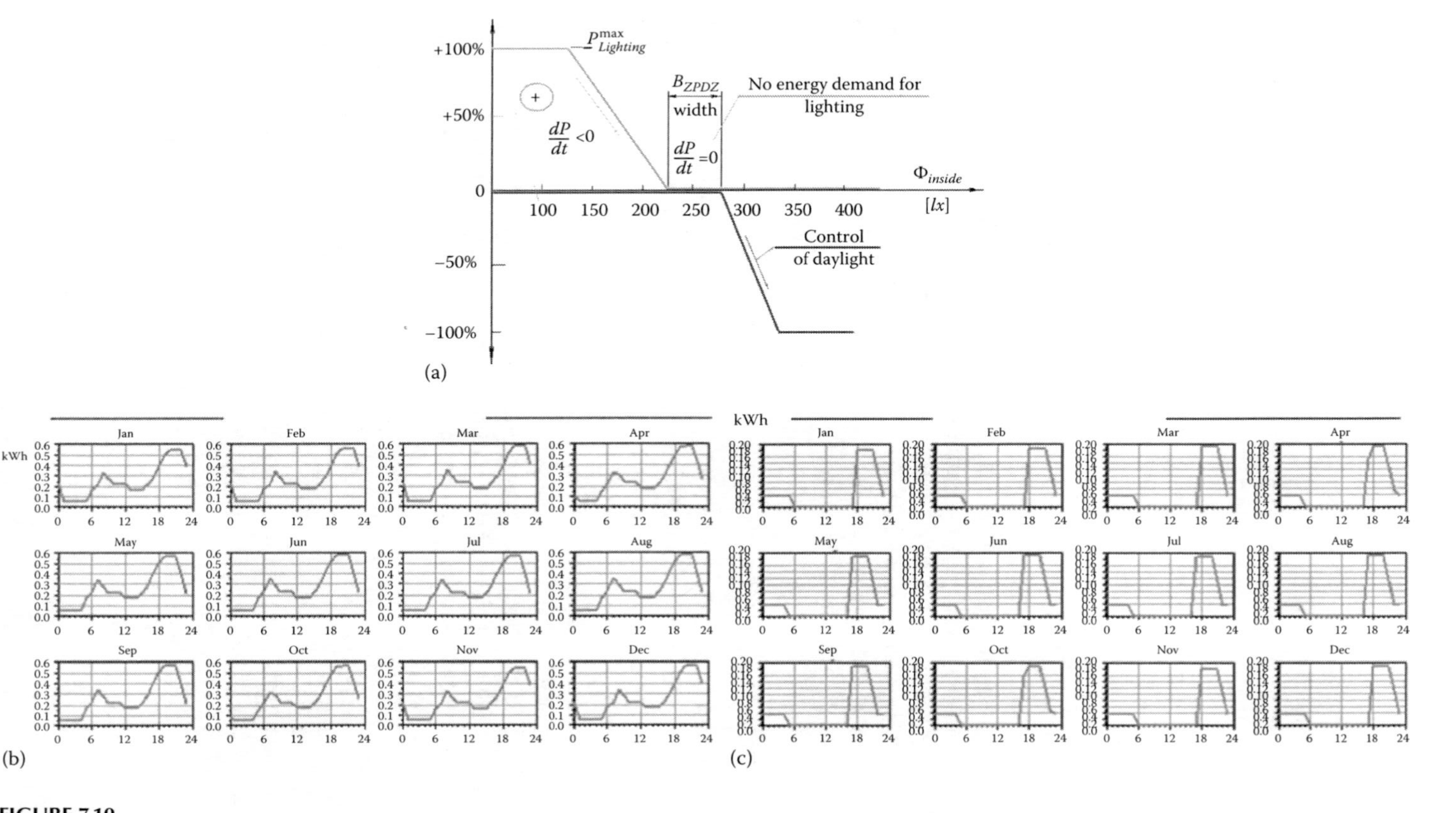

FIGURE 7.10
Regulation of the power of the visual comfort system with respect to the demand for energy. (a) Zero demand zone, (b) indoor lighting, and (c) outdoor lighting.

to provide a visual comfort since the natural illuminance of the interior will decrease. That *bound and ambiguous* effect of the dimensions of the envelope-glazed element *calls for* envelope optimization. Its profit function should comprise the net present value of the initial investments and the future payments for heating and illumination energy. At the same time, the effect of window overall dimensions on the *interior natural illuminance* should also be assessed.

Consider methods employed to calculate the dimensions of apertures, transmitting natural light to the inhabited areas. Providing a visual comfort is an important task, which was induced by previous local studies of premises of buildings erected in the 1970s. Note that some designers were taken up with esthetics or technology at that time. The result was *too large* building glazing, whereas the total glazing coefficient of some buildings amounted to 35% ÷ 40%, and it was significantly higher for some east, south, and west facades (Димитров 2003). Due to unjustified calculations of the window overall dimensions (the surface area of the light transmitting aperture A_w, respectively), a systematic overheating occurred when the solar flux I_i uncontrollably entered the window. In gloomy days, however, there was a sense of excessive cold and thermal discomfort.

7.4.2 Calculation of the Coefficient of Daylight (CDL)

As studies and experience show, the depth of the *periphery area* depends mainly on the window height "*H*" and depth "*B*," considering fixed floor height and module step of the construction. We propose here an *analytical expression* enabling one to assess the illuminance at an arbitrary spot. It can be incorporated in algorithms of stepwise minimization of window dimensions (Димитров 2003).

To conform to the Bulgaria Nation Standard of normal illuminance of a control area,* the coefficient of daylight defined as $CDL = E_P/E_{Sc}$ should satisfy the following inequality (Димитров 1990):

$$CDL \geq [CDL]_{St}, \tag{7.31}$$

where

E_P is the natural illumination at point P of the control line, which is parallel to the floor and at the level of the work visual area (an area with expected visual comfort)

E_{Sc} is the illuminance of the vertical window surface after the last glazing row

$[CDL]_{St}$ is the standardized nominal value of the coefficient of daylight, guaranteeing visual comfort within the inhabited area

* The control line is 1 m away from the wall opposing the window and 0.8 m above the floor.

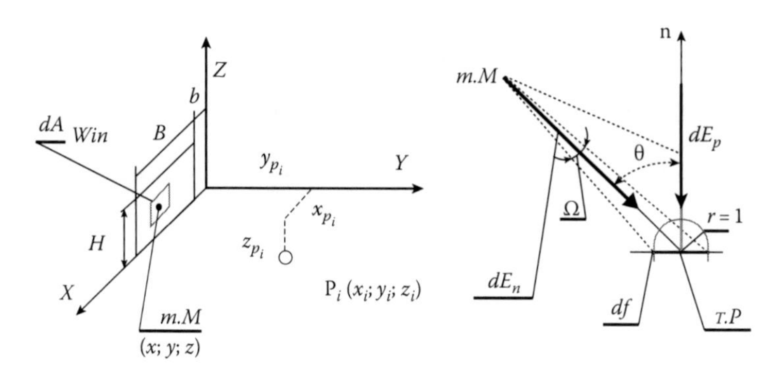

FIGURE 7.11
Stereometry of the interior natural illuminance.

The problem of finding the coefficient of natural illumination is reduced to the projection of the area of the light receiving surface on the semi-spherical surface with radius $r = 1$ and center P (x_P, y_P, z_P) belonging to the control surface. Consider an elementary area dA_{Win} of the light-transmitting aperture with center $M(0, y, z)$ (see Figure 7.11). Assume that the illumination variation at point P obeys the cosine law of Lambert and it can be assessed via

$$dE_P = E_n \cdot \cos(\theta) = L \cdot df \cdot \cos(\theta), \tag{7.32}$$

where
- $E_n = Ldf$ is the light flux incident on an area with an observation angle θ
- θ is the observation angle defined by the normal $\vec{n}_1$ and the straight-line MP (see Figure 7.11)

The coefficient of daylight is found from the definition equality integrating expression (7.31):

$$CDL_P = \frac{1}{E_{Sc}} \int_{f_0} dE_P = \frac{1}{\pi L} \int_{f_0} L \cdot \cos(\theta) \cdot df = \frac{1}{\pi} \int_{A_{Win}} \left(\frac{r}{s}\right)^2 \cos(\delta) \cdot \cos(\theta) dA_{Win}.$$

Taking into account the geometry of the illumination area (the window), it takes the form

$$CDL_P = \frac{x_P}{\pi} \int_{\zeta_1}^{\zeta_2} \left(\frac{1}{2} \int_{\eta_1}^{\eta_2} \frac{d\eta^2}{\chi^2 + \eta^2 + \zeta\chi^2} \right) d\zeta,$$

whose boundaries are

$$\zeta_1 = x_{P_i} - b; \quad \zeta_2 = x_{P_i} - b - B; \quad \eta_1 = h - z_{P_i}; \quad \eta_2 = H + h - x_{P_i}.$$

Then, we find

$$CDL_P = \frac{15.91 y_{P_i}}{\sqrt{y_{P_i}^2 + (-z_{P_i})^2}} \left[arctg \frac{x_{P_i} - B - b}{\sqrt{y_{P_i}^2 + (-z_{P_i})^2}} - arctg \frac{x_{P_i} - b}{\sqrt{y_{P_i}^2 + (-z_{P_i})^2}} \right]$$
$$- \frac{15.91 y_{P_i}}{\sqrt{y_{P_i}^2 + (H - z_{P_i})^2}} \left[arctg \frac{x_{P_i} - B - b}{\sqrt{y_{P_i}^2 + (H - z_{P_i})^2}} - arctg \frac{x_{P_i} - b}{\sqrt{y_{P_i}^2 + (H - z_{P_i})^2}} \right], \tag{7.33}$$

where
- B, m is the window width (it is assumed as a basic width, and it is stepwise varied)
- H, m is the window height
- b, m is the distance between window and wall
- h, m is the sill height
- x_{P_i}, y_{P_i}, z_{P_i}, m are coordinates of a point of the control surface at which CDL is assessed assuming $Z_i = (0.8\ \text{h})$

The present study uses a simplified version of the analytical expression (7.33) to illustrate the stepwise minimization. CDL is assessed at points of a line perpendicular to the window plane and passing through the control point of *standard assessment*. The minimization of the window surface area of a standard student's hostel has been incorporated into a cyclic numerical procedure, whose results are plotted in Figure 7.12.

Plots of CDL are shown in Figure 7.12, and simulation stepwise minimization of the overall dimensions of the light-transmitting aperture is performed following Equation 7.33 and using the DERIVE 5 software. Window width is varied by step $\Delta B = 0.4$ m until the value of CDL at the control point enters the interval $1.02[CDL]_{St} \geq CDL_P \geq [CDL]_{St}$. The overall window dimensions thus found should be taken as minimal pursuant to *sanitary-hygiene criteria*. Windows of the students' hostel with dimensions 1700/1200 mm have been selected as a result. They have been taken into account in a subsequent project of building recovery (Димитров 2003). Additional simulations have been performed to verify the results of the CDL distribution within a plane parallel to the floor and containing the control point. The levels of illumination have also been found (Димитров 2003).

The simulation results are shown in Figure 7.13 as isolines located within the room dimmed out area. The plot concerning a window with dimensions

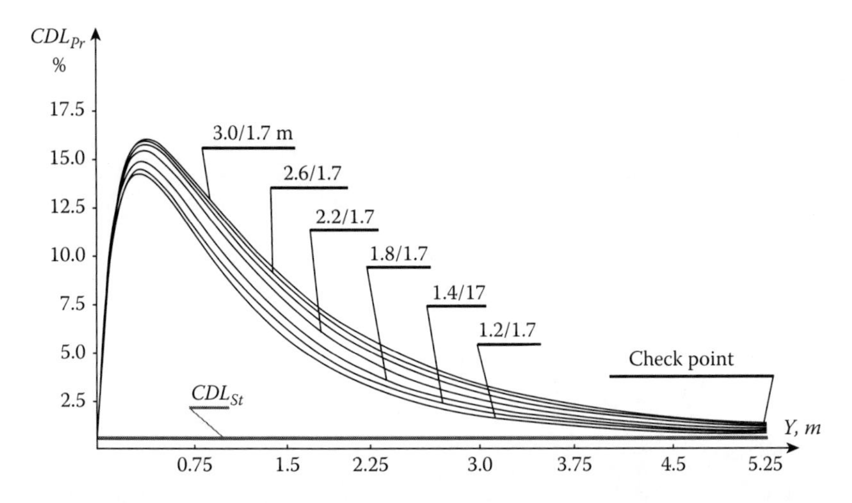

FIGURE 7.12
Stepwise minimization of window overall dimensions. (The distribution of the coefficient of daylight (CDL) in room no 213 of a student's hostel is found using the theoretical model.)

1.7/3.0 (the actual case) indicates that the CDL is five times larger than the standard value proving that the windows are significantly overcalculated. In addition, Figure 7.13b shows the distribution of the CDL isolines within the same plane but with adopted window dimensions 1.70/1.20 m. It is seen that the CDL meets the normative requirement although that the window surface area is 2.5 times smaller. This implies a requirement of reducing the dimensions of the windows of the standard premises to 1.70/1.20 m, which must be included in the project.

Hence, the reduction of the glazed area is an essential part of building recovery. It yields multiple effects:

- Drop of the initial investments (costs of the glazed elements are relatively higher than those of solid elements) and hence, a shorter period of investment reimbursement
- Significant thermal energy savings (the thermal resistance of glazed elements is 2.6–4.2 times smaller than that of the solid elements)
- Improvement of the room thermal comfort (the sense of thermal comfort in the inhabited areas will grow significantly stronger due to the reduced surface area of the *cold* windows)
- Guaranteed normal visual comfort
- Reduction of premises overheating at the east, west, and north facades—note that overheating is caused by solar radiation

The proposed method of minimizing the overall dimensions of the light-transmitting apertures can be easily applied using a PC and appropriate software packages.

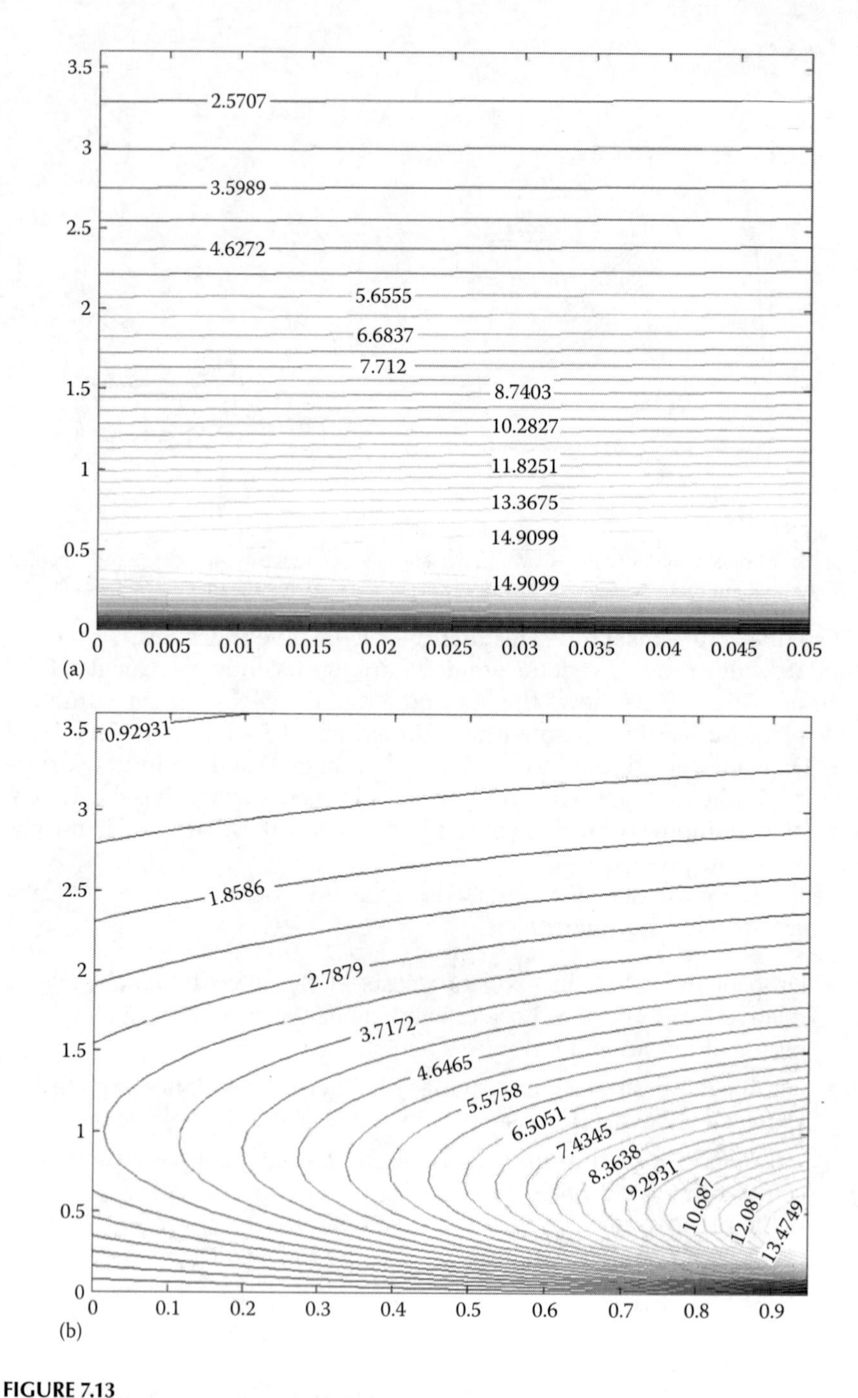

FIGURE 7.13
Distribution of CDL, assessed using the theoretical model: (a) actual situation (a window with dimensions 1.7 × 3.0 m); (b) recovered premises with windows 1.7 × 1.2 m.

7.5 Method of Reducing the Tribute of the Construction and the Thermal Bridges to the Energy Inefficiency

Saving primary energy by a practical introduction of low-energy technologies and highly efficient equipment is the consumer's *most successful energy strategy* on the monopoly energy market, where prices are controlled by social and not market mechanisms. It is in accordance with the requirement of the official *Energy Strategy 2020* with a task to attain 20% rate of energy efficiency. Ratio *PEC/GDP* toe/Eu is adopted to measure the energy efficiency of industrial technologies.* Here *PEC, toe* is the primary energy consumption and *GDP*, Eu is the gross domestic product.

At *consumer's* level, however, that measure is *not useful* and seems to be *senseless. GDP cannot be calculated at* everyday life's level, all the more that household appliances are not involved in any production but are used to satisfy the individual comfort and hygiene needs (bath, laundry), for entertainment (watching TV, listening to the radio), etc. Those arguments are valid for buildings, too, which are intended to guarantee men's habitation, relaxation, entertainment, etc., and where *GDP* is not to be spoken of.

Another basic problem facing customers and related to *energy efficiency* is the involvement of specific terms. To make customer's choice of electric appliances easier, the EU introduced a *system of energy labeling of household appliances* (directive 92/75/EC). The state administration obliges *manufacturers and dealers* to *inform customers* (end consumers of energy) on the energy efficiency† upon purchase. At the same time, *manufacturers and dealers* should also supply the equipment with *a class of energy efficiency (CEE)* depending on the so-called *index of energy efficiency* (see directives 96/57/EU, 92/42/EU and 2000/55/EU). Seven classes of energy efficiency (from *A* to *G*) are specified. Appliances class A are considered to be the most energy efficient ones, while those of class E should not be admitted at all to the European market, as stated by directive 96/57/EU.

Analysis of data shows that energy consumed by *households* amounted to 23.2% of the total energy consumption in 2005‡ (or 2.148 * 10^6 toe, respectively 90.22 * 10^6 GJ). Due to the lack of own statistical data, the energy consumption can be divided into portions using the model§ of energy consumption of residential buildings (Dimitrov 2014a). Table 7.5 gives the following picture of the structure of energy consumption.

---

* The energy efficiency of national economy on macroeconomic scale is also calculated on such a basis.

† Directive 2005/32/EU treats all energy-consuming appliances: boilers, ventilators, pumps, air conditioners, refrigerators, laundry machines and dishwashers, PCs, monitors, scanners, printers, copiers, fluorescent lamps, etc.

‡ The absolute amount of energy consumed by local households in 2005 was 2.148 * 10^6 toe or 90.23 * 10^6 GJ.

§ We use the EPA (2000) and DOE (2005) statistics.

TABLE 7.5

Indicative Targets of Household Energy Savings

Household appliances	$0.104 * 10^6$ toe
DHW	$0.068 * 10^6$ toe
Illumination	$0.038 * 10^6$ toe
Thermal comfort	$0.113 * 10^6$ toe
Others	$0.107 * 10^6$ toe
Total	$0.43 * 10^6$ toe
National indicative target of the branch 20%	$\Rightarrow 0.086* 10^6$ toe

The national indicative task of increasing the energy efficiency by 20% of the end energy consumption, documented in the *Energy Strategy 2020* ($0.086 * 10^6$ toe), cannot be fulfilled if for instance we reduce *only* the fourth article in Table 7.5 (regarding thermal comfort) by 50% (by $0.0565 * 10^6$ toe) applying mass enveloping of all buildings. A useful idea of solving that problem is the use of classes of energy efficiency of all energy-consuming systems in the building (Димитров 2008a) and stimulating the engagement of designers and investments in the manufacture of energy-efficient products. Customers should be provided with subsequent energy characteristics and *class of energy efficiency* when purchasing or renting buildings, offices, or apartments.

7.5.1 Characteristics of Heat Transfer through Solid Inhomogeneous Multilayer Walls

The physical picture of heat transfer within the control volume of a ETS via conductance and convection is idealized and incomplete, although it is used to describe transfer phenomena through solid structural elements, such as floors, roofs, walls, etc. The real physics however is different owing mostly to two groups of reasons:

1. Technological
2. Architectural structural

The first group concerns all deviations from and violations of the constructional requirements (see Figure 7.14).

The second group concerns the incorporation of materials in the building envelope components whose conductivity strongly differs from that of traditional materials. Such components are steel bar–reinforced concrete columns, beams, concrete walls (see Figure 7.15). It also accounts for buildings with varied geometry, for instance, those with bay windows, angles, contra-angles, etc.

To assess the amount of heat passing through such complex inhomogeneous structures, we use thermal bridges (structures) yielding additional

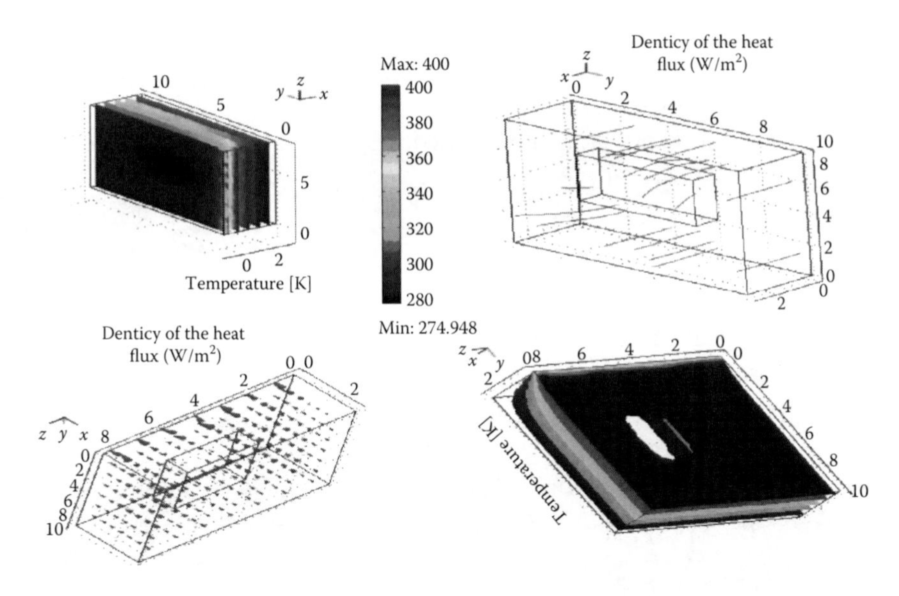

FIGURE 7.14
Structural inhomogeneity of solid walls due to technological errors: simulation of heat transfer in an inhomogeneous physical medium using a specialized software.

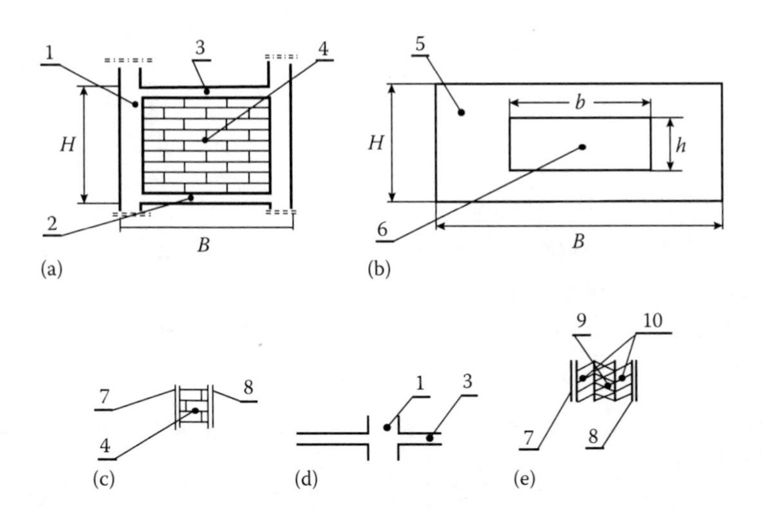

FIGURE 7.15
Structural inhomogeneities due to components with different heat conductivity: (a) brick wall; (b) panel wall; (c) cross section of a brick wall; (d) node of structural system as a thermal bridge; (e) multilayer wall: (1) steel bar–reinforced column (λ = 1.63 W/mK); (2) steel bar–reinforced plate (λ = 1.45 W/mK); (3) steel bar–reinforced beam (λ = 1.63 W/mK); (4) brick wall (λ = 0.52 W/mK); (5) steel bar–reinforced panel, sandwich type (U_w = W/m²K); (6) glazed element (U_{np} = 2.6 W/m²K stuck together wooden window); (7) lime-cement coating (λ = 0.81 W/mK); (8) internal stucco (λ = W/mK); (9) foaming polystyrene (λ = 0.03 W/mK); (10) steel bar–reinforced concrete (λ = 1.63 xW/mK).

transfer of heat through the control surfaces. Note that transfer in this case is more intensive than that through a homogeneous medium (Dimitrov 2013).

We apply the principle of superposition to real structural elements (regarding heat transfer through the homogeneous base and the bridges):

$$\vec{H}_{inhom} = \vec{H}_{hom} + \vec{H}_{bridges}, \tag{7.34}$$

where

$\vec{H}_{inhom}$ J is the heat transferred through a real wall
$\vec{H}_{hom}$ J is the heat transferred through a homogeneous wall
$\vec{H}_{bridges}$ J is the heat transferred through the thermal bridges

Heat flux $\vec{Q}$ (rate of heat transfer) through a structural element with surface area A_0 reads

$$\vec{Q}_{in\,\mathrm{hom}} = \vec{Q}_{\mathrm{hom}} + \vec{Q}_{bridges}$$

where

$\vec{Q}_{hom} = \left(\sum_{1}^{n} \frac{\lambda_i}{\delta_i} + U_{out} + U_{in}\right) A_0.(T_i - T_e) = U_{Hom} A_0.(T_i - T_e)$, W is the heat flux through an element without a thermal bridge

$\vec{Q}_{bridges} = \left(\sum_{j=1}^{k} l_j.\psi_j + \sum_{1}^{m} \chi_k\right)(T_i - T_e) = U_{bridges}(T_i - T_e)$ W is the heat flux though the thermal bridges

$\sum_{j=1}^{k} l_j \psi_j$ W is the additional heat flux through linear thermal bridges with length l_j m and intensity ψ_j W/mK

$\sum_{k} \chi_k$ is the additional heat flux through point thermal bridges with intensity χ W/K

The temperature difference $(T_i - T_e)$ is the operational one between the *cold* and the *hot* sources of the ETS.

Until recently (November 2009), the tribute of the thermal bridges to the thermal losses due to heat conduction* was accounted for via a linear dependence involving the gross surface area of the solid elements A_i (Section 8.5) (Димитров and Бояджиев 2010):

$$Q_{bridges} = 0.1 * A_i (T_i - T_e)\,\mathrm{W}. \tag{7.35}$$

* Near angles, transoms, columns, door and window frames, jetties, balconies, etc.

Bridge tribute to the total heat transfer was considered to be insignificant and not exceeding 8%–10%. A new version of regulation no. 7 (Димитров and Бояджиев 2010) was adopted in the beginning of 2010 without a trial period. Together with all emotional comments on its operation, an unexpected effect was disclosed: the transfer tribute of the thermal bridges *significantly exceeded* 10% of the total amount of energy transferred through the envelope—it is even higher than 60% in some specific cases. This verified the control of thermal bridges emerging during the envelope assembly.

We shall prove the necessity of normative amendment using *method of assessing index (IEE_{Const}) and class (CEE_{Const}) of the energy efficiency of constructions*. The method may be useful in the design of new buildings.

7.5.2 Method Described Step by Step

Select as a first step a quantity, sensitive enough to be used as *an index of energy efficiency adopted to assess the building structural modifications* (Figure 7.16). Then, propose a *correlation table* so that the official building administration would regulate the class of energy efficiency of the construction conforming to Appendix 4 of regulation no. 7.

To assess the effect of different building structures on the transfer of heat through the thermal bridges, emerging within the envelope and owing to floor *plates, columns, beam, and steel bar reinforced walls*, we use a new *standard form*, called *energy standard of the construction (EE_{Const})*.

7.5.3 Description of the Energy Standard of the Construction (EE_{Const})

It is proposed that all structures of newly designed buildings should be *assessed by comparing* them to the structure of an *imaginary cylindrical building*, called *energy standard*, which has the following characteristics:

- *Building volume* V_{ESt} (equal to the volume of the designed building bound to certification $V_{ESt} = V_{Buil}$)
- *Building height* H_{ESt} (equal to the denivelation between the ground components (level 0.00) and the level of the last carrying plate (roof or last floor) that creates a linear thermal bridge

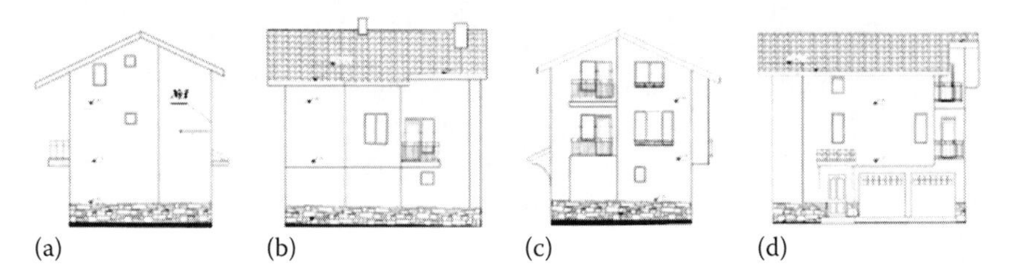

FIGURE 7.16
Facade of an inspected building UPI XXIII, No. 69, Dragitchevo village: (a) northern facade; (b) eastern facade; (c) southern facade; (d) western facade.

- *Glazing degree*—0% (without apertures transmitting natural light)
- *Building system*—a system where all structural components (columns, steel bar–reinforced walls, beams, etc.) are located deep into the building core or far enough from the envelope

The selected *energy standard* of the *building structure* has linear bridges with minimal length L_{ESt}, ($L_{ESt} \Rightarrow$ min), whereas other quantities, such as building volume (V_{Buil}), number of floors (n_{Buil}), and building height H_{ESt}, are kept constant. Knowing the building volume (V_{Buil}), height H_{ESt}, and number of carrying plates (n_{cep}), one can assess the total length of the linear bridges of EE_{Const} as

$$L_{ESt} = n_{Buil} * \sqrt{\frac{4\pi.V_{Buil}}{H_{ESt}}}. \tag{7.38}$$

The value (L_{ESt}) is further used to calculate the index IEE_{Constr} and class CEE_{Const} of the construction energy efficiency. The introduction of the term class of energy efficiency is in agreement with the requirements of the EU normative documents and practice. It facilitates the communications between designers with different qualification and investors, helping also to the formulation of clear methodical regulations of the design of buildings and related systems and components (Димитров 2008a).

Conforming to the methods of assessing the energy efficiency approved by the EU members, *the index of energy efficiency* of constructions (IEE_{Constr}) is defined as (Димитров 2008a)

$$IEE_{Const} = \frac{SEM_{Const}}{EE_{Const}}, \tag{7.39}$$

where

SEM_{Const} is the *specific energy measure* of the structure of the designed (passportized) building:

$$\left(SEM_{const} = \sum_{k} l_k.\psi_k + \sum_{j} \chi_j, k = 1 \div n_1; j = 1 \div n_2\right) \text{W/K}$$

EE_{Const} W/K is the *energy standard* (benchmark)

$$\left[EE_{Const} = L_{ESt} * \psi_{IF3} = n_{Buil}\sqrt{\frac{4\pi.V_{Buil}}{H_{ESt}}} * 1\,\text{W/K}\right],$$

where $\psi_{IF3} = 1$ W/mK for linear thermal bridges at the level of the intermediate floors (Димитров and Бояджиев 2010). Performing transformations, we find the efficiency index of the building structure:

$$IEE_{Const} = \frac{\sum_k l_k^{Const} * \phi_k + \sum_j \chi_j^{Const}}{n_{Ctp} * \sqrt{\dfrac{4\pi * V_{Ctp}}{H_{EEConst}}}}. \tag{7.40}$$

Cases $\sum_k l_k^{Const} * \psi_k$ comprise lengths l_k of the introduced thermal bridges only.*

7.5.4 Employment of the Energy Standard to Assess How the Building Structure Affects the Energy Efficiency

The method of using EE_{Const} consists of seven steps:

Step 1: Compose a map of the facade thermal bridges (see Figure 7.17a).

Step 2: Calculate the length of the linear thermal bridges regarding their type (see the data shown in Figure 7.17a).

Step 3: Calculate SEM_{Constr} (the numerator of Equation 7.39).

Step 4: Calculate the building volume V_{Buil} and height H_{Buil} (using the building documentation).

Step 5: Calculate the minimal admissible length of the linear thermal bridges of the energy standard L_{ESt} (formula 7.38).

Step 6: Calculate the index of energy efficiency IEE_{Constr} of the designed (passportization) building using Equation 7.40).

Step 7: Determine the construction energy efficiency (use the table of correspondence—see Table 7.7).

We use a work project of an individual residential building in order to illustrate the execution of the seven steps. The building facades are shown in Figure 7.16.

A steel bar–reinforced structure is erected (the construction scheme is shown in Figure 7.18a). It comprises seven columns, two steel bar–reinforced walls, and three carrying steel bar–reinforced plates at levels +2.7, +5.5, and +8.7 m. Conforming to *step 1*, we make a map of the linear thermal bridges shown in Figure 7.17a. The map *does not show* bridges

---

* Linear thermal bridges, emerging around envelope glazed elements and jetties, are not accounted for in the estimation of SEM_{Constr}. Yet, they are considered when assessing $SEM_{Evelope}$ — the measure of envelope specific energy. Note that the analysis of $SEM_{Evelope}$ is beyond the scope of this book (Димитров and Н.Бояджиев 2010).

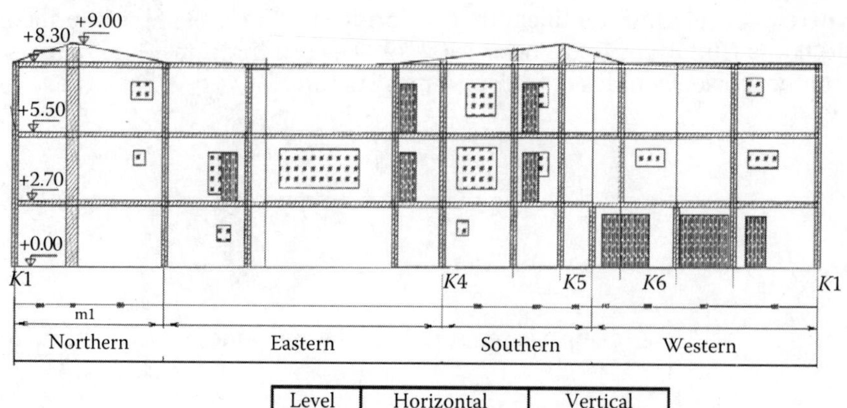

Level	Horizontal	Vertical
+2.70	37.22 m	27.00 m
+5.50	35.70 m	25.20 m
+8.30	8.50 m	8.60 m

(a)

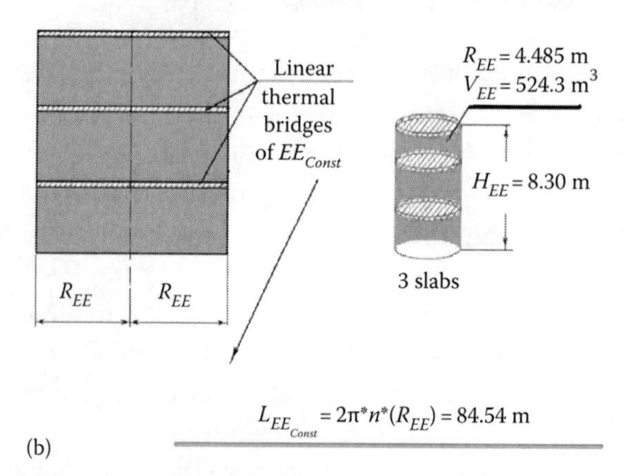

(b)

FIGURE 7.17
Buildings flat pattern: (a) a map of facade thermal bridges located within the envelope; (b) building energy standard.

emerging as a result of building geometry (edges and jetties) and those around glazed components and doors. They should be referred to the envelope affecting its energy efficiency (to find IEE_{Constr} in this particular case we assume a fixed building geometry). The map of the thermal bridges presents a specific clockwise unfolding of the building envelope, taking column 1 in Figure 7.17a as an origin. The map shows all spots where the structure *punches* the envelope. They are specified in a table (*step 2*)—see the table in Figure 7.17a. Apply now *step 3*—calculate the specific energy measure of the construction SEM_{Constr} (the numerator of Equation 7.39) and find that SEM_{Constr} = 184.39 W/K.

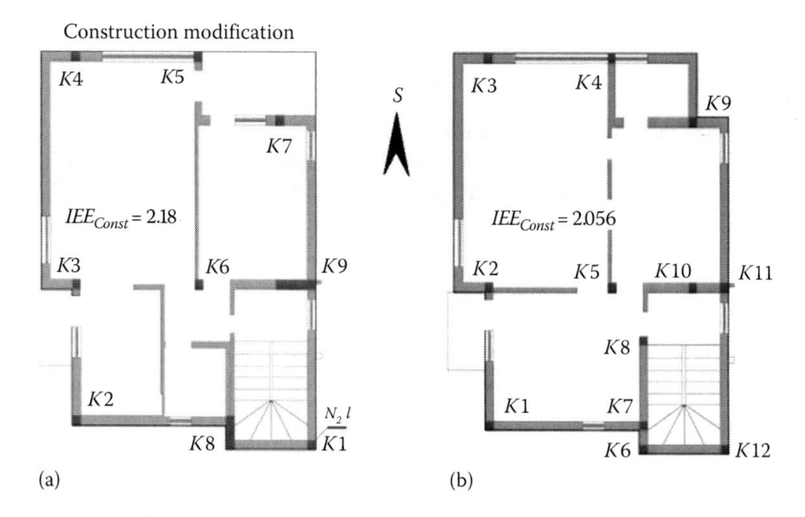

FIGURE 7.18
Decrease of the length of the linear thermal bridges via construction modification. (a) Skeleton structure, and (b) skeleton-beam structure.

The construction under consideration has been modified in order to reduce the length of the linear thermal bridges. The modification consists in the replacement of two steel bar–reinforced walls projected with columns (nos. 6, 7, 10, and 11) and shortening the column total length by the means of beams.

This is illustrated in Figure 7.18b. In fact, the decreased value of SEM_{Constr} amounts to $SEM^{MOD}_{Const} = 173.86$ W/K, which is a reduction by 7.7%. Next, the parameters of the energy standard are found—the *model of the studied building*, equivalent with respect to volume and height (a cylindrical building). *Step 4* comprises calculation of the building volume ($V_{Buil} = 524.33$ m^3) and height ($H_{Buil} = 8.3$ m). *Step 5* consists in the calculation of the linear thermal bridges of the energy standard using the expression (7.38) ($L_{ESt} = 84.54$ m), while the energy standard EE_{Const} is equal to $EE_{Const} = 84.54$ W/K.

Finally, using Equation 7.40, we find the index of the energy efficiency of the construction IEE_{Constr}. In this particular case (see Figure 7.18a), IEE_{Constr} amounts to $IEE_{Constr} = (184.39/84.54) = 2.18$. The result shows that the specific measure of the particular construction exceeds that of the energy standard by 218%.

The modified building structure in Figure 7.18b has an insignificantly different index of energy efficiency. Since the reduction of SEM_{Constr} of the modified construction amounts to 7.8% as compared to the standard, the reduction of IEE_{Constr} is 5.7%. Consider a fixed architectural solution of a building. Then, the similarity between SEM_{Constr} and IEE_{Constr} proves that a *cosmetic* modification of the construction is a weak factor of improving the energy efficiency.

Based on those considerations, more essential mechanisms of modifying the building structure were sought, keeping its geometry. This was achieved by involving other building systems, different from the *skeleton-beamless and skeleton-beam ones* (SBS) (Димитров 2011). The idea was to use constructions where walls carry significant loads while the necessary structural elements have been introduced deeply into the envelope.

The investigation trends are illustrated in Figure 7.19 and systemized in Table 7.6. The results in Table 7.6 yield the following conclusions:

1. The examples given unambiguously illustrate how sensitive is the proposed index of energy efficiency to the modification of the construction, keeping the other conditions.

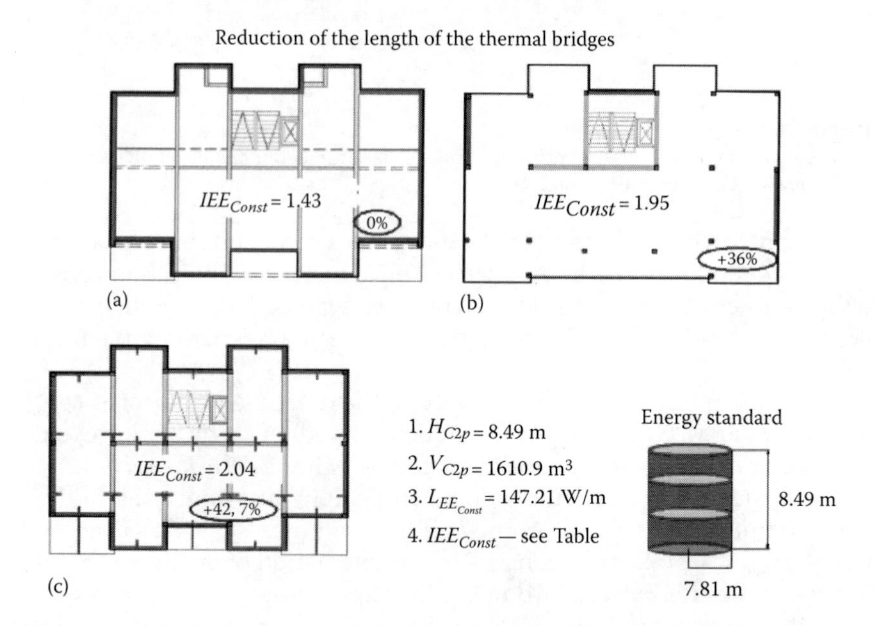

FIGURE 7.19
Effect of the construction on the value of the energy efficiency index IEE_{Const}. (a) Panel structure, (b) PPL-package lifted plates, and (c) skeleton-beam structure.

TABLE 7.6

Influence of Structural System of the Building on Its Energy Efficiency by Index IEE_{Const}

Construction	Horizontal $\sum_1^n l_k\psi_k$ W/K	Vertical $\sum_1^n l_k\psi_k$ W/K	SEM_{Const} W/K	IEE_{Const}	Difference ΔIEE_{Const}	ΔIEE_{Const} (%)
Panel	211.05	0	211.05	1.43	—	—
PLP	211.05	76.8	287.85	1.95	+0.52	+36.0
SBS	211.05	89.6	300.65	2.04	+0.61	+42.7

TABLE 7.7

Table of the Correspondence of the Index and Class of Energy Efficiency for the Building Structural System

IEE_{Const}	Class of *EE* of the Construction	Estimates and Further Steps
$IEE_{Const} \leq 0.75$	A^0	Excellent+
$0.75 \leq IEE_{Const} \leq 1.25$	*A*	Excellent
$1.26 \leq IEE_{Const} \leq 1.5$	*B*	Very good
$1.51 \leq IEE_{Const} \leq 1.75$	*C*	Good
$1.76 \leq IEE_{Const} \leq 2.0$	*D*	Fairly good, but needing modification
$2.01 \leq IEE_{Const} \leq 2.25$	*E*	Should be modified
$2.26 \leq IEE_{Const} \leq 2.5$	*F*	Should be modified
$2.51 \leq IEE_{Const}$	*G*	Should be modified

2. We propose the inclusion of a correspondence table (Table 7.7) into building standards and codes, concerning the building structural system. There, the state building administration should regulate the class of energy efficiency.
3. The quantity proposed to assess construction modifications is the index of energy efficiency (e.g., Equation 7.40), which is suitable for use in the design of constructions and for guaranteeing their methodical independence from energy supervisors.

7.6 Assessment of Leaks in the Building Envelope and the Air-Conditioning Systems

The central air-conditioning systems are widespread in local public and commercial buildings. They are also used in residential buildings in southern Europe, the Mediterranean, and the United States. The experience of that prolonged use shows that efficiency considerably deteriorates due to cracked sections and leaks (Dimitrov 2008). For example, the standard techniques of leaks measurement including leaks in the building envelope and known as *fan pressurization* have been improved and applied in air-conditioning and ventilation installations (HVAC I).

As stated by the ASTM E1554-03 (2003), there are two methods to locate *leaks* in an air-conditioning/ventilation system (all air):

1. Through duct pressurization tests
2. Through blower door subtraction tests

The first provides more precise results and is more frequently used. The second method is compromised due to equipment errors.

TABLE 7.8
Basic Characteristics of the Used Methods

Method	HVAC Location of Leaks in Ducts	Separated Leaks in Supply and Return Ducts	Measurement of Leakage through the Building Envelope	Use in Real Operational Conditions
Duct pressurization	Yes	No	Yes	Yes
Test house pressure	Yes	N	Yes	Yes
Nulling pressure test	Yes	No	Yes	Yes
Tracing gas	Yes	Yes	Yes	No
Blower door subtraction	Yes	No	Yes	No
Test Delta-*Q*	Yes	Yes	Yes	Yes

The general disadvantages of both methods, however, boil down to that

- They cannot be used in real operational conditions
- They cannot distinguish leaks in the supply from those in the return duct and also they cannot locate leaks in the building envelope

Several alternative methods of locating *leaks* in the ducts of air-conditioning systems (all air) have been developed in the last fifteen years. The basic characteristics of the most important methods are shown in a tabular form (see Table 7.8) (Dimitrov 2008). Their common feature is the large set of data yielding calculation errors and uncertainties.

We propose here a methodology, performing sizing of the primary data via so called *internal scales*. They are applicable both in the *online* processing of data collected by automated systems and in the design of statistical models based on *in situ* manual calculations. We use them to estimate leak incorporated into the balance equations of the *modified* ΔQ *method*.

7.6.1 Measuring Equipment of the Method "Delta-*Q*"

It is known that there is a relationship between the fluid delivery (air) Q and the pressured drop Δp in the ducts, which can be expressed by a Darcy exponential functional form:

$$Q = C(\Delta p)^n, \tag{7.41}$$

where
Q is the fluid rate (m^3/s)
C is the leakage coefficient
Δp is the leak pressure drop (Pa)
n is the pressure exponent

Exponent n varies in the interval $Q_{BI} + \Delta Q_e + \Delta Q_s + Q_s + \Delta Q_r - Q_r = 0$, corresponding to a fully developed laminar flow. Yet, it has been experimentally proved that its value is close to 0.66.

A new laboratory-measuring equipment has been designed to apply the Delta-*Q* method (Dimitrov 2008) (see Figure 7.20a). It comprises the following major elements:

- Test chamber (position 1)
- Air-conditioning handler (position 4)
- Measuring equipment (position 5)
- Supply (position 6) and return (position 7) ducts

The chamber is taken as a working volume (59.4 m^3) modeling a thermal zone and undergoing air-conditioning. Using two axial fans (positions 3 and 4)–type *blower door*, one can vary internal pressure in the range −100 to +100 Pa by air pressurization into or subtraction from the test chamber

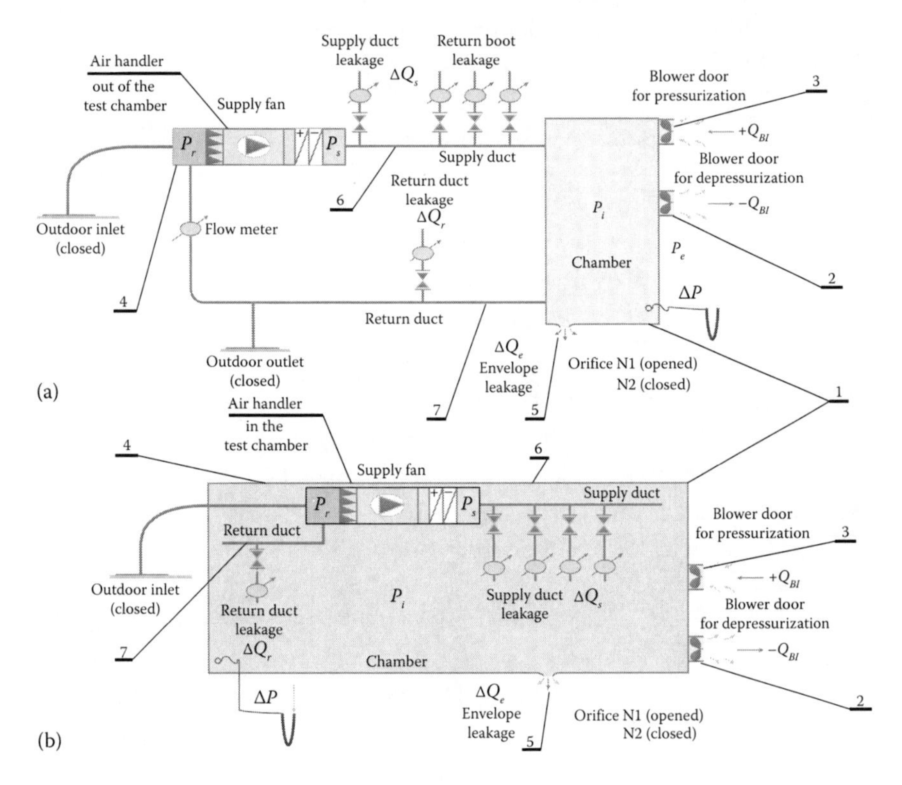

FIGURE 7.20
Equipment arrangement: (a) AH out of the thermal zone; (b) AH inside the thermal zone: (1) test chamber; (2) return fan; (3) supply fan; (4) air handler; (5) measuring equipment; (6) supply ducts; (7) return duct.

(position 1). The arrangement of the equipment allows for the successive supercharge and discharge of the test chamber (the air-conditioned area) through the ducts performing pressure control. In the technical literature, this method is known as the *Delta-Q method* (Dimitrov 2008). The used duct system of air supply is similar to the typical AC (all air) ducts system. It has eight supply ducts and a single return duct. Special devices are also mounted, controlling the air leaks in the experimental stand amounting to 5%–60%. Note that leakage can be described using Equation 7.41, where the value of the exponent n is guaranteed to be 0.6 (Dimitrov 2008). Air leakage through the envelope (infiltration–exfiltration) is controlled within a wide range (0.24 ÷ 2.4 m^3/s at 25 Pa) by six holes with calibrated dimensions.

7.6.2 Modified Balance Equation of Leaks in Air Ducts, Air-Conditioning Station, and Envelope

A new form of ΔQ equations is proposed in the present study, accounting for the balance of air delivery (see Figure 7.20a and b):

At air handler *on*, we have

$$Q_{BI} + \Delta Q_e + \Delta Q_s + Q_s + \Delta Q_r - Q_r = 0 \tag{7.42}$$

(assuming that $Q_s = Q_r$).

$$Q_{BI} + \Delta Q_e + \Delta Q_s + \Delta Q_r = 0, \tag{7.43}$$

where

Q_{BI} m^3/s is the air delivery of the blower door
$Q_s = Q_r$ m^3/s is the air handler supply/return volume rate
ΔQ_e m^3/s is the rate of leakage through the envelope
ΔQ_s m^3/s is the leakage of the supply duct
ΔQ_r m^3/s is the leakage of the return duct

At air handler *off*, ΔQ_e, ΔQ_s, and ΔQ_r can be expressed as (see Figure 7.20a)

$$\Delta Q_e = C_e \left(p_i - p_e\right)^{n_e} \tag{7.44}$$

$$\Delta Q_r = C_r \left(p_r - p_e\right)^{n_r} \tag{7.45}$$

$$\Delta Q_s = C_s \left(p_s - p_e\right)^{n_s}. \tag{7.46}$$

Equations 7.44, 7.45, and 7.46 are put in Equation 7.43, finding that

$$Q_{Bl} + C_e (p_i - p_e)^{n_e} + C_r (p_r - p_e)^{n_r} + C_s (p_s - p_e)^{n_s} = 0 \quad (7.47)$$

At air handler *off*, we have $Q_{Bl} = Q_{off}$ and $p_s = p_r = p_i$.
Then, it follows from Equation 7.47 that

$$Q_{off} + C_e (p_i - p_e)^{n_e} + C_r (p_i - p_e)^{n_r} + C_s (p_i - p_e)^{n_s} = 0$$

Using the substitution: $(p_i - p_e) = \Delta p$, the balance equation becomes

$$Q_{off} + C_e \cdot \Delta p^{n_e} + C_r \cdot \Delta p^{n_r} + C_s \cdot \Delta p^{n_s} = 0. \quad (7.48)$$

At air handler *on*, we have $Q_{Bl} = Q_{on}$ and $p_s \neq p_r \neq p_i$.
Then Equation 7.47 reads

$$Q_{on} + C_e (p_i - p_e)^{n_e} + C_r (p_r - p_e + p_i - p_i)^{n_r} + C_s (p_s - p_e + p_i - p_i)^{n_s} = 0.$$

Transforming the aforementioned relation, we obtain

$$Q_{on} + C_e (p_i - p_e)^{n_e} + C_r \left[(p_r - p_i) + (p_i - p_e)\right]^{n_r} + C_s \left[(p_s - p_i) + (p_i - p_e)\right]^{n_s} = 0, \quad (7.49)$$

and performing the substitution $p_r - p_i = \Delta p_r$ and $p_s - p_i = \Delta p_s$ the balance equation (Equation 7.49) takes the following form:

$$Q_{on} + C_e \Delta p^{n_e} + C_r (\Delta p_r + \Delta p)^{n_r} + C_s (\Delta p_s + \Delta p)^{n_s} = 0. \quad (7.50)$$

Subtract Equation 7.48 from Equation 7.49:

$$(Q_{on} - Q_{off}) + C_r \left[(\Delta p_r + \Delta p)^{n_r} - \Delta p^{n_r}\right] + C_s \left[(\Delta p_s + \Delta p)^{n_s} - \Delta p^{n_s}\right] = 0 \quad (7.51)$$

Coefficients C_r and C_s are found from Equation 7.42:

$$C_r = \frac{\Delta Q_r}{(p_r - p_e)^{n_r}} = \frac{\Delta Q_r}{(p_r - p_e + p_i - p_i)^{n_r}} = \frac{\Delta Q_r}{\left[(p_r - p_i) + (p_i - p_e)\right]^{n_r}} = \frac{\Delta Q_r}{(\Delta p_r + \Delta p)^{n_r}},$$

and from Equation 7.45:

$$C_s = \frac{\Delta Q_s}{(p_s - p_e)^{n_s}} = \frac{\Delta Q_s}{(p_s - p_e + p_i - p_i)^{n_s}} = \frac{\Delta Q_s}{\left[(p_s - p_i) + (p_i - p_e)\right]^{n_s}} = \frac{\Delta Q_s}{(\Delta p_s + \Delta p)^{n_s}}$$

Substitute (Q_{on}–Q_{off}) for ΔQ and rewrite Equation 7.51:

$$\Delta Q = \frac{\Delta Q_r}{(\Delta p_r + \Delta p)^{n_r}}\left[\Delta p^{n_r} - (\Delta p_r + \Delta p)^{n_r}\right] + \frac{\Delta Q_s}{(\Delta p_s + \Delta p)^{n_s}}\left[\Delta p^{n_s} - (\Delta p_s + \Delta p)^{n_s}\right].$$

Then, we obtain a new modified equation for ΔQ:

$$\Delta Q = \Delta Q_r\left[\left(\frac{\Delta p}{(\Delta p_r + \Delta p)}\right)^{n_r} - 1\right] + \Delta Q_s\left[\left(\frac{\Delta p}{(\Delta p_s + \Delta p)}\right)^{n_s} - 1\right], \tag{7.52}$$

where

ΔQ_s are leak rates in the supply duct
ΔQ_r leak rates in the return duct.

The total Delta-Q balance equation, which should be correctly used in practice, takes the following forms:

$$\Delta Q = \alpha \Delta Q_r + \beta \Delta Q_s, \tag{7.53}$$

where α and ß are coefficients, shown in Table 7.9 depending on the location of the air handler (see Figure 7.20a and b). The modified equation (Equation 7.53) does not oppose the basic algorithm and the basic idea of the method ΔQ, but expresses only one of its improvements.

7.6.3 Delta-Q Procedure: Data Collection and Manipulation

The Delta-Q test is based on measuring the change of the air rate through the duct leaks under different pressure (Dimitrov 2008). Pressure difference has been created by pressurizing and depressurizing the whole building (including the ducts) by means of two fans–type blower doors (positions 2

TABLE 7.9

Coefficients α and ß of the Modified Delta-Q Balance Equations

Location of the Air Handler	α	ß
On top of the roof or in the basement (outside of the thermal zone)	$\left[\left(\frac{\Delta p}{(\Delta p_r + \Delta p)}\right)^{n_r} - 1\right]$	$\left[\left(\frac{\Delta p}{(\Delta p_s + \Delta p)}\right)^{n_s} - 1\right]$
Inside the thermal zone	$\left[\left(\frac{\Delta p}{(\Delta p_r)}\right)^{n_s} - 1\right]$	$\left[\left(\frac{\Delta p}{(\Delta p_s)}\right)^{n_s} - 1\right]$

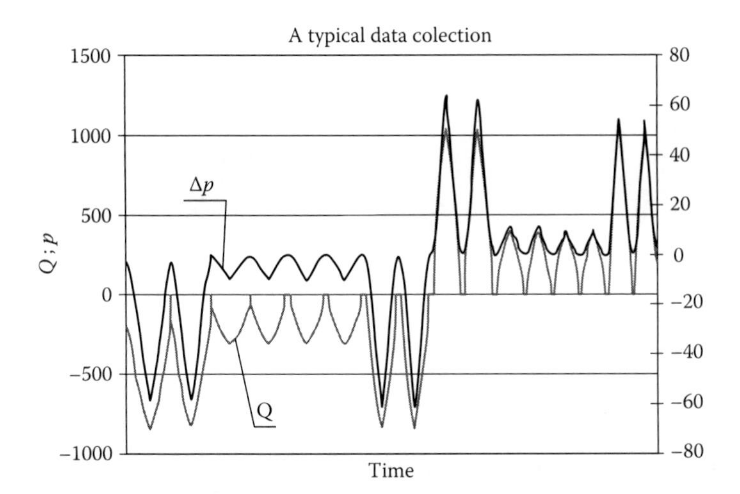

FIGURE 7.21
The primary data obtained by the measuring station according to the time.

and 3; Figure 7.20). They generate pressure within the range $-50 \leq p \leq +50$ Pa by step of 10 Pa. The procedure operates under two regimes—*handler off* and *handler on*.

One and the same pressure-measuring station is used to measure pressure difference at two points. The difference between the two flow rates gives the Delta-Q of the corresponding pressure difference Δp between both sides of the envelope—see more details in (Dimitrov 2008). An example illustrating the distribution of the data collected for leakage rates Q within a full measurement cycle (overpressure and underpressure) is given in Figure 7.21, corresponding to mode *AC handler on*. The data set comprises about 10,000 values of pressure difference Δp and leakage rate Q. These data are used in the balance Equation 7.53, applying statistical averaging. Thus, the total leakage rate ΔQ and leaks in the supply ΔQ_s and return ΔQ_r ducts, respectively, are found.

Primary data, plotted in Figure 7.22, show a data set comprising a full cycle *depressurization–pressurization*. *Quasi-periodicity* in the distribution of the collected data for $Q = Q(t)$ and $\Delta p = p(t)$ is observed: from zero to the extrema of Q_{max} and p_{max} at t_{max} and then back to zero.

Considering the observed *quasi-periodicity*, an idea about the use of Q_{max}, p_{max}, and t_{max} *as internal scales* of each cycle emerged.

The task was to obtain similar distributions for each cycle (so-called modeling) expressed in the coordinate system $\bar{Q} - \bar{t}$. Here, $\bar{Q}$ and $\bar{t}$ are dimensionless delivery and time $\bar{Q} = Q/Q_{max}$ and $\bar{t} = t/t_{max}$, at points of each cycle (from t_b to t_e).

Calculations over 4230 points were made using specialized software for statistical processing of experimental data, named *Stata* (Dimitrov 2008). The results for $3 \leq m \leq 8$ are shown in Figure 7.23 (similar results were also found

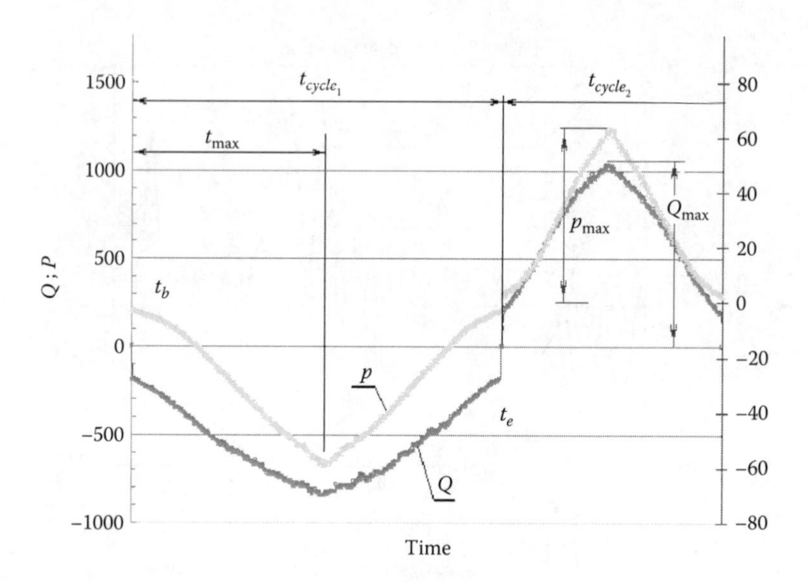

FIGURE 7.22
A full cycle, consisting of depressurization and pressurization.

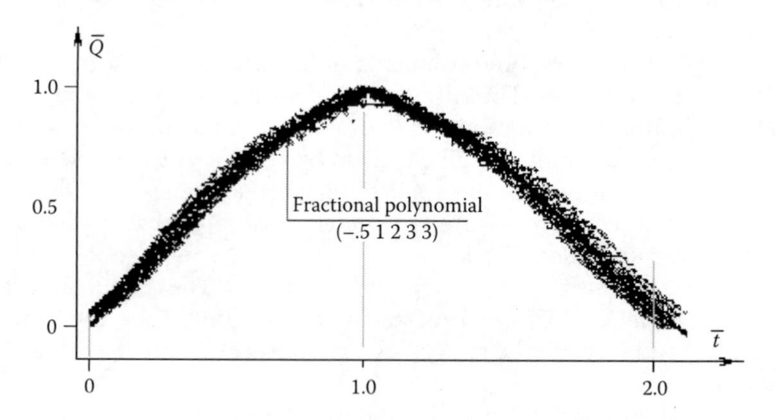

FIGURE 7.23
Distribution Stata plot, corresponding to 4230 points.

for the distribution of $3 \le m \le 8$). Function *Stata*, called Fracpoly, was used to describe that type of profiles.

The designed model has a coefficient of reliability $3 \le m \le 8$ within the range $0.9834 \le R \le 0.9889$, thus proving its high degree of reliability. Hence, the final form of the statistical model reads

$$\bar{Q} = 0.33 - 0.0000741\bar{t}^{-0.5} + 0.5126124\bar{t} + 2.431693\bar{t}^{2} - 2.210527\bar{t}^{3}$$
$$+ 1.25671\bar{t}^{3} \cdot \log \bar{t} \;\; Integral = 1.3686. \tag{7.54}$$

Function "POLY_INTERPOLATE" of ACC DERIVE 6 was used as an alternative option of model design:

$$\bar{Q} = 0.3833733t^4 - 1.55\bar{t}^4 + 1.2237\bar{t}^2 + 0.6997\bar{t}^2 - 2.210527\bar{t} + 0.206.$$
$$Integral = 1.34909. \qquad (7.55)$$

The difference between the two models is negligible, and it can be ignored as proved by the difference Δ between of the integrals of Equations 7.54 and 7.55. It amounts to

$$\Delta = \frac{1.3686 - 1.34909}{1.3686} 100 = 0.03\%.$$

This fact was encouraging in carrying the next step of finding an automodeled distribution of the collected experimental data.

7.6.4 Normalization of the Collected Data

Normally, the experimental data are presented in the coordinate system Q-p using dimensional data. There are certain difficulties in designing an appropriate statistical model due to the occasional disturbances of the collected data. Different types of physical or mathematical filters are used to create statistically reliable models. We propose an alternative approach, which uses dimensionless data normalized by internal scales Q_{max} and p_{max} belonging to the same cycle. The advantage of that data presentation is that it minimizes the effects of random factors and false measurements during the experiment. The collected data are inspected visually or by using an appropriate algorithm, cycle by cycle, looking for extreme values of Q_i and p_i in each jth cycle. Thus, we select the internal scales Q^j_{max} and p^j_{max}. Then, data from cycles Q_i and p_i are normalized by means of the already selected internal scales Q^j_{max} and p^j_{max}:

$$\bar{Q}_i = \left(\frac{Q_i}{Q^j_{max}} \right)_{j=const} \quad u \ \bar{p}_i = \left(\frac{p_i}{p^j_{max}} \right)_{j=const} \quad (i = 1 - n_j). \qquad (7.56)$$

The results are presented in the ranges $-1 \le \bar{Q}_i \le +1; -1 \le \bar{p}_i \le +1$. Here, n_j is the number of data in the jth cycle, as $j = 1 \div m$ (m is the number of cycles in the series). In our case (see Figure 7.21), we have $m = 16$ cycles, but n_j acquires different values (about 250 ÷ 300 ps), and it depends on the settings of the measuring equipment.

The normalized data are in the range [–1, 1]. All collected data were processed using the Stata software and plots are shown in Figure 7.24. Note that data are compactly distributed, proving the effectiveness of the method of data presentation.

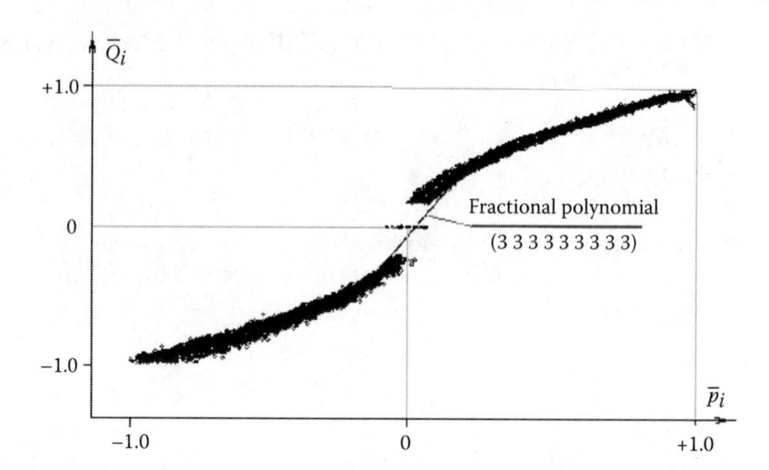

FIGURE 7.24
Stata graph of the normalized *Q*-*p* data distribution.

Results: Design of a *step-by-step* statistical model of delivery prediction.

The first step of the proposed statistical method is the design of the $\overline{Q}-\overline{p}$ model. The regression procedure incorporated in the Stata Fracpoly could aid the selection of an appropriate model with a fairly high degree of reliability ($0.98 \leq R \leq 0.993$). The imported DERIVE model has the following structure:

$$\overline{Q} = 1.8\overline{p} - 0.02\overline{p}^2 - 2.47\overline{p}^3 + 0.04\overline{p}^4 + 2.8\overline{p}^5 - 0.02\overline{p}^6 - 1.17\overline{p}^7, \quad (7.57)$$

which is more convenient for use (see Figure 7.25).

The second step in the suggested method consists in determining the *inner scale* Q_{max} and p_{max}. A fourth degree polynomial is suitable for that purpose,

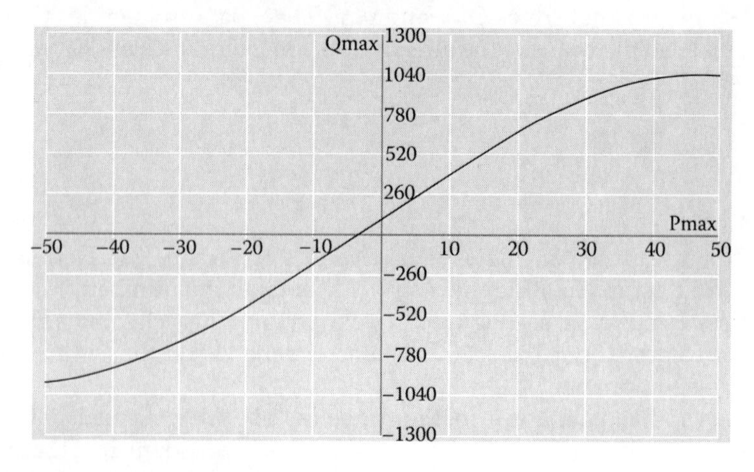

FIGURE 7.25
Graph of Q_{max}–p_{max} relation.

requiring the introduction of the coordinates of at least four points. We must consider also both ends of the measured interval since the relationship is nonlinear (if belongs to the recommended values would be).

Doubling the point number would increase the reliability of the *Stata* model. We found the following form of the *Stata* model:

$$Q_{max} = 30.3 + 1954.91p_{max}^3 - 5175.171p_{max}^3 \log p_{max}. \tag{7.58}$$

The improvement by means of ACC "DERIVE" yields

$$\bar{Q}_{max} = -10^{-5} p_{max}^4 - 0.004p_{max}^3 + 30p_{max} + 100, \tag{7.59}$$

and the plot is shown in Figure 7.25.

The third step of the proposed method is the design of model $Q = f(p)$, which would enable one to predict the leakage rate Q. For this purpose, the definition expression for $\bar{Q} = Q/Q_{max}$ is put in Equation 7.57:

$$Q = Q_{max} * \bar{Q} = Q_{max} * \left(1.8\bar{p} - 0.02\bar{p}^2 - 2.47\bar{p}^3 + 0.04\bar{p}^4 + 2.8\bar{p}^5 - 0.02\bar{p}^6 - 1.17\bar{p}^7\right).$$

So, using Equation 7.59 and $\bar{p} = p/p_{max}$ we obtain the following structure of the model:

$$Q = \left(-10^{-5}\ p_{max}^4 - 0.004\ p_{max}^3 + 30\ p_{max} + 100\ \right) * \tag{7.60}$$

$$\left(1.8\frac{p}{p_{max}} - 0.02\left(\frac{p}{p_{max}}\right)^2 - 2.47\left(\frac{p}{p_{max}}\right)^3 + 0.04\left(\frac{p}{p_{max}}\right)^4\right.$$

$$\left. + 2.8\left(\frac{p}{p_{max}}\right)^5 - 0.02\left(\frac{p}{p_{max}}\right)^6 - 1.17\left(\frac{p}{p_{max}}\right)^7\right).$$

Performing transformation via ACC DERIVE, we may use Equation 7.60 to predict the total leak rates. This is illustrated in Figure 7.26 in the coordinate system Q-p, considering two limit cases: $p_{max} = -50$ Pa and $p_{max} = +50$ Pa. It is important to note that this model works properly only in the particular interval [–50 ÷ +50 Pa] (extrapolations should be avoided).

The model introduced by Equation 7.60 should be designed accounting for *handler-off* mode. The difference Delta-Q can be determined easily by subtracting two similar models valid for *handler-off* and *handler-on* regimes. That difference is plotted in Figure 7.27.

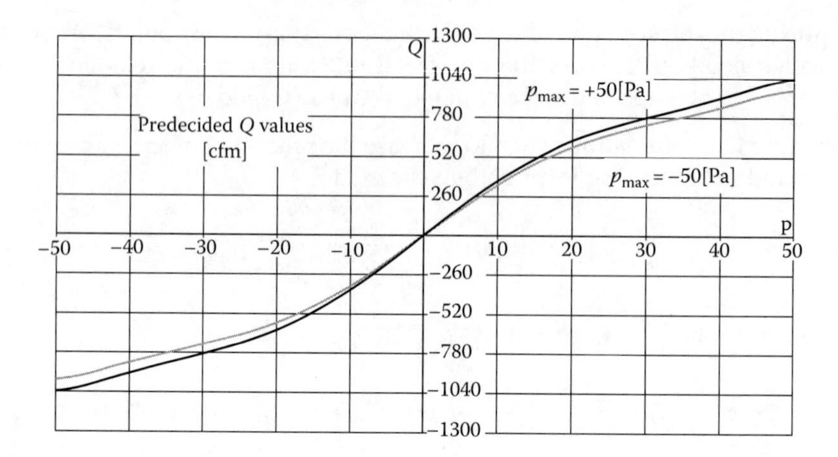

FIGURE 7.26
Plot of the model for the prediction of Q-p, obtained by "DERIVE 6."

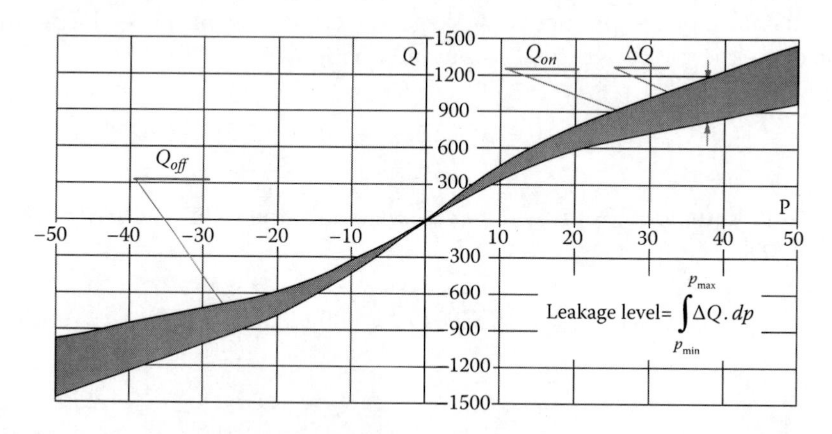

FIGURE 7.27
"DERIVE 6" plot of the ΔQ subtraction.

7.7 Mathematical Model of the Environmental Sustainability of Buildings

7.7.1 General Structure of the Model

We apply in this paragraph the principle of *superposition of ecological effects* over the building ecological sustainability in order to describe formally building environmental sustainability. Conforming to that principle, building ecological harmonicity presents a sum of sustainability shares introduced by the following independent ecological factors: used building materials $IES_{Buil} = ES_{Buil} / ES_{Buil}^{ESt} \Rightarrow 0 \le IES_{Buil} \le 1.0$; landscape and settlement environment $IES_{Buil} = ES_{Buil}/ES_{Buil}^{ESt} \Rightarrow 0 \leq IES_{Buil} \leq 1.0$; internal comfort and healthy environment $IES_{Buil} = ES_{Buil}/ES_{Buil}^{ESt} \Rightarrow 0 \leq IES_{Buil} \leq 1.0$; energy and atmosphere $IES_{Buil} = ES_{Buil}/ES_{Buil}^{ESt} \Rightarrow 0 \leq IES_{Buil} \leq 1.0$; water management $IES_{Buil} = ES_{Buil}/ES_{Buil}^{ESt} \Rightarrow 0 \leq IES_{Buil} \leq 1.0$; employed innovation technologies ($IES_{Buil} = ES_{Buil}/ES_{Buil}^{ESt} \Rightarrow 0 \leq IES_{Buil} \leq 1.0$); and local priorities $IES_{Buil} = ES_{Buil}/ES_{Buil}^{ESt} \Rightarrow 0 \leq IES_{Buil} \leq 1.0$.

To assess the building environmental sustainability $IES_{Buil} = ES_{Buil}/ES_{Buil}^{ESt} \Rightarrow 0 \leq IES_{Buil} \leq 1.0$, it seems + useful to introduce it as a profit function, formally presented in the form

$$IES_{Buil} = \frac{ES_{Buil}}{ES_{Buil}^{ESt}} \Rightarrow 0 \leq IES_{Buil} \leq 1.0, \tag{7.61}$$

called *building general ecological model.* Here $IES_{Buil} = ES_{Buil}/ES_{Buil}^{ESt} \Rightarrow 0 \leq IES_{Buil} \leq 1.0$ is the estimation of the partial tribute of the jth ecological factor.

In its character, the profit function $IES_{Buil} = ES_{Buil}/ES_{Buil}^{ESt} \Rightarrow 0 \leq IES_{Buil} \leq 1.0$ cannot be quantitatively expressed, but it has a qualitative estimating character. There are no quantitative measurers of the factors determining the environmental sustainability of the whole building, but only subjective assessments. For instance, factor *internal comfort and healthy environment* attends to the subjective assessment, and it operates at *comfort and healthy* level and *discomfort and unhealthy* level. Factor *local priorities* depends on political and regional strategic subjective tendencies and ideas, and it also operates at two levels—*approval* and *disapproval.*

To treat qualitatively those *quantitative* arguments, we use selected *standards* for comparison. Consider some characteristics that determine building ecological sustainability. Consider also *an actual building* and that are called *a standard of ecologically harmonic building* or *ecological standard* in brief. Then, the *measurement procedure* consists in *juxtaposition* of those characteristics. The assessment of the ecological sustainability of the actual building is based on the deviation of the characteristics of the object under study from the standard ones. Mathematically, this is written as a ratio between the *profit functions* $IES_{Buil} = ES_{Buil}/ES_{Buil}^{ESt} \Rightarrow 0 \leq IES_{Buil} \leq 1.0$ and $IES_{Buil} = ES_{Buil}/ES_{Buil}^{ESt} \Rightarrow 0 \leq IES_{Buil} \leq 1.0$ of the building ecological sustainability, on the one hand, and the *ecological standard,* on the other hand. We will call it an *index of building ecological sustainability*:

$$IES_{Buil} = \frac{ES_{Buil}}{ES_{Buil}^{ESt}} \Rightarrow 0 \leq IES_{Buil} \leq 1.0. \tag{7.62}$$

Buildings with $ES_{Buil}^{ESt} = \sum_{1,7} PES_j^{ESt} = \sum_{1,7} a_j^{ESt} ES_j^{ESt} \Rightarrow Mat, W, EE, IC, L, Inn,$ RP are *ecologically harmonious* to a degree corresponding to the respective *ecological standard* (buildings, whose index is significantly lower, significantly deviate from the standard. *They should be renovated or simply destroyed as hazardous*).

We use as an *ecological standard* a building, which brings admissibly *low ecological damage to the landscape, atmosphere, material and energy sources, global climate, and ozone layer.* Admissible levels are regulated conforming to local regulations of environment preservation and to the European and interstate regulation norms. The mathematical expression of the ecological standard is as follows:

$$ES_{Buil}^{ESt} = \sum_{1,7} PES_j^{ESt} = \sum_{1,7} a_j^{ESt} ES_j^{ESt} \Rightarrow Mat, W, EE, IC, L, Inn, RP. \qquad (7.63)$$

It is found by applying the definition Equation 7.61 to appropriately chosen levels* of the factorial standard arguments. We use the following notations:

- ES_{Buil}^{ESt}—assessment value of the profit function of the ecological standard.
- PES_j^{ESt}—assessment of the partial tribute of the *j*th factor of the ecological building standard.
- $a_j^{Est} = \dfrac{PES_j^{ESt}}{ES_{Buil}^{ESt}} = \dfrac{PES_j^{ESt}}{\sum_{j=1,7} PES_j^{ESt}}$—values of the *j*th factor accounting for its tribute to the total ecological sustainability of the standard. Note that Equation 258 yields the important conclusion that $\sum_{j=1,7} a_j^{ESt} = 1.0$.
- ES_j^{ESt}—ecological sustainability of the standard generated by the *j*-th factor.

In what follows, we will call expression (Equation 7.63) a model of the building ecological standard, while the weighting ratios $a_j^{ESt} \Rightarrow j = 1,7$ form a *basic set of weighting factors of the ecological standard.*

The general index of the ecological sustainability

$$IES_{Buil} = \frac{ES_{Buil}}{ES_{Buil}^{ESt}} = \frac{\sum_{j=1,7} PES_j}{\sum_{j=1,7} PES_j^{ESt}} = \frac{PES_M}{\sum_{j=1,7} PES_j^{ESt}} + \ldots + \frac{PES_{EE}}{\sum_{j=1,7} PES_j^{ESt}}$$

$$+ \frac{PES_{RP}}{\sum_{j=1,7} PES_j^{ESt}} \Rightarrow j = Mat, W, EE, IC, L, Inn, RP$$

defined by Equation 7.62,

* Depending on building's function, they can be empirically found.

can now be presented as

$$IES_{Buil} = \frac{ES_{Buil}}{ES_{Buil}^{ESt}} = \frac{\sum_{j=1,7} PES_j}{\sum_{j=1,7} PES_j^{ESt}} = \frac{PES_M}{\sum_{j=1,7} PES_j^{ESt}} + \ldots + \frac{PES_{EE}}{\sum_{j=1,7} PES_j^{ESt}}$$

$$+ \frac{PES_{RP}}{\sum_{j=1,7} PES_j^{ESt}} \Rightarrow j = Mat, W, EE, IC, L, Inn, RP. \quad (7.64)$$

Performing transformation of the *j*th term $\left(PES_j \Big/ \sum_j PES_j^{ESt} \right) * \left(PES_j^{ESt} / PES_j^{ESt} \right) = a_j^{ESt} * IES_j$, Equation 7.64 takes the following *final form*:

$$IES_{Buil} = \sum_{j=1,7} a_j^{ESt} * IES_j \Rightarrow j = Mat, W, EE, IC, L, Inn, RP. \quad (7.65)$$

In an expanded form, the *general index* of building ecological sustainability IES_{Buil}, presenting also the building ecological model, reads

$$IES_{Buil} = a_{Mat}^{ESt} * IES_{Mat} + a_W^{ESt} * IES_W + a_{EE}^{ESt} * IES_{EE} + a_{IC}^{ESt} * IES_{IC}$$

$$+ a_L^{ESt} * IES_L + a_{Inn}^{ESt} * IES_{Inn} + a_{RP}^{ESt} * IES_{RP}. \quad (7.66)$$

There IES_{Buil} comprises seven different *factor indices* $IES_j = PES_j / PES_j^{ESt}$. (The tributes of the seven factors are *used building materials, landscape, energy and atmosphere, waters; internal comfort and healthy environment, innovations, local priorities*). It is important for the integral estimation of the whole building.

The proposed expression (7.66) for the calculation of the *index of building ecological sustainability* $a_j^{ESt} = PES_j^{ESt} / ES_{Buil}^{ESt} \Rightarrow j = Mat, W, EE, IC, L, Inn, RP$ comprises seven weighting factors carrying the *genetic code of the ecological standard* since they express *the tribute* of the *j*th factor to the sustainability of the building standard

$$a_j^{ESt} = \frac{PES_j^{ESt}}{ES_{Buil}^{ESt}} \Rightarrow j = Mat, W, EE, IC, L, Inn, RP.$$

More details on the discussed issues on ecological sustainability can be found in references (Димитров and Назърски 2011; Димитров et al. 2011; Димитров 2011b; Димитров 2012, 2013a,b; Назърски et al. 2011a; Назърски et al. 2011b; Dimitrov 2014).

7.7.2 Selection of an Ecological Standard: Table of Correspondence

Buildings designed pursuant to ecology requirements have strongly reduced overhead expenses for energy and fresh water, their market price rises, emissions of carbon dioxide resulting from energy generation drop by average 352 $1.0 \geq IES_{Buil} \geq 0.725$, inhabitants' life grows better and healthier and taxes are reduced raising the image of investors, owners, and builders. The adequate selection of a building ecological standard is of essential importance to reach such results. The standard should reflect the specific national characteristics, local conditions and resources, function of the building under consideration, and good world practices.

As is known (Димитров and Назърски 2011), two systems of rating the building ecological sustainability grew popular in the recent 15 ÷ 20 years, being in use in foreign and national architectural practice.* These are *BREEAM* (building research establishment environmental assessment method) and *LEED* (leadership in energy and environmental design). They are user-oriented and employed by state and communal authorities to manage constructional investments. Their applicability to the certification of constructional projects and in pursuing judicious state policy is proved by the increasing number of buildings, certified and supported by the state and communal administrations of more than *90 countries* worldwide (including EU members). The growing tendency of using the LEED method in the period 1998 ÷ 2010 is illustrated in Figure 7.28, while the number of buildings certified in conformity with BREEAM is more then 10^5, and buildings awaiting conclusion of the procedure are about half a million.

Regardless of the fact that the *two rating methods* emerged 6 years apart from each other, they share qualitatively identical ecological factors, which are compared in Figure 7.29.

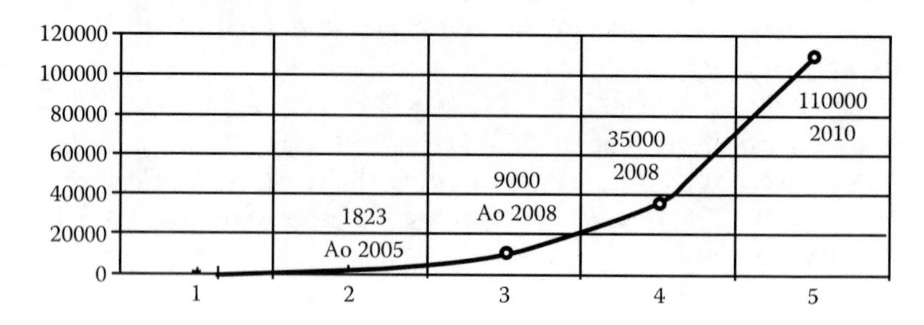

FIGURE 7.28
Number of buildings certified in the United States according to the LEED method, period 1998–2010.

* Known but not so popular are also GreenStar (Au), CASBEE (Jap), HQE (Fr), (PEARLS—*The Pearls Rating System for Estidata),* and DGNB (BRG established in 2008). See a comparison between the two systems in the Appendix.

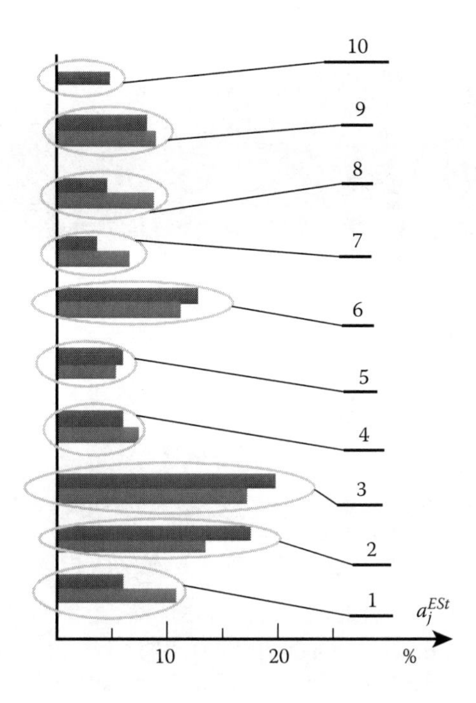

FIGURE 7.29
Weighting coefficients accounting for the tribute of the *j*th factor to the total ecological sustainability of the standard, employing methods BREEAM and LEED: (1) management; (2) indoor; (3) energy; (4) transport; (5) waters; (6) materials; (7) wastes; (8) earth and ecology; (9) pollutants; (10) innovations.

Plots show that 5 out of 10 factors have close weighting coefficients, while those of the remaining factors significantly differ from each other. Note that *differences exceed* 20%, and for three factors—1 (*management*), 7 (*wastes*), and 8 (*earth and ecology*)—the differences amount to 100%. Factor 10 (*innovation*) lacks at all in *BREEAM* since it is newly introduced in the *LEED* method.

The availability of those differences naturally results in differences between the ecological ratings of identical buildings (as is with the Exhibition Hall *Herman Miller* or the *Bakery of Van de Kamp*—part of the historical heritage—see Figure 7.30. Besides, it is established that ratings pursuant to *LEED* are lower than those pursuant to *BREEAM*.

At the end of the last decade, however, *the ecological standards* of both methods approached each other. As is seen in Table 7.10, they practically coincide. This means that eventual difference between the ecological ratings of the two methods may be due to auditors' subjective errors. At the same time, the last version of the *LEED* method from 2009 (*V*3) displays a tendency of accounting for the building function— buildings were divided into eight categories. New ecological standards concerning various types of buildings

FIGURE 7.30
Building assessments performed using old versions of methods LEED and BREEAM differ from each other (the higher points are in LEED).

TABLE 7.10
New Versions of the Systems LEED 3.0 (LEED 2009), BREEAM 2008 (2009) Make Standards Closer—They Use Identical Ecological Factors with Close Weighting Coefficients

LEED		BREAM	
Energy	33%	Energy	33%
Materials	13.5%	Materials	13.5%
Site selection	24.5%	Site selection and ecology	20.5%
Indoor environmental quality	14%	Indoor environmental quality	13%
Water	5.5%	Water	2.5%
Innovation	6.5%	Innovation	6.5%
Regional priority	4.0%	Facility management	12%

Source: Modified from Elgendy, K., Comparing Estidama's Perl rating system to LEED and BREEAM, *Carbon*, Middle East Sustainable Citees, 2010.

also emerged—for residential and commercial buildings, schools, existing buildings, and clinics. Table 7.11 presents the maximal number of points assigned to the buildings in the list, and the respective ecological model corresponding to the building function is implicitly accounted for. For instance, factor *indoor environmental quality* is larger in the school ecological model—it gets 1.27 times more points than the similar factor valid for existing and new buildings, but of another type (the third and the fifth columns). The situation with factor *material and resources* is similar. Points for commercial and new buildings are 1.4 times more than those for existing buildings.

TABLE 7.11

Ecological Models Used by the Method LEED, *V*3 for Different Types of Building Activity—Maximal Number of Points

Factor of Ecological Sustainability	Existing Buildings *V*3	Commercial Buildings *V*3	New Buildings *V*3	School Buildings *V*3
Sustainable sites	26	21	26	24
Water efficiency	14	11	10	11
Energy and atmosphere	35	37	35	33
Material and resources	10	14	14	13
Indoor environmental quality	15	17	15	19
Innovation and design	6	6	6	6
Regional policy	4	4	4	4
Total	110	110	110	110

The weighting coefficients of the ecological standards are calculated as ratios between points assigned to buildings as conforming to the subsequent factor, on the one hand, and the total number of points assigned to buildings certified pursuant to the *LEED* method, on the other hand. Those coefficients are given in Table 7.12. They prove the established tendency of accounting for the building-specific function participating in the ecological model.

The design of a full list of adopted and approved ecological models is beyond the scope of this book. Hence, we shall concentrate only on those factors, which concern the operation of the building energy system and its components. These factors are *energy and atmosphere* and *indoor environmental quality*. The analysis of their share shows that they are of crucial importance for building ecological sustainability—their share is $0.3 \div 0.34$ and $0.14 \div 0.17$, respectively, for different buildings (see Table 7.13).

However, we submit mathematical ecological models, which have been checked in building practical certifications similar to those whose weighting coefficients are specified in Table 7.12. On that basis, we use the mathematical models in Table 7.13 introducing building ecological sustainability via expressions (7.67 through 7.71).

Regarding the building type, we calculate the *index of the ecological sustainability of a certain building* $1.0 \geq IES_{Buil} \geq 0.725$ after applying a procedure of finding *the seven factorial indices*—$1.0 \geq IES_{Buil} \geq 0.725$ discussed in what follows and using the mathematical expressions in Table 7.17. To assign a class of ecological sustainability to a building, however, one must use an appropriate rating scale (we use in our study the so-called *table of correspondence*—see Table 7.14).

TABLE 7.12

Weighting Coefficients of Ecological Models Used by the LEED, *V*3 Method to Rate Different Buildings

	Factor of Ecological Sustainability	Weighting Coefficients			
		Existing Buildings *V*3	Commercial Buildings *V*3	New Buildings *V*3	School Buildings *V*3
1	Sustainable sites	0.23	0.19	0.23	0.22
2	Water efficiency	0.13	0.1	0.09	0.11
3	Energy and atmosphere	0.32	0.34	0.32	0.3
4	Material and resources	0.09	0.13	0.13	0.12
5	Indoor environment quality	0.14	0.15	0.14	0.17
6	Innovation and design	0.05	0.05	0.05	0.05
7	Regional policy	0.04	0.04	0.04	0.03
	Total	1.0	1.0	1.0	1.0

TABLE 7.13

Mathematical Ecological Models of Buildings with Different Structure, Derived from the General Ecological Model, which Employs Indices of Ecological Sustainability

(7.67) For existing buildings

$$IES_{Buil} = 0.079 * IES_{Mat} + 0.13 * IES_W + 0.32 * IES_{EE} + 0.14 * INS_{IC} + 0.23 * IES_L + 0.05 * IES_{Inn} + 0.04 * IES_{RP}$$

(7.68) For commercial buildings

$$IES_{Buil} = 0.13 * IES_{Mat} + 0.1 * IES_W + 0.34 * IES_{EE} + 0.15 * INS_{IC} + 0.19 * IES_L + 0.05 * IES_{Inn} + 0.04 * IES_{RP}$$

(7.69) For new buildings

$$IES_{Buil} = 0.13 * IES_{Mat} + 0.09 * IES_W + 0.32 * IES_{EE} + 0.14 * INS_{IC} + 0.23 * IES_L + 0.05 * IES_{Inn} + 0.04 * IES_{RP}$$

(7.70) For schools

$$IES_{Buil} = 0.12 * IES_{Mat} + 0.11 * IES_W + 0.3 * IES_{EE} + 0.17 * INS_{IC} + 0.22 * IES_L + 0.05 * IES_{Inn} + 0.04 * IES_{RP}$$

(7.71) For building core and envelope

$$IES_{Buil} = 0.12 * IES_{Mat} + 0.09 * IES_W + 0.34 * IES_{EE} + 0.11 * INS_{IC} + 0.25 * IES_L + 0.05 * IES_{Inn} + 0.04 * IES_{RP}$$

TABLE 7.14

Table of Correspondence

Index of Environmental Sustainability IES_{Buil}	Class of Environmental Sustainability CES_{Buil}	Comparison with US-GBE
$IES_{Buil} \geq 0.725$	A	Platinum
$0.60 \leq IES_{Buil} \leq 0.72$	B	Gold
$0.50 \leq IES_{Buil} \leq 0.59$	C	Silver
$0.4 \leq IES_{Buil} \leq 0.49$	D	LEED certified
$0.3 \leq IES_{Buil} \leq 0.39$	E	
$0.2 \leq IES_{Buil} \leq 0.29$	F	
$IES_{Buil} \leq 0.195$	G	

Note: The index of building ecological sustainability IES_{Buil} is calculated using expressions ($40 \div 44$).

To successfully assign ratings, the proposed rating scale is specifically structured. This means that the high levels of ecological assessment correspond to those found by using well-known and recognized rating systems. The use of the *correspondence table* is shown in Table 7.14. Note that it uses a *system* outlining the degree of building ecological sustainability, which is distinguishable in the EU. For that purpose, seven classes are introduced and denoted by the first letters of the Latin alphabet—*A, B, … G* (the highest class is *A*).

The values of the building index of ecological sustainability $1.0 \geq IES_{Buil} \geq 0.725$ for class *A* are chosen to be within the $1.0 \geq IES_{Buil} \geq 0.725$. They correspond to the *platinum* rating of the LEED *V*3 system. Those for classes *B, C,* and *D* correspond to ratings *gold, silver,* and *LEED certificated*. Three additional classes of ecological sustainability are specified in Table 7.14—*E, F,* and *G*. Adopting them, one can perform a more detailed building classification. The proposed system of building ecological rating is directed to the *European users,* and it is similar to that rating other industrial products offered within the EU. It is clear for investors, designers, and constructors, guaranteeing easy communication between them.

At a design level, the proposed rating system defines the responsibilities of specialized teams of designers in tackling complex ecology-related problems. For example, the satisfaction of the need for friendly *indoor environment* requires successful efforts of designers with different qualification—power engineers, installation engineers (electrical, H&AC, ventilation and plumbing), and designers of indoor planting and specialists in building materials.

At certification level, the proposed methods require a thorough preparation of the building passport and verification of the registered building characteristics. The *methods* of specifying the *factorial indexes of ecological sustainability* comprise the following stages:

1. Calculation of the factorial index *landscape* IES_L, where $0 \leq IES_L \leq 1.0$ (Назърски et al. 2011a)

   This is done by an audit accounting for seven indexes (checks), with a task to register: significant traces on the landscape due to construction activities or overbuild, as well as lack of infrastructure; availability of public transport; change of the variety of species and inhabitants' behavior; usage of rain waters and preservation of the soil humus; steady house painting; pollution of the *night sky*.
2. Calculation of the factorial index *water efficiency* IES_W ($0 \leq IES_W \leq 1.0$)

   This index is bound to three checks establishing: lack of soil pollution, usage of innovative water technologies, and economy of fresh water.
3. Calculation of the factorial index *indoor environmental quality* IES_{IC} ($0 \leq IES_{IC} \leq 1.0$)

   That index is discussed in (Назърски et al. 2011b). We define two requirements that should be met in order to attain optimal parameters of the indoor environment—cleanness and no smoking. We should check the quality of the air supplied from outside; the correspondence between the supplied clean air and the number of inhabitants of the controlled area; availability of a management plan of sustaining clear outdoor environment; usage of building materials with low emission of organic pollutants (*VOC's*); lack of external sources of chemical and physical emissions; control of the system of thermal comfort and that of daylight illumination and sunlight.
4. Calculation of the factorial index *energy and atmosphere* IES_{EE} ($0 \leq IES_{EE} \leq 1.0$)

   Details on the calculation of that index are given in (Димитров 2011b, 2013a,b). We only note that the algorithm consists of two checks and it reflects three requirements for verification of the construction correspondence to the building documentation; performance of functional tests of whether the designed goals have been achievement; implementation of innovations; attaining zero annual energy consumption (links of the building cogeneration system with the C&DES network); and environmentally friendly management of the refrigerant agents of thermopumps and cooling systems. Checks should verify that energy is supplied via innovative technologies and by on-site generation and find whether the consumption of energy from primary sources is optimal.*

5. Calculation of the factorial index *materials and resources* IES_{Mat} (here $0 \leq IES_{Mat} \leq 1.0$)

   An issue requiring supply and use of recyclable materials arises here. We should check the use of local traditional materials preserving the cultural identity; the preparation of a management plan for building postdestruction wastes; the use of recyclable materials including wood; the use of building components fabricated from recycled materials; and the use of materials fabricated at a remote site (1000 km away from the construction site) Назърски et al. (2011a).

6. Calculation of the factorial index *innovation and design* IES_{Inn} ($0 \leq IES_{Inn} \leq 1.0$)

   Two checks should be performed: a check of innovative employment of nontraditional energy technologies and that of designers' professional capability of ecological treatment of construction issues.

7. Calculation of the factorial index *regional priorities* IES_{RP} ($0 \leq IES_{RP} \leq 1.0$).

It should be performed after the issue of a documented act with a three-stage resolution of the local authorities on the construction significance—*useful, necessary, of vital (inevitable) necessity.*

We shall describe in what follows two sets of methods proposed to find the values of the factorial indexes *indoor environmental quality* and *energy and atmosphere*. Those methods are linked with the mathematical models of building ecological sustainability—expressions (Equations 7.67 and 7.71) and Table 7.13.

7.7.3 Comparison of Systems Rating the Ecological Sustainability in Conformity with the General Criteria

Together with the discussed systems *BREEAM* (Great Britain) and *LEED* (the United States), there are a number of other systems rating ecological sustainability of buildings. These are, for instance, GB Tool (Canada), CASBE (2003) (Japan), REKOS and Hakinen (Finland), Green Star (Australia), HQE (France), Pearls (UAE), and DGNB (Germany). All are of regional importance and limited applicability due to two basic reasons:

1. Complex algorithms of application
2. Limited capabilities of communication between end users and auditors due to the involvement of different ecological standards and rating scales

* In the last version of BG_LEED (Димитров 2013b), there is an additional requirement for a zero balanced energy demand by means of on-site generation.

We propose here an analysis of the applicability of four of the most famous methods. In addition, we shall discuss the method proposed in the present study and called metaphorically *Bulgarian landscape, energy and environmental dreams—BG_LEED* using six common criteria for comparison:

1. Distinguish ability from builders' and users' viewpoint
2. Capability of rating transfer
3. Normative commitment with the EU
4. Applicability to different types of constructions
5. Employment of nonecological rating factors, including economical and political
6. Degree of complication from a technological view

First of all, we shall comment data specified in Table 7.15 where the ratings of ecological sustainability employed by different methods are given in a tabular form. It is seen that the highest ecological rating (*outstanding, platinum, five stars, gold, class A*) is assigned to buildings exceeding 70% of the possible maximal rating. The lowest rating (*LEED certificated*) is assigned to buildings exceeding 40%. Besides, the rating scale of US_GBC is the most liberal one and that of DGNB is the most conservative one (its highest ratings are assigned to buildings exceeding 85% of the possible maximum and the lowest ratings to buildings amounting to 45% of the possible maximum). The system with the widest span is that of *BREEAM*, where the maximal rating (*outstanding*) is assigned to buildings exceeding 85% of the possible maximum and the minimal rating (*pass*) to buildings exceeding 30% of the possible maximum. The system proposed here avoids that *disadvantage*. It practically comprises the all possible ratings within the range 20%–100%.

The results of the analysis are shown in Table 7.16. Most of all, note that the systems of ecological rating are important marketing instruments. Hence,

TABLE 7.15

Comparison of the Levels in the Rating Scales of Well-Known Methods

Method	BREEAM (%)	LEED (%)	GreenStar (%)	NGNB (%)	BG_LEED
Maximum	над 85	над 72	75–100	над 80	над 0.725
	70	55–71	60–74	65	0.6–0.71
	55	36–54	45–59	50	0.5–0.59
	45	27–35			0.4–0.49
	30				0.3–0.39
					0.2–0.29
Total	105	110	100	855	1.0

TABLE 7.16

Table of Comparison between the Methods Involving General-Functional Criteria

	Criterion/Version (Year of Introduction)	BREEAM (2008)	US-GBC LEED (*V*3) (2009)	DGNB (2008)	BG_LEED Proposed (*V*2012)
1	Distinguish ability from the viewpoint of the world construction community	Yes	Yes	No	No
2	Rate transfer between the systems	No	No	No	Yes
3	Commitment of EU documents	No	NO	Yes	Yes
4	Usability of buildings of different function	Yes	Yes	No	Yes
5	Use of economical indexes	No	No	Yes	Yes
6	Social and regional indexes	Yes	Yes	Yes	Yes
7	Use friendliness	No	Yes	No	Yes
8	Degree of communicability with EU users	No	No	No	Yes

their approval/disapproval by the end users (investors, owners, moderators) is of crucial importance. It should be a landmark for civil servants preparing normative documents and auditors.

The analysis of the advantages and disadvantages of the existing methods proves that our method of rating the building ecological sustainability is a promising instrument of the design, audit, and management of local investments. That statement is tabulated and verified by the sixth column in Table 7.16. First of all, the method uses notations that local users know from other products, and they are in conformity with the notations approved by the EU. Secondly, the adopted ecological models correspond to the building types, cultural *aura* and functionality. Thirdly, the methods used to find the factorial indices of ecological sustainability are based on national building practices and standards treating the preparation of respective documentation and building passports.

A starting point of the use of a national *rating system* in the assessment of building environmental sustainability can be the *idea of transforming university and school buildings, together with the adjacent areas, into areas of ecological harmony.* It may stimulate teaching and training, scientific research, and the intellectual maturity of children and young people (Димитров and Назърски 2011).

The proposed system may be useful for *state and communal educational administration* in *investments management.* It can also be a unique assistant of architects, designers, and investment auditors in project preparation and passportization of new and reconstruction of existing buildings.

7.8 Conclusion

This book proposes the *author's generalized idea* on the nature of the energy interactions running in the building envelope at microscopic thermodynamic level. It is a basis of the design of a representative list of macroscopic instruments needed to assess the state of the *building envelope* and the running energy transfer.

We formulate a *hypothesis* explaining the mechanism of scatter of solar energy flux within the envelope. It is assumed that solar photons excite oscillation and internal polarization or ionization of the atomic structures of the envelope material. The described mechanism of energy transfer within the envelope is called *model of lagging temperature gradient* (see Figure 2.13). Various macroscopic energy transformations within the envelope are analyzed on that basis. Energy models of *energy exchange within the envelope* have been designed, passing through the following stages:

- An integral form of the gross potential of the field of a three-functional *ETS* is defined as a sum of the internal energy U_{1-2} and the work L_{1-2}^{Gross} done by the generalized force F_{Env} of the surroundings. $L_{1-2}^{Gross} = -Q_{1-2} + L_{M_{1-2}} + L_{E_{1-2}}$ is the final work done when the system passes from state 1 into state 2, both characterized by respective state parameters (V_1, T_1, p_1) and (V_2, T_2, p_2):

$$U_{1-2} - Q_{1-2} + L_{M_{1-2}} + L_{E_{1-2}} = U_{1-2} + L_{1-2}^{Gross} = Const = -\Omega_{Env}. \tag{3.15}$$

- The free energy function is defined as (Ψ_{1-2}),* and it is assessed in the form $\Psi = pv - Ts_0 + Ve_0$ using 6 *macroscopic* thermodynamic characteristics (Equation 3.43). It is treated as a *generalized characteristic of transfer* running in the envelope, instead of treating separately the variation of temperature T, electric potential V, and pressure p at the TDS control borders.
- A new parametric differential equation is designed including the variation of temperature T, electric potential V, pressure p, and entropy reduced to 1*kg* of system mass:

$$Tds_0 - Vde_0 - c_v dT - pdv = 0 \tag{3.18}$$

* We sued the substitution $(\Psi = \Psi_{1-2})$.

The *important steps* of designing the mathematical model of transfer within solid walls are as follows:

- A *differential equation describing the state of the work medium* in *tri-functional* systems is composed (a link between the partial derivatives of the generalized potential difference ψ and the basic state parameters is found; Equation 3.39):

$$\frac{\left(\frac{\partial\psi}{\partial v_0}\right)_{T,V}\left(\frac{dv_0}{de_0}\right)_T+\left(\frac{\partial\psi}{\partial s_0}\right)_{V,p}\left(\frac{ds_0}{de_0}\right)_p}{\left(\frac{\partial\psi}{\partial e_0}\right)_{T,p}}=-1.$$

- A *differential equation describing the state of the work medium* in bi-functional systems is composed (a link between the partial derivatives of the free energy function ψ and the basic state parameters is found):
  - Considering systems undergoing *electric and thermal impacts* Equation 3.39 is modified as

$$\frac{M}{P}\frac{ds_0}{de_0}=-1 \quad \text{or} \quad \frac{de_0}{\left(\frac{\partial\psi}{\partial e_0}\right)_{T,p}}+\frac{ds_0}{\left(\frac{\partial\psi}{\partial s_0}\right)_{V,p}}=0 \tag{3.40}$$

  - While in systems undergoing *electric and mechanical impacts*, only, Equation 3.39 reads

$$\frac{N}{P}\frac{dv_0}{de_0}=-1 \quad \text{or} \quad \frac{de_0}{\left(\frac{\partial\psi}{\partial e_0}\right)_{T,p}}+\frac{dv_0}{\left(\frac{\partial\psi}{\partial v_0}\right)_{T,V}}=0. \tag{3.41}$$

A *new form* of the generalized law of transfer is found—Equation 4.5 ÷ Equation 4.7:

$$h=-[\Gamma]\{\nabla\psi\}=-\begin{bmatrix}\Gamma_\lambda & 0 & 0\\ 0 & \Gamma_\beta & 0\\ 0 & 0 & \Gamma_{\bar{\Omega}}\end{bmatrix}\begin{Bmatrix}(\nabla\psi)_S\\ (\nabla\psi)_v\\ (\nabla\psi)_e\end{Bmatrix},$$

where formulas of Fourier, Fick, and Ohm participate as special cases. The *new forms* of the generalized law of transfer $\vec{h}=-[\Gamma]\vec{n}_\psi\nabla(pv-s_0T+e_0V)$

(Equation 4.7) enable one to use commercial software, supplied with graphical CAD editors, automatic discretizers, and solvers of *FEM* algorithms. It can be upgraded with modules libraries of the physical properties of envelope structural materials (coefficients of heat, mass, and electric conductivity_ λ, β, and $\bar{\Omega}$). Thus, time and efforts are saved, increasing the efficiency of the work of architects and designers of integrated building envelopes;

- New forms of the differential equation of transfer are found (Equation 5.13):

$$\frac{\partial}{\partial \tau}\left[C_{\psi}\rho\int_{V}\left(pv - s_0T + e_0V\right)\right] - \nabla\left[\Gamma\vec{n}_{\psi}\nabla\left(pv - s_0T + e_0V\right)\right] - G_h = 0$$

- A new form of the integral equation of transfer is found (Equation 5.22):

$$\frac{\partial}{\partial \tau}\int_{v}\left(Ve_0 - Ts_0 + pv\right)dv - \frac{[\Gamma]}{C_{\psi}\cdot\rho}\int_{A_0}\frac{\partial}{\partial \vec{n}_{\psi}}\left(Ve_0 - Ts_0 + pv\right)dA - \frac{1}{C_{\psi}\cdot\rho}\int_{v}G_h dv = 0.$$

It is written in Cartesian, cylindrical, and spherical coordinates.

- Other forms of the equation of convective transfer read
  - Vector form (Equation 6.16):

$$\frac{\partial}{\partial \tau}\int_{v}\psi\, dv + \frac{1}{C_{\psi}\cdot\rho}\int_{A_0}\vec{n}\cdot\left(\psi\cdot\vec{V} - [\Gamma].\nabla\psi\right)dA - \frac{1}{C_{\psi}\cdot\rho}\int_{v}G_h dv = 0,$$

  - Gradient form (Equation 6.17) in Cartesian (Equation 6.18), cylindrical (Equation 6.19), and spherical (Equation 6.20) coordinates.

Boundary conditions of the envelope components are defined and physical process running in the vicinity of the interface *air envelope* are described. The importance of specific building features is analyzed and their effect on transfer is disclosed—note, for instance, the general architectonics, landscape and building location, wind rose orientation, openness of the building site, proximity and overshadowing of large 3D objects:

- FE equations in a matrix form are derived and examples of particular FE structures are given.
- Three functions of the envelope are outlined: barrier, filter, and intelligent membrane opposing external impacts.

- A set of author's *methods of engineering calculation* of envelope components is proposed, enabling the envelope action as an energy barrier to external impacts.
- Calculation of the thermal resistance of solid structural elements (Equation 7.8 ÷ 7.18).
- Assessment of the tribute of structure and thermal bridges to the increase of building energy efficiency (Equation 7.40).
- Calculation of solar shading devices with complex stereometry (Equation 7.16).
- Modeling of the heat exchange between a solar shading devices, a window and the surroundings.
- Specification of envelope light-transmitting apertures by means of CDL (Equation 7.33).
- Discussion of leaks in the envelope and the air ducts (exfiltration and infiltration)—the Delta-*Q* method (Equation 7.60).
- Design of a mathematical model of an environmentally sustainable building adopted by a general engineering method called BG_LEED (Equation 7.63), correlation table (Table 7.14), and equations for the calculation of IES_{Buil} of different buildings (Table 7.13).
- Priorities of the study and development of building components are formulated, so that they would act as intelligent membranes opposing external impacts.

The research and practical experience shared so far enables one to put forward the following recommendations to future activities:

Problems that are to be solved *in the short term*:

1. Creation of a general database of envelope materials, reflecting the three basic transfer properties—electro, heat, and moisture conductivity. The database should be incorporated into FE software.
2. Performance of simulation calculations for comparative analysis of the envelope functioning as *an energy filter* and *an intelligent energy membrane* with a task of efficient exploitation of the energy resources. An important issue at present is the active *research and development* of semiconducting films with controllable conductivity to be used as envelope components. Note also their integration with traditional and new nanomaterials and usage in envelopes with a composite structure.
3. Design of envelope energy standards that should be normatively approved and open to actualization.
4. Preparation and periodical actualization of tables of correspondence of indices and classes of envelope energy efficiency.

Problems that are to be solved *in the medium term*:

1. Development of structural materials and products with *intelligent* transfer properties. They should actively react to changes in the surroundings, controlling external energy impacts to conform to indoor needs.
   a. Windows with changeable transparency
   b. Solid elements with controllable conductivity
2. Design of scheme solutions using structural materials and products with *flexible* conductivity.
3. Design of building envelopes adjusting to surroundings changes.

The creation of intelligent building envelopes initially used in official buildings but later introduced in regular constructions is an enormous challenge to the entire construction community (from scientific research units, companies, and laboratories engaged in the design/erection/assembly of constructions to management teams).

Acknowledgments

I would like to thank the associate professors Dr. eng. Robert Kazangiev (IM-BAN) and Dr. math. Elena Varbanova (TU-Sofia) for their help of the English variant of this book. Finally, I thank my wife, First-Degree Res. Fellow eng. Zita Dimitroff, for the help and patience in my writing activities.

8

Applications (Solved Tasks and Tables)

Example 8.1

Design a matrix equation of the facial element N = "1," belonging to the linear area L_0 shown in Figure 5.11 from Section 5.3.3.1. The cross section is a rectangle with a side $a = 0.1$ m or a circle with diameter $d = 0.113$ m (the area of the cross section is $A_O = A_\square = 0.01\ m^2$). A detailed notation of the boundary conditions is shown in Figure 8.1.

The substance flows from the *interior* into node "0" as a direct flux h_i = 1000 [tq/s m²]* and as a convective flux, for $\alpha_i = 0.25$ [tq/s m² °] and ψ_i = 250°. The element loses substance through the surface that bounds the perimeter by means of direct flux $h_{A,h} = 10$ [tq/s m²] and convective flux through the same surface, for $\alpha_i = 0.25$ [tq/s m² °] and $\bar{\psi}_i = 40°$. Matter is not recuperated within the element ($G_h = 0$). Substance is lost through the *bottom surface (node "1"), too, by means of diffusion.*

Solution

Initially, one should design the matrix of conductivity $[K^{(1)}]$, the matrix of the surface properties $[F^{(1)}]$, and the generalized matrix of the element conductivity $[G^{(1)}] = [K^{(1)}] + [F^{(1)}]$.

8.1 Matrix of Conductivity $[K^{(1)}]$

Consider a 1D simple element, as is the facial element "1." Consider also the value of the generalized coefficient of conductivity $\Gamma = 1.2$ [tq/s m °], the element length $L_0 = 0.2$ m, and the area of the cross-section $A_0 = 10 \cdot 10^{-3}\ m^2$. Then the matrix of conductivity for both cross sections will take the following form in accordance with Equation 5.69:

$$[K^{(1)}] = 6A_0\begin{bmatrix} 1 & -1 \\ -1 & 1 \end{bmatrix} = 60 \cdot 10^{-3}\begin{bmatrix} 1 & -1 \\ -1 & 1 \end{bmatrix}.$$

* [*tq/s m²*] is the [*transfered quantity/s m²*]: [*J/s m²*] [*C/s m²*] [*kg/s m²*]. It is a measure for density of transfer matter flux.

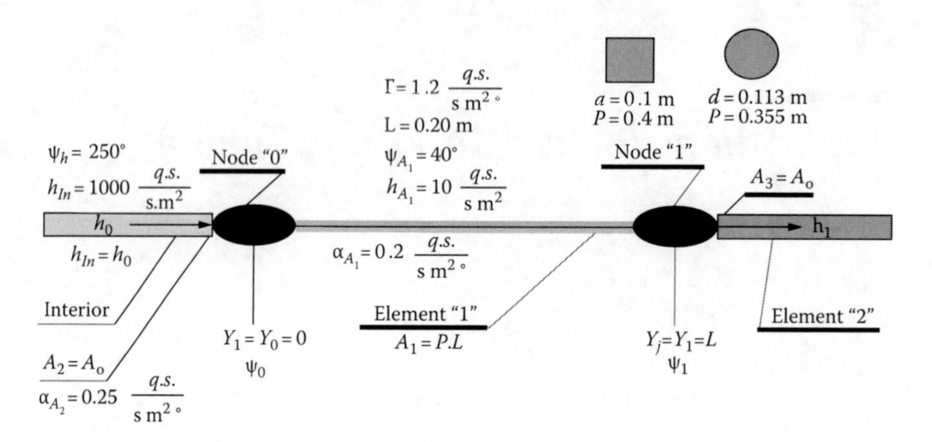

FIGURE 8.1
Boundary condition in a 1D simple finite element.

8.2 Matrix of Surface Properties $[F^{(1)}]$

As for the example discussed herein, matrix $[F^{(1)}]$ consists of two terms only—see Equation 5.70 (according to the problem formulation, there is no convective flux on A_3 and hence $\psi_{A,3} = 0$).

Taking into account that the absolute coordinates of nodes y_i and y_j that form the element are $y_{i=0} = 0$ and $y_{j=1} = L_0$, and that $\alpha_{A_{1,\varphi}} = \alpha_i = 0.2[\text{tq/sm}^2\,°]$ and $\alpha_{A_{2,\varphi}} = \alpha_i = 0.25[\text{tq/sm}^2\,°]$, the matrix of surface properties $[F^{(1)}]$ takes the form

$$[F^{(1)}] = \frac{0.2P}{L_0^2}\int_0^L \begin{bmatrix} (L_0 - y)^2 & (L_0 - y)(-0 + y) \\ (-0 + y)(L_0 - y) & (-0 + y)^2 \end{bmatrix} dy - 0.25A_0 \begin{bmatrix} 1 & 0 \\ 0 & 0 \end{bmatrix}$$

$$= \frac{0.2 \cdot PL}{6}\begin{bmatrix} 2 & 1 \\ 1 & 2 \end{bmatrix} - 0.25A_0 \begin{bmatrix} 1 & 0 \\ 0 & 0 \end{bmatrix}.$$

As for the *circular cross section,* we have $Po = 0.355$ m $Ao = 0.01$ m²:

$$[F^{(1)}]_o = \frac{0.2 \cdot 0.355 \cdot 0.2}{6}\begin{bmatrix} 2 & 1 \\ 1 & 2 \end{bmatrix} - 0.25 \cdot 10 \cdot 10^{-3}\begin{bmatrix} 1 & 0 \\ 0 & 0 \end{bmatrix} = 10^{-3}\begin{bmatrix} 2.23 & 2.37 \\ 2.37 & 4.74 \end{bmatrix},$$

while for a *square cross section*, $P_{\square} = 0.4$ m $A_{\square} = 0.01$ m^2:

$$[F^{(1)}]_{\Pi} = \frac{0.2 \cdot 0.4 \cdot 0.2}{6}\begin{bmatrix} 2 & 1 \\ 1 & 2 \end{bmatrix} - 0.25 \cdot 10 \cdot 10^{-3}\begin{bmatrix} 1 & 0 \\ 0 & 0 \end{bmatrix} = 10^{-3}\begin{bmatrix} 2.83 & 2.67 \\ 2.67 & 5.33 \end{bmatrix}.$$

8.3 Generalized Matrix of the Element Conductivity $[G^{(1)}] = [K^{(1)}] + [F^{(1)}]$

For a *circular cross section*,

$$[G^{(1)}] = 60 \cdot 10^{-3}\begin{bmatrix} 1 & -1 \\ -1 & 1 \end{bmatrix} + 10^{-3}\begin{bmatrix} 2.23 & 2.37 \\ 2.37 & 4.74 \end{bmatrix} = 10^{-3}\begin{bmatrix} 62.23 & -57.63 \\ -57.63 & 64.74 \end{bmatrix}.$$

For a *square cross section*,

$$[G^{(1)}] = 60 \cdot 10^{-3}\begin{bmatrix} 1 & -1 \\ -1 & 1 \end{bmatrix} + 10^{-3}\begin{bmatrix} 2.83 & 2.67 \\ 2.67 & 5.33 \end{bmatrix} = 10^{-3}\begin{bmatrix} 62.83 & -57.33 \\ -57.63 & 65.33 \end{bmatrix}.$$

The load vector of element "1" should be calculated in accordance with the boundary conditions specified and the defining equation (Equation 5.69), that is,

$$\{f^{(1)}\} = \left\{f_G^{(1)}\right\} + \left\{f_C^{(1)}\right\} + \left\{f_{Dr}^{(1)}\right\}.$$

8.4 Vector of a Load Due to Recuperation Sources $\{f_C^{(1)}\}$

In compliance with Equation 5.74 and the geometrical data and physical conditions specified in the problem (i.e., the absolute coordinates of nodes y_i and y_j that form the element and have values $y_{i=0} = 0$ and $y_{j=1} = L_0 = 0.20$ m, and the condition that there is no recuperation of matter within the element—$G_h = 0$), the vector of a load due to internal sources $\{f_G^{(1)}\}$ will take the form

$$\left\{f_G^{(1)}\right\} = G_h A_0 \int_0^L \begin{Bmatrix} \left(1 - \dfrac{1}{L_0} y\right) \\ \left(0 + \dfrac{1}{L_0} y\right) \end{Bmatrix} dy = \frac{0.10 \cdot 10^{-3} \cdot 0.2}{2}\begin{Bmatrix} 1 \\ 1 \end{Bmatrix} = \begin{Bmatrix} 0 \\ 0 \end{Bmatrix}.$$

8.5 Vector of a Load due to Convection to the Surrounding Matter $\{f_C^{(1)}\}$

Considering Equation 5.76 and the values of $\overline{\varphi}_{A_{1,\psi}} = \overline{\varphi}_e = 40°$ $\overline{\psi}_{A_{2,\psi}} = \overline{\psi}_{In} = 250°$, we get

$$\left\{f_C^{(1)}\right\} = 0.2 \cdot 40P \int_0^L \left\{ \begin{matrix} \left(1 - \frac{1}{L_0} y\right) \\ \left(0 + \frac{1}{L_0} y\right) \end{matrix} \right\} dy - 0.25 \cdot 250 \int_{A_0} \left\{ \begin{matrix} 1 \\ 0 \end{matrix} \right\} dA$$

$$= \frac{8P \cdot L}{2} \left\{ \begin{matrix} 1 \\ 1 \end{matrix} \right\} - 62.5 A_0 \left\{ \begin{matrix} 1 \\ 0 \end{matrix} \right\}.$$

For a *circular cross section,* we have: $Po = 0.355$ m $Ao = 0.01$ m^2:

$$\left\{f_C^1\right\} = \frac{8.40 \cdot 0.355 \cdot 0.2}{2} \left\{ \begin{matrix} 1 \\ 1 \end{matrix} \right\} - 62.5 \cdot 10 \cdot 10^{-3} \left\{ \begin{matrix} 1 \\ 0 \end{matrix} \right\}$$

$$= \left\{ \begin{matrix} 0.284 \\ 0.284 \end{matrix} \right\} - \left\{ \begin{matrix} 0.625 \\ 0 \end{matrix} \right\} = \left\{ \begin{matrix} -0.341 \\ 0.284 \end{matrix} \right\}$$

For a *square cross section,* $P\square = 0.4$ m $A\square = 0.01$ m^2:

$$\left\{f_C^1\right\} = \frac{8 \cdot 40 \cdot 0.4 \cdot 0.2}{2} \left\{ \begin{matrix} 1 \\ 1 \end{matrix} \right\} - 62.5 \cdot 10 \cdot 10^{-3} \left\{ \begin{matrix} 1 \\ 0 \end{matrix} \right\} = \left\{ \begin{matrix} 0.32 \\ 0.32 \end{matrix} \right\} - \left\{ \begin{matrix} 0.625 \\ 0 \end{matrix} \right\} = \left\{ \begin{matrix} -0.305 \\ 0.32 \end{matrix} \right\}$$

8.6 Vector of a Load due to a Direct Flux $\{f_{Dr}^e\}$

Considering vector $\{f_{Dr}^e\}$ for element "1" in compliance with Equation 5.78 and the boundary conditions specified, it follows that $\{f_{Dr}^e\}$ has three components:

$$\left\{f_{Dr}^{(1)}\right\} = -10P \int_0^L \left\{ \begin{matrix} \left(1 - \frac{1}{L_0} y\right) \\ \left(-0 + \frac{1}{L_0} y\right) \end{matrix} \right\} dy + 1000 \int_{A_0} \left\{ \begin{matrix} 1 \\ 0 \end{matrix} \right\} dA - h_i \int_{A_0} \left\{ \begin{matrix} 0 \\ 1 \end{matrix} \right\} dA$$

$$= -\frac{10LP}{2} \left\{ \begin{matrix} 1 \\ 1 \end{matrix} \right\} + 1000 \cdot A_0 \left\{ \begin{matrix} 1 \\ 0 \end{matrix} \right\} - h_1 \cdot A_0 \left\{ \begin{matrix} 0 \\ 1 \end{matrix} \right\}.$$

For a *circular cross section,* we have $Po = 0.355$ m and $Ao = 0.01$ m^2:

$$\left\{f_{Dr}^{(1)}\right\} = -0.355\begin{Bmatrix}1\\1\end{Bmatrix} + 10\begin{Bmatrix}1\\0\end{Bmatrix} - 0.01h_1 = \begin{Bmatrix}9.645\\-0.355-0.01\cdot h_1\end{Bmatrix}.$$

For a *square cross section,* $P\square = 0.4$ m and $A\square = 0.01$ m^2:

$$\left\{f_{Dr}^{(1)}\right\} = -0.4\begin{Bmatrix}1\\1\end{Bmatrix} + 10\begin{Bmatrix}1\\0\end{Bmatrix} - 0.01h_1 = \begin{Bmatrix}9.4\\-0.4-0.01\cdot h_1\end{Bmatrix}.$$

We find the following relations for the vector of a load, which is the sum of vectors $\{f_G^{(1)}\}$, $\{f_C^{(1)}\}$, and $\{f_{Dr}^{(1)}\}$:

- For a circular cross section,

$$\{f^{(1)}\}o = \left\{f_G^{(1)}\right\} + \left\{f_C^{(1)}\right\} + \left\{f_{Dr}^{(1)}\right\} = \begin{Bmatrix}0\\0\end{Bmatrix} + \begin{Bmatrix}-0.341\\0.284\end{Bmatrix} + \begin{Bmatrix}9.645\\-0.355-0.01h_1\end{Bmatrix}$$

$$= \begin{Bmatrix}9.304\\-0.071-0.01h_1\end{Bmatrix}.$$

- For a square cross section,

$$\left\{f^{(1)}\right\}_{\Pi} = \left\{f_G^{(1)}\right\} + \left\{f_C^{(1)}\right\} + \left\{f_{Dr}^{(1)}\right\} = \begin{Bmatrix}0\\0\end{Bmatrix} + \begin{Bmatrix}-0.305\\0.32\end{Bmatrix} + \begin{Bmatrix}9.4\\-0.4-0.01\cdot h_1\end{Bmatrix}$$

$$= \begin{Bmatrix}9.095\\-0.08-0.01\cdot h_1\end{Bmatrix}.$$

8.6.1 Design and Solution of the Matrix Equation

Having designed the general matrix $[G^{(1)}]$ and the load vector $\{f^{(1)}\}$ for both forms of the cross section of the finite element "1," we can substitute them in Equation 5.61, by solving, we get

$$\{\overline{\psi}^{(1)}\} = [G^{(1)}]^{-1}\cdot\{f^{(1)}\},$$

and for a *circular cross section,* we find that

$$10^3\begin{Bmatrix}\overline{\psi}_0\\\overline{\psi}_1\end{Bmatrix} = \begin{bmatrix}62.23 & -57.63\\-57.63 & 64.74\end{bmatrix}^{-1}\cdot\begin{Bmatrix}9.304\\-0.071-0.01\cdot h_1\end{Bmatrix}.$$

TABLE 8.1

Influence of the Cross-Sectional Shape of 1D Final Element

	h_0	h_1	$\Delta h = h_0 - h_1$	Δh (%)	$\bar{\psi}_1$ (°)	$\bar{\psi}_1 - \bar{\psi}_0$
Ao = 0.01 m^2	1000	730.2	−269.8	−26.98	108.7	141.4
A□ = 0.01 m^2	1000	671.7	−328.3	−32.8	115.4	134.6
Without peripheral losses, $A_0 = 1$ m^2	1000	1000	0	0	83.3	166.7

Performing a similar operation squared cross section FE, we find the following matrix equation:

$$10^3\begin{Bmatrix}\bar{\psi}_0\\ \bar{\psi}_1\end{Bmatrix}=\begin{bmatrix}62.83 & -57.33\\ -57.33 & 65.33\end{bmatrix}^{-1}\cdot\begin{Bmatrix}9.095\\ -80-10\cdot h_1\end{Bmatrix}.$$

Assume that h_0 and $\bar{\psi}_0$ are defined as boundary conditions ($h_0 = 1000$ [tq/s m^2] and $\bar{\psi}_0 = 250°$—see Figure 8.1). Then, we can easily find that the values of h_1 and $\bar{\psi}_1$ for square and circular cross sections differ significantly from each other—see Table 8.1.

Values of $\bar{h}_1$ and $\bar{\psi}_1$ calculated using the classical matrix equation for a 1D finite element (for instance, Equation 5.64) are given for comparison.

The reason for these differences is due to the different enveloping surfaces of the two finite elements (Po = 0.355 m, P□ = 0.4 m) and the geometrical shapes, despite their identical area, Ao = A□ = 0.01 m^2.

The square profile *loses* transferred substance, which is larger than the circular profile by 18%. Hence, it follows that inequality $\bar{h}_{1\square} < \bar{h}_{1\circ}$ holds when the substance *leaves* the finite element at point 1. This in its turn yields smaller *consumption* of the driving potential of field $\bar{\psi}$, and the potential value at the same end point after the *square* finite element is greater than that for the *circular* one, that is, $\bar{\psi}_{1\square} > \bar{\psi}_{1\circ}$.

The example discussed illustrates the capabilities of the finite element method to model various boundary conditions, which account for the geometry of the medium where diffusion processes take place.

Example 8.2

Consider a disc with dimensions R_1 = 0.2 m and R_2 = 1.0 m and height b_0 = 0.1 m. The coefficient of conductivity of the material is Γ_r = 1.2 [tq/m s °]. The internal potential is $\bar{\psi}_{in} = 250°$, and the external one is $\bar{\psi}_{out} = 40°$. The new boundary conditions are specified in Figure 8.2, and they are similar to the boundary conditions of Example 8.1, but are described in Section 5.3.3.3 formulated in Cylindrical coordinates. Design the equation of the finite element.

Solution

Divide the disc into four finite elements, for $r_j - r_i = 0.2$ m (see Figure 8.2). The consecutive steps of finding the solution are the same as those needed to solve the problem in Cartesian coordinates, that is,

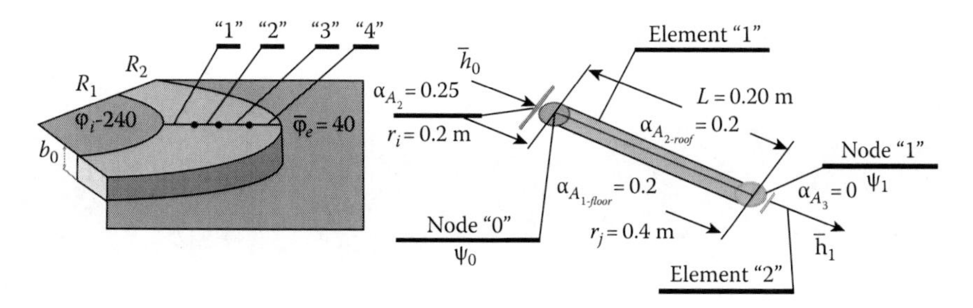

FIGURE 8.2
Boundary conditions in a 1D simple finite element specified in cylindrical coordinates.

1. Design of the matrix of conductivity $[G^e]_C = [K^e]_C + [F^e]_C$:
   - Conductivity matrix:

$$\left[K^e\right]_C = 3.14 \cdot 0.1 \cdot 1.2 \frac{0.2+0.4}{0.4-0.2}\begin{bmatrix} 1 & -1 \\ -1 & 1 \end{bmatrix} = \begin{bmatrix} 1.13 & -1.13 \\ -1.13 & 1.13 \end{bmatrix};$$

   - Matrix of the surface properties:

$$[F^e]_C = (6.28 \cdot 0.2 + 6.28 \cdot 0.2)\frac{0.4-0.2}{12}\begin{bmatrix} 3 \cdot 0.2 + 0.4 & 0.2+0.4 \\ 0.2+0.4 & 0.2+3 \cdot 0.4 \end{bmatrix}$$

$$-6.28 \cdot 0.1 \cdot 0.2 \cdot 0.25\begin{bmatrix} 1 & 0 \\ 0 & 0 \end{bmatrix} = -\begin{bmatrix} 0.0104 & 0.0251 \\ 0.0251 & 0.0585 \end{bmatrix}$$

   Then, we find for the general matrix of conductivity $[G^e]_C$, that is,

$$[G^e]_C = \begin{bmatrix} 1.13 & -1.13 \\ -1.13 & 1.13 \end{bmatrix} + \begin{bmatrix} 0.0104 & 0.0251 \\ 0.0252 & 0.0585 \end{bmatrix} = \begin{bmatrix} 1.1404 & -1.1049 \\ -1.1049 & 1.1885 \end{bmatrix}$$

2. ***Finding the load vector*** $\{f^e\}_C$:
   - Load vector $\{f^e_G\}_C$ due to the recuperation of matter: should be zero, since $G_{h_e} = 0$.
   - Load vector $\{f^e_C\}_C$ due to convection is estimated by

$$\left\{f^e_C\right\}_C = 6.28 \cdot (02 \cdot 40 + 0.2 \cdot 40)\frac{0.4-0.2}{6}\begin{Bmatrix} 2 \cdot 0.2 + 0.4 \\ 0.2 + 2 \cdot 0.4 \end{Bmatrix}$$

$$-6.28 \cdot 0.1 \cdot 0.2 \cdot 250\begin{Bmatrix} 1 \\ 0 \end{Bmatrix} = \begin{Bmatrix} -5.84 \\ 1.675 \end{Bmatrix}$$

- Load vector $\{f_{Dr}^e\}_C$ due to direct flux:

$$\{f_{Dr}^e\}_C = -6.28(0+0)\frac{0.4-0.2}{6}\begin{bmatrix}2\cdot 0.2+0.4\\0.2+2\cdot 0.4\end{bmatrix}+6.28\cdot 0.2\cdot 0.1\cdot 1000\begin{Bmatrix}1\\0\end{Bmatrix}$$

$$-6.28\cdot 0.4\cdot 0.1\bar{h}_1\begin{Bmatrix}0\\1\end{Bmatrix}=\begin{Bmatrix}125.6\\-0.2512\bar{h}_1\end{Bmatrix}.$$

Then, we find the following expression for the load vector $\{f^e\}_C$:

$$\{f^e\}_C = 0+\begin{Bmatrix}-5.84\\1.675\end{Bmatrix}+\begin{Bmatrix}125.6\\-0.2512\bar{h}_1\end{Bmatrix}$$

Put $[G^e]_C$ and $\{f^e\}_C$ in Equation 5.81. Then, the matrix equation of element "1" gets the following form in cylindrical coordinates:

$$\begin{bmatrix}1.1404 & -1.1049\\-1.1048 & 1.1885\end{bmatrix}\begin{Bmatrix}250\\\psi_1\end{Bmatrix}=\begin{Bmatrix}119.76\\1.675-0.2512\bar{h}_1\end{Bmatrix}$$

and it is ready for further application.

Example 8.3

Design an equation of a 2D simple finite element "1" with coordinates that are given in Figure 8.3, considering the following boundary conditions:

$$\bar{\psi}_{A_2}=250°,\quad \alpha_{A_2}=0.25\,[\text{tq/sm}^2\,°],\quad \alpha_{A_1}=\alpha_{A_2}=0,\quad \bar{h}_{A_2}=1000\,[\text{tq/sm}^2]$$

Solution

1. Finding the coefficient of the approximating functions:
   One should undertake this step before the start of the calculation of the components of the matrix equation. The values of the coefficients are given in Table 8.2.
2. Specification of the components of the matrix equation:
   General matrix of conductivity $[G^e]$:
   - The conductivity matrix—according to relation (Equation 5.79):

$$[K^{(1)}]=0.2\cdot 1.2\cdot 0.005\begin{Bmatrix}-10 & -10\\10 & 0\\0 & 10\end{Bmatrix}\begin{bmatrix}-10 & 10 & 0\\-10 & 0 & 10\end{bmatrix}$$

$$=0.12\begin{bmatrix}2 & -1 & -1\\-1 & 1 & 0\\-1 & 0 & 1\end{bmatrix}$$

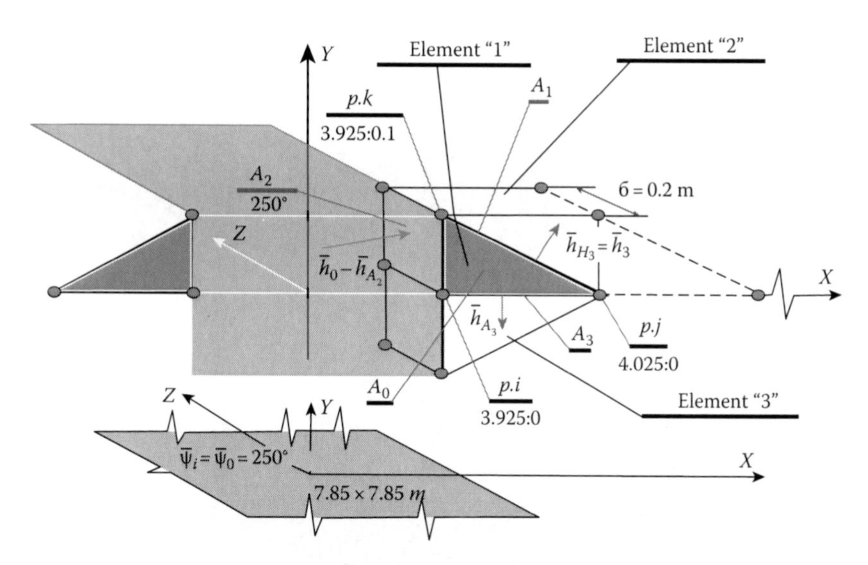

FIGURE 8.3
Boundary conditions in a 2D simple finite element in Cartesian coordinates.

TABLE 8.2

Coefficients of the Approximating Functions

Approximating Functions	a_i	b_i	c_i
P_1	40	−10	−10
P_2	−39	10	0
P_3	0	0	10

- Matrix of the surface properties—according to Equation 5.89:

$$[F^{(1)}] = \frac{0 \cdot 0.5 \cdot 10^{-2}}{12}\begin{bmatrix} 2 & 1 & 1 \\ 1 & 2 & 1 \\ 1 & 1 & 2 \end{bmatrix} + \frac{0 \cdot 0.14 \cdot 0.2}{6}\begin{bmatrix} 0 & 0 & 0 \\ 0 & 2 & 1 \\ 0 & 1 & 2 \end{bmatrix}$$

$$-\frac{0.25 \cdot 0.1 \cdot 0.2}{6}\begin{bmatrix} 2 & 0 & 1 \\ 0 & 0 & 0 \\ 1 & 0 & 2 \end{bmatrix} + \frac{0 \cdot 0.1 \cdot 0.2}{6}\begin{bmatrix} 2 & 1 & 0 \\ 1 & 2 & 0 \\ 0 & 0 & 0 \end{bmatrix}$$

$$= -10^{-3}\begin{bmatrix} 1.6 & 0 & 0.8 \\ 0 & 0 & 0 \\ 0.8 & 0 & 1.6 \end{bmatrix}$$

The generalized matrix of conductivity gets the following form:

$$[G^{(1)}]=[K^{(1)}]+[F^{(1)}]=0.12\begin{bmatrix}2 & -1 & -1\\ -1 & 1 & 0\\ -1 & 0 & 1\end{bmatrix}-10^{-3}\begin{bmatrix}1.6 & 0 & 0.8\\ 0 & 0 & 0\\ 0.8 & 0 & 1.6\end{bmatrix}$$

$$=\begin{bmatrix}0.2384 & -0.12 & -0.1208\\ -0.12 & 0.12 & 0\\ -0.1208 & 0 & 0.1184\end{bmatrix}$$

Vector of the load of the finite element:

- Load due to generation:

$$\left[f_G^{(1)}\right]=\frac{0.2\cdot 0\cdot 0.5\cdot 10^{-2}}{3}\begin{Bmatrix}1\\1\\1\end{Bmatrix}=\begin{Bmatrix}0\\0\\0\end{Bmatrix};$$

- Load due to convection—according to Equation 5.90:

$$\left[f_C^{(1)}\right]=+\frac{0\cdot 0\cdot 0.5\cdot 10^{-2}}{3}\begin{Bmatrix}1\\1\\1\end{Bmatrix}+\frac{0\cdot 0\cdot 0.2\cdot 0.141}{2}\begin{Bmatrix}0\\1\\1\end{Bmatrix}$$

$$-\frac{0.25\cdot 250\cdot 0.2\cdot 0.1}{2}\begin{Bmatrix}1\\0\\1\end{Bmatrix}+\frac{0\cdot 0\cdot 0.2\cdot 0.1}{2}\begin{Bmatrix}1\\1\\0\end{Bmatrix}=\begin{Bmatrix}-0.625\\0\\-0.625\end{Bmatrix};$$

- Load due to direct flux:

$$\left[f_{Dr}^{(1)}\right]=-\frac{0\cdot 0.5\cdot 10^{-2}}{3}\begin{Bmatrix}1\\1\\1\end{Bmatrix}-\frac{\bar{h}_1\cdot 0.2\cdot 0141}{2}\begin{Bmatrix}0\\1\\1\end{Bmatrix}+\frac{1000\cdot 0.2\cdot 0.1}{2}\begin{Bmatrix}1\\0\\1\end{Bmatrix}$$

$$-\frac{\bar{h}_3\cdot 0.2\cdot 0.1}{2}\begin{Bmatrix}1\\1\\0\end{Bmatrix}=\begin{Bmatrix}10-0.0141\bar{h}_{A_3}\\ -0.0141\left(\bar{h}_{A_1}+\bar{h}_{A_3}\right)\\ 10-0.0141\bar{h}_{A_1}\end{Bmatrix}$$

The vector of the finite element load is the sum of the three types of loads, according to Equation 5.91.

3. Design of the matrix equation of the finite element:
   The final form of the matrix equation is determined after putting the matrix components found into relation:

$$\begin{bmatrix} 0.2384 & -0.12 & -0.1298 \\ -0.12 & 0.12 & 0 \\ -0.1208 & 0 & 0.1184 \end{bmatrix} \begin{Bmatrix} \overline{\psi}_i \\ \overline{\psi}_j \\ \overline{\psi}_k \end{Bmatrix} = \begin{Bmatrix} 9.375 - 0.0141\overline{h}_{A3} \\ -0.0141\left(\overline{h}_{A1} + \overline{h}_{A_3}\right) \\ 9.375 - 0.0141\overline{h}_{A_1} \end{Bmatrix}$$

   The equation thus derived can be used to design the matrix equation of the global area and to introduce the boundary conditions.

Example 8.4

Design a matrix equation of the 2D finite element 1 with coordinates that are shown in Figure 8.4, considering the following boundary conditions:

$$\overline{\psi}_{A2} = \overline{\psi}_{In} = \overline{\psi}_i = \overline{\psi}_j = 250°, \quad \overline{h}_0 = \overline{h}_{A2} = 1000(\text{tq/sm}^2),$$

$$\alpha_{A2} = (\alpha_{k-i}) = 0.25(\text{tq/sm}^{2\,\circ}), \quad \alpha_{A1} = \alpha_{A3} = 0\,(\alpha_{j-k} = \alpha_{i-j} = 0).$$

Solution

The equation of element "1" is sought in the form

$$\begin{Bmatrix} \overline{\psi}_i \\ \overline{\psi}_j \\ \overline{\psi}_k \end{Bmatrix} = [G^e]_C^{-1}\{f^e\}_C,$$

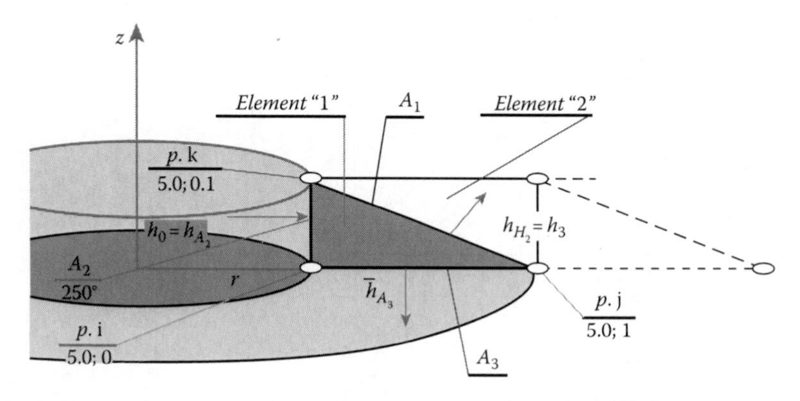

FIGURE 8.4
Boundary conditions in a 1D simple finite element in cylindrical coordinates.

where $\bar{\psi}_i$, $\bar{\psi}_j$, and $\bar{\psi}_k$ are potentials of the thermodynamic field at the nodes of the 2D finite element. In cylindrical coordinates, it has the shape of a rotational body with a triangular generant profile:

1. Determination of the coefficient of the approximating functions: Before calculating the generalized conductivity matrix, one should first of all calculate the coefficients of the approximating functions ($a_i, b_i, c_i \ldots a_k, b_k, c_k$).

   This calculation is performed using *Derive 5* software (Dimitrov 2013). The absolute coordinates of the nodes of finite element 1 are specified in Figure 8.4, and the values of the coefficients found are given in Table 8.3.
2. Components of the matrix equation:
   Generalized conductivity matrix:
   - Conductivity matrix (Equation 5.94):

$$[K^e]_C = 18.97\begin{bmatrix} 2 & -1 & -1 \\ -1 & 1 & 0 \\ -1 & 0 & 1 \end{bmatrix};$$

   - Matrix of the surface properties (Equation 5.95):

$$[F^e]_C = -\frac{6.28\cdot 0\cdot 0.141}{12}\begin{bmatrix} 0 & 0 & 0 \\ 0 & 3\cdot 5.1+5 & 5.1+5 \\ 0 & 5+5.1 & 5.1+3.5 \end{bmatrix}$$

$$+\frac{6.28\cdot 0.25\cdot 0.1}{12}\begin{bmatrix} 3.5+5 & 0 & 5+5 \\ 0 & 0 & 0 \\ 5+5 & 0 & 5+3.5 \end{bmatrix}$$

$$-\frac{6.28\cdot 0\cdot 0.1}{12}\begin{bmatrix} 2 & 1 & 0 \\ 1 & 2 & 0 \\ 0 & 0 & 0 \end{bmatrix} = \begin{bmatrix} 0.262 & 0 & 0.131 \\ 0 & 0 & 0 \\ 0.131 & 0 & 0.262 \end{bmatrix};$$

TABLE 8.3

Coefficients of the Approximating Functions

Approximating Functions	a_i	b_i	c_i
P_1	50	−10	−10
P_2	−50	10	0
P_3	0	0	10

- Generalized conductivity matrix:

$$[G^e]=[K^e]+[F^e]=18.97\begin{bmatrix}2 & -1 & -1\\ -1 & 1 & 0\\ -1 & 0 & 1\end{bmatrix}+\begin{bmatrix}0.262 & 0 & 0.131\\ 0 & 0 & 0\\ 0.131 & 0 & 0.262\end{bmatrix}$$

$$=\begin{bmatrix}38.16 & -18.97 & -18.84\\ -18.97 & 18.97 & 0\\ -18.83 & 0 & 19.23\end{bmatrix}$$

Vector of the finite element load:

- Load due to matter recuperation:

$$\left[f_G^e\right]_C=\frac{6.28\cdot 0\cdot 0.5\cdot 10^{-2}}{6}\begin{Bmatrix}2\cdot 5+5.1+5\\ 5+2\cdot 5.1+5\\ 5+5.1+2\cdot 5\end{Bmatrix}=\begin{Bmatrix}0\\ 0\\ 0\end{Bmatrix};$$

- Load due to convection (Equation 5.96):

$$\left[f_C^e\right]_C=-\frac{3.14\cdot 0\cdot 0\cdot 0.141}{3}\begin{Bmatrix}0\\ 2\cdot 5.1+5\\ 5\cdot 1+2\cdot 5\end{Bmatrix}+\frac{3.14\cdot 250\cdot 0.25\cdot 0.1}{3}\begin{Bmatrix}2.5+5\\ 0\\ 5+2.5\end{Bmatrix}$$

$$-\frac{3.14\cdot 0\cdot 0\cdot 0.1}{3}\begin{Bmatrix}2\cdot 5+5.1\\ 5+2\cdot 5.1\\ 0\end{Bmatrix}=\begin{Bmatrix}98.125\\ 0\\ 98.125\end{Bmatrix}.$$

- Load due to direct flux:

$$\left[f_{Dr}^e\right]_C=-\frac{3.14\bar{h}_{A_1}\cdot 0.141}{3}\begin{Bmatrix}0\\ 2\cdot 5.1+5\\ 5.1+2\cdot 5\end{Bmatrix}+\frac{3.14\cdot 1000\cdot 0.1}{3}\begin{Bmatrix}2\cdot 5+5\\ 0\\ 5+2\cdot 5\end{Bmatrix}$$

$$-\frac{3.14\bar{h}_{A_3}\cdot 0.1}{3}\begin{Bmatrix}2\cdot 5+5.1\\ 5+2\cdot 5.1\\ 0\end{Bmatrix}=\begin{Bmatrix}2213.7-2.22\bar{h}_{A_3}\\ -2.23\left(\bar{h}_{A_1}+\bar{h}_{A_3}\right)\\ 2213.7-2.22\bar{h}_{A_1}\end{Bmatrix}.$$

Summing its components, the vector of the finite element load gets the forms

$$\{f^e\}_C = 0 + \begin{Bmatrix} 98.125 \\ 0 \\ 98.125 \end{Bmatrix} + \begin{Bmatrix} 2213.7 - 2.22\bar{h}_{A_3} \\ -2.23(\bar{h}_{A_1} + \bar{h}_{A_3}) \\ 2213.7 - 2.22\bar{h}_{A_1} \end{Bmatrix} = \begin{Bmatrix} 2311.8 - 2.22\bar{h}_{A_3} \\ -2.23(\bar{h}_{A_1} + \bar{h}_{A_3}) \\ 2311.8 - 2.22\bar{h}_{A_1} \end{Bmatrix}$$

3. Design of the matrix equation of the finite element:
The final form of the matrix equation of finite element 1 is found after putting the calculated components in Equation (5.58). Then by solving, we get:

$$\begin{Bmatrix} \bar{\psi}_i \\ \bar{\psi}_j \\ \vec{\psi}_k \end{Bmatrix} = \begin{bmatrix} 38.19 & -18.97 & -18.83 \\ -18.97 & 18.97 & 0 \\ -18.83 & 0 & 19.23 \end{bmatrix}^{-1} \begin{Bmatrix} 2311.8 - 2.22\bar{h}_{A_3} \\ -2.23(\bar{h}_{A_1} + \bar{h}_{A_3}) \\ 2311.8 - 2.22\bar{h}_{A_1} \end{Bmatrix}$$

It will be used to design the matrix equation of the global area, together with the matrix equations of the other finite elements, considering the corresponding boundary conditions.

Example 8.5

Design the matrix equation of diffusion into a 3D simple finite element with node coordinates that are given in Figure 8.5, considering the following boundary conditions:

$$\bar{\psi}_{A_2} = \bar{\psi}_{A_3} = \bar{\psi}_{A_4} = 250°; \quad \alpha_{A_2} = \alpha_{A_3} = \alpha_{A_4} = 0.25(\text{tq/sm}^2\,°); \quad \alpha_{A_1} = 0,$$

$$\Gamma_x = \Gamma_y = \Gamma_z = \Gamma = 1.2\,(\text{tq/sm}\,°), \quad \bar{h}_{A_2} = 1200(\text{tq/sm}^2), \quad \bar{h}_{A_3} = 500(\text{tq/sm}^2),$$

$$\bar{h}_{A_4} = 100(\text{tq/sm}^2).$$

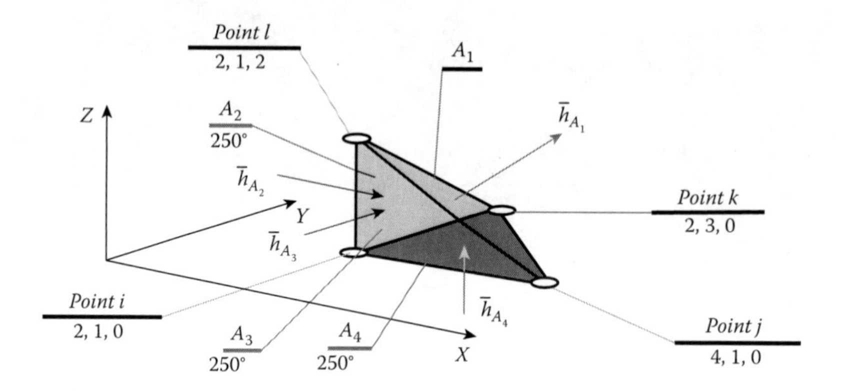

FIGURE 8.5
Boundary conditions in a 3D simple finite element in Cartesian coordinates.

Solution

Before finding the components of the matrix equation, one should find the coefficients of the approximating functions. Hence, one should apply adjugate matrices relations (Equation 5.43):

1. Finding the coefficients of the approximating functions (Table 8.4):

$$V_0 = 1.333 \cdot 10^{-3}\ \mathrm{m}^3; \quad A_1 = 0.0347\,\mathrm{m}^2; \quad A_2 = A_3 = A_4 = 0.02\ \mathrm{m}^2$$

2. Finding the components of the matrix equation:
   Generalized conductivity matrix:
   Conductivity matrix (Equation 5.99):

$$[K^{(1)}] = 10^{-3}\begin{bmatrix} 0.897 & 0.449 & -0.199 & -0.299 \\ 0.449 & 0.224 & -0.099 & -0.149 \\ -0.199 & -0.099 & 0.399 & -0.199 \\ -0.299 & -0.149 & -0.199 & 0.498 \end{bmatrix};$$

- Matrix of surface properties

$$[F^{(1)}] = -\frac{0 \cdot 3.47 \cdot 10^{-2}}{12}\left[M_1^0\right] + \frac{0.25 \cdot 0.02}{12}\left[M_2^0\right] + \frac{0.25 \cdot 0.02}{12}\left[M_3^0\right]$$

$$+\frac{0.25 \cdot 0.02}{12}\left[M_4^0\right] = 10^{-4}\begin{bmatrix} 25.02 & 8.34 & 8.34 & 8.34 \\ 8.34 & 16.68 & 4.17 & 4.17 \\ 8.34 & 4.17 & 16.68 & 4.17 \\ 8.34 & 4.17 & 4.17 & 16.68 \end{bmatrix}.$$

Generalized conductivity matrix:

$$[G^{(1)}] = \begin{bmatrix} 33.99 & 12.83 & 6.34 & 5.34 \\ 12.83 & 18.92 & 3.17 & 2.67 \\ 6.34 & 3.17 & 20.67 & 2.17 \\ 5.34 & 2.67 & 2.17 & 21.66 \end{bmatrix}.$$

TABLE 8.4

Coefficients of the Approximating Functions

	a_i	b_i	c_i	d_i
P_1	1.75	−0.5	−0.25	−0.5
P_2	2.125	−0.25	−0.125	−0.25
P_3	0.5	0	0.5	0
P_4	0.25	0	−0.25	0.5

Vector of the finite element load:

- Load due to recuperation of matter (Equation 5.100):
- For $G_{h_e} = 0$, load $\{f_G^e\}$ is equal to the zero vector.
- Load due to convection (Equation 5.101):

$$\left\{f_C^{(1)}\right\} = -0 \cdot \frac{0.0347}{3}\vec{t}_1 + \frac{0.25 \cdot 250 \cdot 0.02}{3}\vec{t}_2 + \frac{0.25 \cdot 250 \cdot 0.02}{3}\vec{t}_3$$

$$+\frac{0.25 \cdot 250 \cdot 0.02}{3}\vec{t}_4 = \begin{Bmatrix} 1.25 \\ 0.833 \\ 0.833 \\ 0.833 \end{Bmatrix}$$

- Load due to direct flux (Equation 5.102).
  After considering the boundary conditions for $\bar{h}_{A_k}, k = 2,3,4$, we find

$$\left\{f_{Dr}^1\right\} = \begin{Bmatrix} 0.0067\left(h_{A_2} + h_{A_3} + h_{A_4}\right) \\ -0.0116 \cdot \bar{h}_{A_1} + 0.0067\left(h_{A_3} + h_{A_4}\right) \\ -0.0116 \cdot \bar{h}_{A_1} + 0.0067\left(h_{A_2} + h_{A_4}\right) \\ -0.0116 \cdot \bar{h}_{A_1} + 0.0067\left(h_{A_2} + h_{A_3}\right) \end{Bmatrix} = \begin{Bmatrix} 12.06 \\ -0.116 \cdot \bar{h}_{A_1} + 4.02 \\ -0.0116 \cdot \bar{h}_{A_1} + 8.71 \\ -0.0116 \cdot \bar{h}_{A_1} + 11.39 \end{Bmatrix}$$

- Load vector

$$\left\{f^{(1)}\right\} = \begin{Bmatrix} 1.25 \\ 0.833 \\ 0.833 \\ 0.833 \end{Bmatrix} + \begin{Bmatrix} 12.06 \\ -0.0116\bar{h}_{A_1} + 4.02 \\ -0.0116\bar{h}_{A_1} + 8.71 \\ -0.0116\bar{h}_{A_1} + 11.39 \end{Bmatrix} = \begin{Bmatrix} 13.31 \\ -0.0116\bar{h}_{A_1} + 4.85 \\ -0.0116\bar{h}_{A_1} + 9.54 \\ -0.0116\bar{h}_{A_1} + 12.22 \end{Bmatrix}$$

3. Design of the matrix equation of the finite element:
   The forms of the generalized matrix $[G^{(1)}]$ found and the load vector $\{f^{(1)}\}$ are put in Equation 5.103.
   Thus, we find the matrix equation of the finite element valid for the problem under consideration:

$$\begin{Bmatrix} \bar{\psi}_i \\ \bar{\psi}_j \\ \bar{\psi}_k \\ \bar{\psi}_l \end{Bmatrix} = \left[10^{-4}\begin{bmatrix} 3.99 & 12.83 & 6.34 & 5.34 \\ 12.83 & 18.92 & 3.17 & 2.67 \\ 6.34 & 3.17 & 20.67 & 2.17 \\ 5.34 & 2.67 & 2.17 & 21.66 \end{bmatrix}\right]^{-1} \begin{Bmatrix} 13.31 \\ -0.0116\bar{h}_{A_1} + 4.85 \\ -0.0116\bar{h}_{A_1} + 9.54 \\ -0.0116\bar{h}_{A_1} + 12.22 \end{Bmatrix}$$

It will be used for the design of the equation of the global area—an object of study of the next part of this book.

Example 8.6: (Transfer in 1D Global Area)

Considering the material discussed in Sections 5.3 through 5.5, we use here the linear approximating function (5.36), which describes the change of the ETS characteristics (see Figure 5.11). The general solution of the area is sought in the form

$$\psi(x) = \sum_{i=0}^{n} P_i \psi_i = P_0 \psi_0 + P_1 \psi_1 + \cdots + P_n \psi_n$$

Note that Equations 5.60 through 5.62 are valid for each finite element. However, the different (N in number) finite elements *operate* under different boundary conditions. They can be divided into two groups: boundary and internal elements. The first group comprises elements with numbers 1 and N, where there is inflow or outflow of substance diffusing from the surrounding medium. Hence, load vector $f_i^{(e)}$ has nonzero components. The second group comprises all elements with numbers from 2 to $N-1$. Their characteristic feature is that they are balanced with respect to substance, that is, inflowing and outflowing fluxes are equalized, and hence, load vectors $\{f^{(e)}\}$ are zero ones.

We shall successively design the matrix equation of the *facial* element "1" and that of the last element N, being representatives of the first group but operating under different initial conditions.

Then, we shall design the respective matrix equation of an arbitrary ith element, which is a representative of the second group.

Start with element "1." We have the following relations for fluxes of substance transferred from node "0" to node "1," and from node "1" to node "0":

$$I_{0-1} = h_{0-1} A_0 = -\frac{\Gamma}{\Delta x} A_0 (\bar{\psi}_1 - \bar{\psi}_0) = \frac{\Gamma}{\Delta x} A_0 (\bar{\psi}_0 - \bar{\psi}_1),$$

$$h_{0-1} = h_0 = \frac{\Gamma}{\Delta x} (\bar{\psi}_0 - \bar{\psi}_1), \quad A_0 = 1$$

the area of the cross section of the finite element:

$$I_{1-0} = h_{1-0} A_0 = -\frac{\Gamma}{\Delta x} A_0 (\bar{\psi}_0 - \bar{\psi}_1), \quad \left(h_{1-0} = -\frac{\Gamma}{\Delta x} (\bar{\psi}_0 - \bar{\psi}_1) \right)$$

These two equations can be written in a matrix form:

$$\begin{Bmatrix} h_0 \\ -h_1 \end{Bmatrix} = -\frac{\Gamma}{\Delta x} E \begin{Bmatrix} \bar{\psi}_0 \\ \psi_1 \end{Bmatrix}, \quad \text{where } E = \begin{bmatrix} 1 & -1 \\ -1 & 1 \end{bmatrix} \tag{8.1}$$

Values $\bar{\psi}_0$ or $h_0 = h_{1-0} = -h_{0-1}$ are often specified in applied engineering problems as boundary conditions. This will be illustrated in the numerical example that follows.

The last finite element N of the linear area consists of nodes with numbers $n-1$ and n, and it is subjected to the impact of a potential difference $(\bar{\psi}_{n-1}-\bar{\psi}_n)$. The equation of this finite element is similar to Equation 8.1 and it has the following form:

$$\begin{Bmatrix} h_{(n-1)-(n)} \\ -h_{(n)-(n-1)} \end{Bmatrix} = -\frac{\Gamma}{\Delta x} E \begin{Bmatrix} \bar{\psi}_{n-1} \\ \bar{\psi}_n \end{Bmatrix}.$$

(Note that values of $\bar{\psi}_n$ or $h_n = h_{(n-1)-(n)} = -h_{(n)-(n-1)}$ are also specified in the boundary conditions.)

An arbitrary finite element of the internal area, the ith one, for instance, consists of nodes numbered i and $i+1$. In this case for them, the next equation is valid:

$$\begin{Bmatrix} h_{(i)-(i+1)} \\ -h_{(i+1)-(i)} \end{Bmatrix} = -\frac{\Gamma}{\Delta x} E \begin{Bmatrix} \bar{\psi}_{i} \\ \bar{\psi}_{i+1} \end{Bmatrix}$$

The last three equations describe the diffusion processes that take place in all N elements of the global area. They cannot be separately solved, and they should be joined in a system where the *redundant* algebraic equations should be removed. This process is schematically shown in Figure 8.7. for an area consisting of four finite elements numbered 1, 2, 3, and 4 (see Figure 8.6). The matrix equations in Figure 8.7 correspond to these finite elements. As a result, the system of equations of the global area gets the form

$$-\frac{\Gamma}{\Delta x}\begin{bmatrix} 1 & -1 & 0 & 0 & 0 \\ -1 & 2 & -1 & 0 & 0 \\ 0 & -1 & 2 & -1 & 0 \\ 0 & 0 & -1 & 2 & -1 \\ 0 & 0 & 0 & -1 & 1 \end{bmatrix} \cdot \begin{Bmatrix} \bar{\psi}_0 \\ \bar{\psi}_1 \\ \bar{\psi}_2 \\ \bar{\psi}_3 \\ \bar{\psi}_4 \end{Bmatrix} = \begin{Bmatrix} -\frac{h_0}{\Gamma}\cdot\Delta x \\ 0 \\ 0 \\ 0 \\ -\frac{h_4}{\Gamma}\Delta x \end{Bmatrix}$$

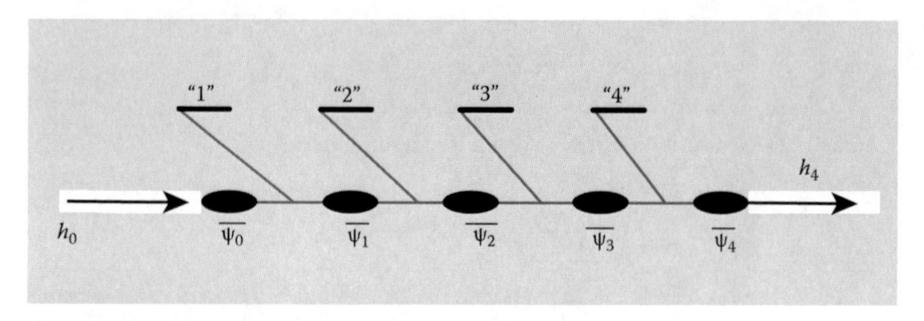

FIGURE 8.6
One-dimensional global area.

$$-\frac{\Gamma}{\Delta x}\begin{bmatrix}1 & -1\\ -1 & 1\end{bmatrix}\cdot\begin{Bmatrix}\varphi_0\\ \varphi_1\end{Bmatrix}=\begin{Bmatrix}h_0\\ -h_1\end{Bmatrix}$$
$$-\frac{\Gamma}{\Delta x}\begin{bmatrix}1 & -1\\ -1 & 1\end{bmatrix}\cdot\begin{Bmatrix}\varphi_1\\ \varphi_2\end{Bmatrix}=\begin{Bmatrix}h_1\\ -h_2\end{Bmatrix}$$
$$-\frac{\Gamma}{\Delta x}\begin{bmatrix}1 & -1\\ -1 & 1\end{bmatrix}\cdot\begin{Bmatrix}\varphi_2\\ \varphi_3\end{Bmatrix}=\begin{Bmatrix}h_2\\ -h_3\end{Bmatrix}$$
$$-\frac{\Gamma}{\Delta x}\begin{bmatrix}1 & -1\\ -1 & 1\end{bmatrix}\cdot\begin{Bmatrix}\varphi_3\\ \varphi_4\end{Bmatrix}=\begin{Bmatrix}h_3\\ -h_4\end{Bmatrix}$$

$$\rightarrow\ 1.\psi_0 - 1.\psi_1 = -\frac{h_0}{\Gamma}.\Delta x$$
$$\rightarrow\ -1.\psi_0 + 1.\psi_1 + 1.\psi_1 - 1.\psi_2 = \frac{(-h_1+h_1)}{\Gamma}.\Delta x$$
$$\rightarrow\ -1.\psi_1 + 1.\psi_2 + 1.\psi_2 - 1.\psi_3 = \frac{(-h_2+h_2)}{\Gamma}.\Delta x$$
$$\rightarrow\ -1.\psi_2 + 1.\psi_3 + 1.\psi_3 - 1.\psi_4 = \frac{(-h_3+h_3)}{\Gamma}.\Delta x$$
$$\rightarrow\ -1.\psi_3 + 1.\psi_4 = -\frac{h_4}{\Gamma}.\Delta x$$

FIGURE 8.7
Design of the global matrix of the matrix equations.

It stands clear that the solution depends on the values of $\bar{\psi}_0$ or $\bar{\psi}_4$ (i.e., on the boundary conditions), as well as on the medium physical properties such as conductivity and overall dimensions.

Example 8.7: (Transfer in a 2D Area)

We shall demonstrate herein a technique of designing the general conductivity matrix $[G^{(0)}]$ and the load vector $\{f^{(0)}\}$ of the global area (Fagan 1984).

Consider a global area shaped as a rectangle with dimensions 0.2 × 0.2 m (see Figure 8.8), with discretization that is performed using eight simple finite elements ($m = 8$). The sequence of ordering the nodes of each finite element is given in the following address (Table 8.5).

The nodes of each finite element are outflanked in positive direction (counterclockwise) with a formal origin—at the node denoted by a star at scheme (see Figure 8.8). For instance, the finite element numbered 2 is defined by nodes 4, 1, and 5 according to that sequence (see Table 8.5). Hence, the discrete analogue of that global area is *presented* by 9 nodes.

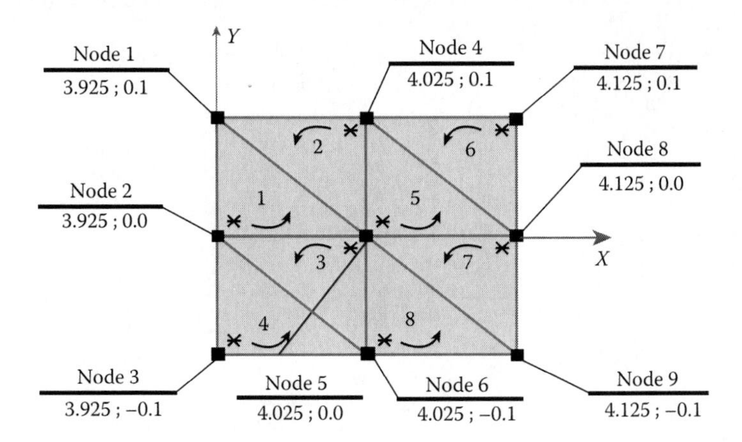

FIGURE 8.8
Two-dimensional global area.

TABLE 8.5

Sequence of Ordering the Nodes of Each Final Element in 2D Area

Number of the Finite Element	Successive Nodes (*i*–*j*–*k*)
1	2–5–1
2	4–1–5
3	5–2–6
4	3–6–2
5	5–8–4
6	7–4–8
7	8–5–9
8	6–9–5

Using the software package Derive 5, we calculate the approximating functions P_i (i = 1–3) using relation (Equation 5.43), and then, using relation (Equation 5.45), we find the conductivity matrices

$$[G^{(1)}]=[G^{(4)}]=[G^{(5)}]=[G^{(8)}]=1.44\begin{bmatrix} 2 & -1 & -1 \\ -1 & 1 & 0 \\ -1 & 0 & 1 \end{bmatrix}$$

$$[G^{(2)}]=[G^{(3)}]=[G^{(6)}]=[G^{(7)}]=1.44\begin{bmatrix} 1 & 0 & -1 \\ 0 & 1 & -1 \\ -1 & -1 & 2 \end{bmatrix}$$

Each component of the conductivity matrix $[G^{(e)}]$, corresponding to a specific finite element, has two addresses (see Figure 8.9):

- A local address indicating the position (row and column) of the matrix component
- An absolute address specified by the number of nodes constructing the finite element and ordered in compliance with the address table

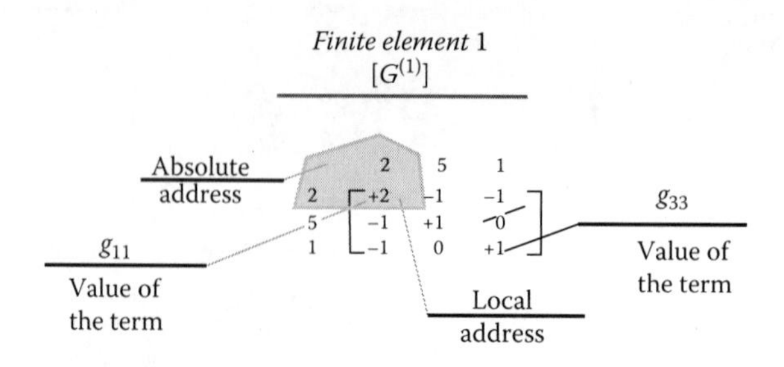

FIGURE 8.9
Addressing the conductivity matrix of a finite element.

For instance, the finite element numbered "1," composed of nodes 2–5–1, has a conductivity matrix $[G^{(1)}]$ with term numbers $g_{i,j}$. For example, the term number g_{11} has a value +2, that is, $g_{11} = +2$, having a local address (1, 1) and an absolute address (2, 2), and the term number g_{33} (with value $g_{33} = 1$) has a local address (3, 3) and an absolute address (1, 1) (see Figure 8.9).

Since the model of the global area considered contains nine nodes, the global conductivity matrix $[G^{(0)}]$ formed has dimensionality 9 × 9, and it should be zero at the beginning. The next step which is shown in Figure 8.10 is the transfer of terms of the conductivity matrices $[G^{(e)}]$, $1 \leq e \leq 8$ to the newly formed matrix $[G^{(0)}]$, and they should be transferred to their absolute addresses starting with matrix $[G^{(1)}]$.

We shall describe the process of preaddressing the already considered term g_{33} of matrix $[G^{(1)}]$ of the finite element 1 with a local address (3, 3) and absolute address (1, 1). The value of g_{33} ($g_{33} = 1$) is put into matrix $[G^{(0)}]$ in a cell with a local address (1, 1), corresponding to the absolute address of term g_{33} in matrix $[G^{(1)}]$. It is added to the value inherited in the same cell—it is zero at this stage. This procedure is applied to all remaining terms $g_{i,j}$ of matrix $[G^{(1)}]$.

We shall follow the process of preaddressing the other aforementioned term of matrix $[G^{(1)}]$, that is, g_{11} ($g_{11} = +2$), having a local address (1, 1) and an absolute address (2, 2). It is put into the global matrix in a cell with a local address (2,2), and its value (+2) is added to the available value, which is zero. The preaddressing of the conductivity matrix $[G^{(2)}]$ of finite element 2 is performed in the same way. For instance, term g_{33} ($g_{33} = 2$) with a local address (3, 3) has an absolute address (5, 5), and value (+2) will be added to the value available in cell (5, 5) of the global matrix $[G^{(0)}]$.

After performing this operation to all finite elements of the area, we find the final global conductivity matrix $[G^{(0)}]$ (see Figure 8.10). As seen in Figure 8.10, some of the matrix terms remain equal to zero. They correspond to nodes of the finite elements between which no diffusion takes place. For instance, Figure 8.10 shows that the terms of $[G^{(0)}]$ $g_{71} - g_{76}$ *and* g_{79}, with addresses from (7, 1) to (7, 6) and (7, 9) are zero in the global matrix.

This is so since node No. 7 does not participate in combinations with nodes 1, 2, 3, 5, 6, and 9 to form finite elements of the global area

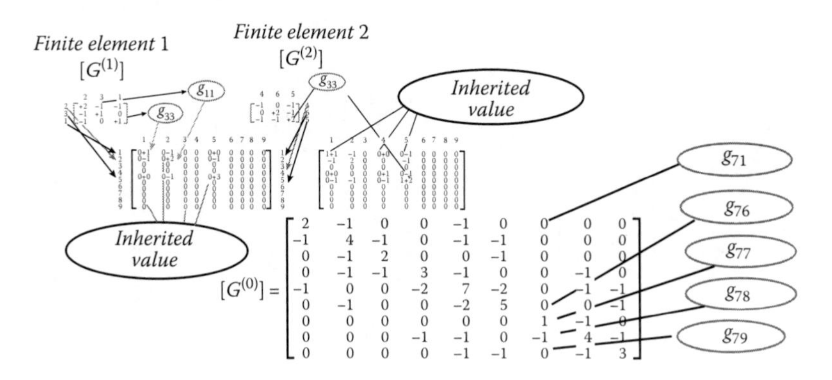

FIGURE 8.10
Preaddressing of the conductivity matrix of the global area.

(see Figure 8.8). Besides, as Figure 8.10 shows, the term $g_{7\,8}$ of the global conductivity $[G^{(0)}]$ that has a local address (7, 8) has a value $g_{7\,8} = -1$, which proves that diffusion would take place between nodes 7 and 8 only. Hence, the link between the characteristics of the diffusion process in the global and those valid in the neighborhood of node 7 (excluding node 8) is very weak.

The next step of designing the global matrix equation is a set up of the area load vector.

Since the matrix equations of the particular finite elements are already derived, their load vectors are defined and the procedure of their summation is to be performed. Table 8.6 shows the load vectors of all finite elements for the case under consideration. They are found after assuming that the convective heat exchange at the boundaries and the generation of substance can be disregarded. The set up of the load vector of the global area can be performed in the same manner as that of designing the global conductivity matrix.

Initially, the global load vector $\{f^{(0)}\}$ is to be a matrix column with rank 9, which contains only zero terms—see Figure 8.11. During the *second step,* $\{f^{(0)}\}$ is filled by the components of the load vector $\{f^{(1)}\}$, belonging to the finite element. The way of filling $\{f^{(0)}\}$ is shown in Figure 8.11

TABLE 8.6

Vectors of the Load of the Finite Elements of the 2D Area

Finite Element	$\{f^{(e)}\}_1$	$\{f^{(e)}\}_2$	$\{f^{(e)}\}_3$	$\{f^{(e)}\}_4$
Load Vector	$\begin{Bmatrix} h_{0-1} - h_{1-3} \\ -1.41h_{1-2} - h_{1-3} \\ -1.41h_{1-2} + h_{0-1} \end{Bmatrix}$	$\begin{Bmatrix} -h_{2-5} - h_{2-0} \\ 1.41h_{1-2} - h_{2-0} \\ 1.41h_{1-2} - h_{2-5} \end{Bmatrix}$	$\begin{Bmatrix} -h_{3-8} + h_{1-3} \\ -1.41h_{3-4} - h_{1-3} \\ -1.41h_{3-4} - h_{3-8} \end{Bmatrix}$	$\begin{Bmatrix} h_{0-4} - h_{4-0} \\ 1.41h_{3-4} - h_{4-0} \\ 1.41h_{3-4} + h_{0-4} \end{Bmatrix}$
Finite Element	$\{f^{(e)}\}_5$	$\{f^{(e)}\}_6$	$\{f^{(e)}\}_7$	$\{f^{(e)}\}_8$
Load Vector	$\begin{Bmatrix} h_{2-5} - h_{1-7} \\ -1.41h_{5-6} - h_{1-7} \\ 1.41h_{5-6} + h_{2+5} \end{Bmatrix}$	$\begin{Bmatrix} -h'_{6-0} - h'''_{6-0} \\ 1.41h_{5-6} - h''_{6+0} \\ 1.41h_{5-6} - h'_{6-0} \end{Bmatrix}$	$\begin{Bmatrix} -h_{7-0} + h_{5-7} \\ 1.41h_{8-7} + h_{5-7} \\ 1.41h_{9-7} - h_{7-0} \end{Bmatrix}$	$\begin{Bmatrix} h_{3-8} - h_{8-0} \\ -1.41h_{8-7} - h_{8-0} \\ -1.41h_{8-7} + h_{3-8} \end{Bmatrix}$

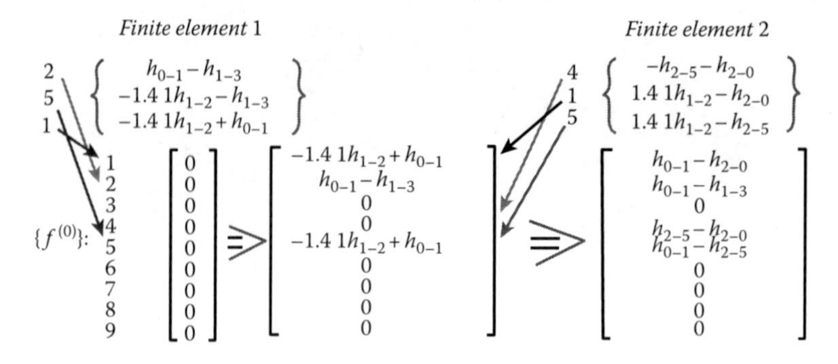

FIGURE 8.11
Preaddressing the load vector.

This procedure is repeated for all finite elements until finishing their full list valid for the area.

After completing the preaddressing of all components of $\{f^{(0)}\}$, it gets the following final form:

$$\{f^{(0)}\} = \begin{Bmatrix} h_{0-1} - h_{2-0} \\ h_{0-1} - h_{0-4} \\ h_{0-4} - h_{4-0} \\ -h_{2-0} - h''_{6-0} \\ 0 \\ -h_{40} - h_{80} \\ -h'_{60} - h''_{60} \\ -h'_{60} - h_{70} \\ -h_{70} - h_{80} \end{Bmatrix}.$$

It is seen in this relation that the fifth component of the vector is zero only. This means that only the fifth finite element is quantitatively balanced, and inflowing and outflowing fluxes are equated. The processes that develop in all other finite elements will depend on the condition of diffusion exchange between these elements.

The conductivity matrix $[G^{(0)}]$ and the load vector $\{f^{(0)}\}$ found are put in Equation 5.59, thus obtaining a system of equations valid for the global area:

$$\begin{bmatrix} 2 & -1 & 0 & 0 & -1 & 0 & 0 & 0 & 0 \\ -1 & 4 & -1 & 0 & -1 & -1 & 0 & 0 & 0 \\ 0 & -1 & 2 & 0 & 0 & -1 & 0 & 0 & 0 \\ 0 & -1 & -1 & 3 & -1 & 0 & 0 & -1 & 0 \\ -1 & 0 & 0 & -2 & 7 & -2 & 0 & -1 & -1 \\ 0 & -1 & 0 & 0 & -2 & 5 & 0 & 0 & -1 \\ 0 & 0 & 0 & 0 & 0 & 0 & 1 & -1 & 0 \\ 0 & 0 & 0 & -1 & -1 & 0 & -1 & 4 & -1 \\ 0 & 0 & 0 & 0 & -1 & -1 & 0 & -1 & 3 \end{bmatrix} \begin{Bmatrix} \phi_1 \\ \phi_2 \\ \phi_3 \\ \phi_4 \\ \phi_5 \\ \phi_6 \\ \phi_7 \\ \phi_8 \\ \phi_9 \end{Bmatrix} = \begin{Bmatrix} h_{0-1} - h_{2-0} \\ h_{0-1} - h_{0-4} \\ h_{0-4} - h_{4-0} \\ -h_{2-0} - h''_{6-0} \\ 0 \\ -h_{40} - h_{80} \\ -h'_{60} - h''_{60} \\ -h'_{60} - h_{70} \\ -h_{70} - h_{80} \end{Bmatrix}$$

We shall consider in what follows a technique of adding boundary conditions to the matrix form of the system of equations found, which describes diffusion processes developing in the global area.

Example 8.8

Determine the class of energy efficiency of the building of residential type shown in Figure 7.16 and used in Section 7.5.3.

Solution

1. *Spherical standard value*:
   In this case, similarly to examples in Section 7.1, we apply Equation 226 (Dimitrov 2013) to determine the spherical standard value, but here,
   - The building volume is V_{Buil} = 9.0 * 63.00 = 567.00 m^3
   - The gross area of the envelope is

$$A_0 = 2(54 + 108 + 63) = 450 \text{ m}^2;$$

   - The referred value of the spherical standard is the same:

$$U_{ref} = 0.5 \text{W}/(\text{m}^2 \text{K})$$

   Thus, we have

$$U_{Sp.St} = 4.84 U_{Ref} \frac{V_{Buil}^{0.666}}{A_0} = 4.84 * 0.5 \frac{567.0^{0.666}}{450.0} = 0.369 \text{ W}/(\text{m}^2 \text{K}).$$

2. *U-value of the wall* (see Figure 8.12)
   How it is known from the theory (see Down 1969), in the case of multilayer envelope structure, the *R*-value (thermal resistivity) of wall is $R_{Wall} = \sum_3 R_i = \sum_3 (\delta_i/\lambda_i) = 0.89 \text{ m}^2 \text{K/W}$ (see Figure 8.12). Here, δ_i(m) and λ_i(W/(Km)) are the thickness and coefficient of conductivity of the *i*th layer. After calculations (see the table in Figure 8.12): $U_{Wall} = 1/R_{Wall}$ = 1.123 W/m^2K. We accept

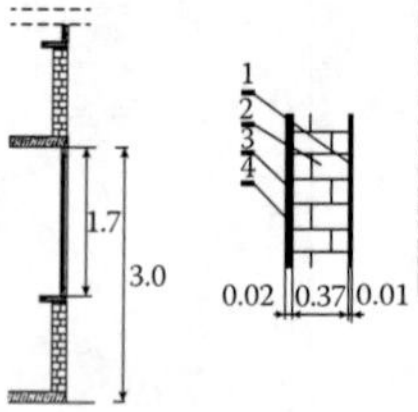

	δ_i (m)	λ_i (W/mK)	$R_i = \delta_i/\lambda_i$ (m^2K/W)
1	0.01	0.72	0.0139
2	0.37	0.52	0.71
3	0.02	0.12	0.167

$R_{Wall} = \sum_3 R_i = 0.89$ m^2K/W

$U_{Wall} = \frac{1}{R_{Wall}} = 1.123$ W/m^2K

1 – Cement plaster, sand
2 – Air-bricks
3 – Lime plaster, perlite
4 – Paint

FIGURE 8.12
Structure of the building envelope—before.

that coefficients of transitivity U_3 of walls, floor, and roof are equal of 0.336 W/m² K; the coefficients of transitivity of outside door U_1 and windows U_2 are taken from *the catalogs of manufacturers* (see Tables 10 and 11, Dimitrov A.V., 2013):

- For outside door, $U_1^{OD} = 4.00$ W/m² K (wooden, single, no glazing—Table 11)
- For garage door, $U_1^{GD} = 4.6$ W/m² K (wooden, double, no glazing—Table 11)
- For balcony door, $U_1^{GD} = 3.2$ W/m² K—(wooden, single, 80% glazed—Table 11)
- For windows, $U_2 = U_{window} = 2.4$ W/m² K—(double, glazed *e*-0.1; 12.7 mm—Table 10)

3. *U-value of the building*

 3.1. Unfolding of the envelope: See Figure 8.13 (similar to Figure 7.17a).
 You should calculate the next areas:
 A_F, m, components of the envelope gross area; A_1, m², outside door area; A_2, m², *windows and glazing elements* area; A_3, m², solid-wall facade area

 3.2. Table with data (Table 8.7)
 The total coefficient of total heat transitivity of the building U_{Buil} is determined by an application of Equation 224 (Dimitrov 2013):

$$U_{Buil} = \frac{\sum_{1-3}\sum_{1-6} U_i A_i}{A_0} = \frac{(101.72 + 54.05 + 417.3)}{450} = \frac{573.07}{450}$$

$$= 1.273\ \text{W}/(\text{m}^2\,\text{K})$$

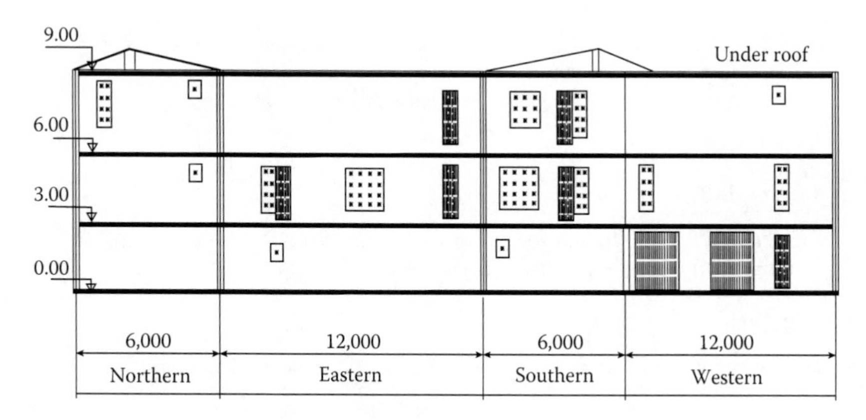

FIGURE 8.13
Unfolding of the building facade, estimated building.

TABLE 8.7

Description of the Elements of the Envelope and Their Contribution to Heat Transfer of the Building

Facade	A_F(m²)	A_1(m²)	A_2(m²)	A_3(m²)	U_1A_1 (W/K)	U_2A_2 (W/K)	U_3A_3(W/K)
N	54	—	2.76	51.24	0	6.62	57.54
E	108	4.41	5.44	98.15	3 * 1.46 * 3.2	13.06	110.22
S	54	2.94	9.88	41.18	2 * 1.46 * 3.2	23.71	46.24
W	108	17.31	4.44	86.25	1 * 2.3 * 4.0 + 2 * 7.5 * 4.6	10.65	96.86
Roof	63	—	0	63	0	0	70.94
Floor	63	—	0	63	0	0	70.94 * 0.5
					$\sum_6 U_1A_1 = 101.72$	$\sum_6 U_2A_{22} = 54.05$	$\sum_6 U_3A_3 = 417.5$

4. *Index of energy efficiency:*
   After processing, we employ Equation 7.7 to calculate the index of energy efficiency value:

$$IEE_{Buil}^{Envelope} = \frac{U_{Buil}}{U_{Sp.S.}} = \frac{1.273}{0.369} = 3.45$$

5. *Class of energy efficiency:*
   You can determine the class of energy efficiency of the building envelope from Table 8.8 (a sample from Table 8.9). See just a part from this table.
   For a value $IEE_{Buil}^{Envelope} = 3.45$, the class of energy efficiency of the building envelope **is F**.

The recommendation *is to make an additional insulation*

Example 8.9

Calculate the thickness of additional insulation (see position 5 in Figure 8.14) if *polyfoam material is used for this.*

TABLE 8.8

Determination of the Class of Energy Efficiency of a Building Envelope

Index of Energy Efficiency	Class of Energy Efficiency	Recommendation *What to Do*
$2.0 < IEE_{Env} \leq 4.0$	→ *F*	→ To make an additional insulation

TABLE 8.9

Correlation Table between the Index and Class of Energy Efficiency of Building Envelope: IEE_{Env} and CEE_{Env}

Index of Energy Efficiency of a Building Envelope $IEE_{Env} = U_m/U_{SpSt}$	Class of Energy Efficiency of a Building Envelope, $-CEE_{Env}$	Recommendation (Target Year 2010)	Recommendation (Target Year 2020)
Less than 0.25	A^0	Hyper-insulated	Hyper-insulated
$0.25 < IEE_{Buil} \leq 0.5$	$A+$	Excellently insulated	Excellently insulated
$0.5 < IEE_{Buil} \leq 0.75$	A	Excellently insulated	Very well insulated
$0.75 < IEE_{Buil} \leq 1.0$	B	Very well insulated	Well insulated
$1.0 < IEE_{Buil} \leq 1.25$	C	Very well insulated	Well insulated
$1.25 < IEE_{Buil} \leq 1.5$	D	Well insulated	Additional thermal insulation
$1.5 < IEE_{Buil} \leq 2.0$	E	Well insulated	Additional thermal insulation
$2.0 < IEE_{Buil} \leq 4.0$	F	Additional thermal insulation	Additional thermal insulation
$IEE_{Buil} > 4.0$	G	Additional thermal insulation	Additional thermal insulation

1 2 3 6 4 5

0,02 0,37 δ 0,02

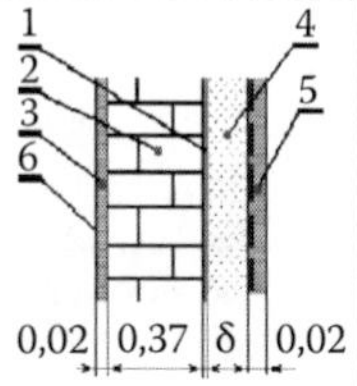

	δ_i (m)	λ_i (W/mK)	$R_i = \delta_i/\lambda_i$ (m²K/W)
1	0.01	0.72	0.0139
2	0.37	0.52	0.71
3	0.02	0.16	0.125
4	? 0.07	0.028	2.5
5	0.02	0.87	0.022

1–Cement plaster, sand;
2–Air-bricks;
3–Lime plaster, perlite;
4–Additional insulation –Polyfoam;
5–Cement plaster, sent
6–Paint

$$R_{Wall} = \sum_5 R_i = 3.37 \text{ m}^2\text{K/W}$$

$$U_{Wall} = \frac{1}{R_{Wall}} = 0.297 \text{ W/m}^2\text{K}$$

FIGURE 8.14
Structure of the building envelope—after.

Solution

The thickness of *the new additional layer* of insulation is determined by the next expression (see Equation 7.2):

$$\delta_{ec} = \frac{\lambda_{Ins}}{U_{ref}} * IEE^{-0.5} = \frac{0.03}{0.5} * 0.75^{-0.5} = 0.07\,\text{m},$$

where

- $U_{ref} = 0.5\,\text{W/K}\,\text{m}^2$, is the referred value of the energy standard, suggested by the Buildings Department
- λ_{Ins}, W/mK, is the coefficient of conductivity of the used insulation material
- IEE_{Env} is the index of energy efficiency, determined by the investor, according to his/her accepted class of energy efficiency (Table 8.9), for example for Class *A*, $IEE_{Env} = 0.5$; for Class B, $IEE_{Env} = 0.75$, and for Class C, $IEE_{Env} = 1.0$

The obtained value for the thickness amounts to 0.07 m and ensures new value $U_{Wall} = 0.297$ W/(m²K). Overall, this is a 70% decrease compared to the initial case (see Figures 8.12 and 8.14).

We could achieve the same effect in combination of the optimization of the size of the glass components of the building envelope, using the daylight coefficient methodology (Dimitrov 2011).

See Tables 8.10 through 8.13.

TABLE 8.10

Correspondence between the Class and Index of Energy Efficiency and Assessments of R_{ec}^{Env} and δ_{ec}

Class of Envelope Energy Efficiency CEE_{Env}	Index of Envelope Energy Efficiency $IEE_{Env} = \frac{SEM_{Env}}{EE_0^{Env}}$	Economically Profitable Thermal Resistance R_{ec}^{Env}, m² K/W ($R_{Sh.S_{ec}} = 2.0$ m² K/W)	Economically Profitable Envelope Thickness δ_{ec}, m $\left(\begin{array}{l} R_{Sh.S_{ec}} = 2.0\ \text{m}^2\ \text{K/W} \\ \lambda_{ins} = 0.03\ \text{W/mK} \end{array}\right)$
A^+	0.25	4	0.12
A	0.5	2.8	0.085
B	0.75	2.3	0.07
C	1.0	2.0	0.06
D	1.25	1.79	0.05
E	1.5	1.63	0.049
F	2.0	1.41	0.042
G	4.0	1	0.03

TABLE 8.11

Physical Properties of the Materials

	ΘD	ν_D	Tm	ρ	Cp	λ	α	M	cm	nvz
	K	THz	K	kg/ m^3	J/kg K	W/mK	106 m^2/s	kg mol	m/s	rp
Al	418	8.700452	933	2,702	903	237	97.13489	26.97	0.357191	3
Ag	225	4.683258	1235	10,500	235	429	173.86018	107.9	0.194142	1
An	165	3.434389	1336	19,300	129	317	127.32458	197	0.142052	1
Be	1160	24.1448	1550	1,850	1825	200	59.23732	9	0.780083	2
Bi	117	2.435294	545	9,780	122	7.86	6.587549	209	0.128859	5
Cd	300	6.244344	594	8,650	231	96.8	48.444811	112.4	0.279918	2
Cr	402	8.367421	2118	7,160	449	93.7	29.146085	52	0.308969	6
Co	445	9.262443	1769	8,862	421	99.2	26.588744	26.6	0.254763	8
Cu	339	7.056109	1358	8,933	385	401	116.59671	63.54	0.258747	1
Ge	366	7.6181	1211	5,360	322	59.9	34.706128	72.5	0.346097	4
Fe	467	9.720362	1810	7,870	447	80.2	22.797757	55.8	0.356065	8
Pb	117	2.435294	601	11,340	129	35.3	24.130812	207.2	0.122303	4
Mg	406	8.450679	923	1,740	1024	156	87.553879	24.3	0.388035	2
Mo	425	8.846154	2894	10,240	251	138	53.691484	96	0.355663	6
Ni	456	9.491403	1728	8,900	444	90.7	22.952728	59	0.339973	8
Nb	252	5.245249	2741	8,570	265	53.7	23.64545	92.9	0.221346	5
Pd	275	5.723982	1827	12,020	244	71.8	24.481056	106.6	0.225914	8
Pl	229	4.766516	2045	21,450	133	71.6	25.097709	195	0.189684	8
Si	658	13.69593	1685	2,330	712	148	89.212519	12	0.450983	4
Ta	231	4.808145	3269	16,600	140	57.5	24.741824	180.6	0.203146	5
Ti	278	5.786425	505	7,310	227	66.6	40.135714	47.9	0.206464	4
W	379	7.888688	3660	19,300	132	174	68.299576	183.8	0.318831	6
V	273	5.682353	2192	6,100	489	30.7	10.291998	50.9	0.219761	5
Zn	308	6.41086	693	7,140	389	116	41.764778	65.2	0.255501	2
Zr	270	5.61991	2125	6,570	278	22.7	12.428413	91.2	0.257533	4

TABLE 8.12

Physical Properties of the Crystal Materials

	Mass	n_{vz}	$K_{t=300K}$	$n_{vz}\frac{\theta_D^3}{\theta_F}$	T_m	$\rho \cdot c_p$	α
Crystal	K mol	Number of Electrons	W/m K	$\times 10^3$	K	$J/m^3K \times 10^6$	$m^2/s \times 10^6$
Ag	107.9	1	429	1.71	1235	2.47	174
Al	26.97	3	237	1707	933	2.46	97.1
Pb	207.9	4	36.7	42.5	601	1.46	251

TABLE 8.13

Coefficients of the Mathematical Model of the Shading Devices of Second Degree

Type of Solar Shading Surface	a_0	a_1	a_2	a_3	a_4	a_5	a_6
Plane, parallel to Oz.	V	$tg\alpha$	l	—	—	—	—
Plane, parallel to Oy.	$-W$	$-tg\alpha$	—	l	—	—	—
Cylinder, axis parallel to Oz: $K = 1$, circle, $K < 1$, and $K > 1$ ellipse.	$K^2(V^2 - l^2)$	—	$2K^2V$	—	l	K^2	—
Cylinder, axis parallel to Oy: $K = 1$, circle, $K < 1$, and $K > 1$ ellipse.	$K^2(W^2 - l^2)$	—	—	$-2K^2W$	l	—	K^2
Parabolic cylinder, axis parallel to Oz.	$(V^2 - l^2)$	l K	$-2V$	—	—	l	—
Parabolic cylinder, axis parallel to Oy.	$(W^2 - l^2)$	l K	—	$-2W$	—	—	l
Ellipsoids: $l = \sigma = 1$—*Spherical.* $\sigma = 1; \tau > 1$—*Stretched circular.* $\sigma < 1; \tau = 1$—*Collapsed circular.*	$\sigma^2(V^2 + \tau^2 \cdot W^2 - \tau^2 l^2)$	—	$2 \cdot \sigma^2$ V^2	$-2 \cdot W$ $\tau^2 \cdot V^2$	τ^2	σ^2	$\tau^2\sigma^2$

References

Банов, И., 1999, *Топлинен режим на сгради при периодично прекъсната работа на отоплителната инсталация*, [Thermal regime of buildings in periodically interrupted operation of the heating system]. Doctor of Philosophy, SSC "Energy technologies and machineries." TU-Sofia, Sofia (Bul).

Богословский, В.Н., 1982, Строительная теплофизика [Building thermophysics]. Moscow (Russ).

Бъчваров, С., 2006, *Теоретична механика-част първа* [Theoretical mechanics—First part]. TU of Sofia, Sofia (Bul).

Власов, О.Е., 1927, *Плоские тепловые волны* [Flat heat waves], Reports of Heat Technical Institute, N3/26. Moskow (Russ).

Димитров, А.В., 1986, *Оптимизация на топлинното съпротивление на плътните архитектурно-строителни елементи* [The thermal resistance of solid constructional element optimization], Annual reports of UCAG, XXXII, 1985–1986, part 8. Sofia, (Bul).

Димитров, А.В., Я.Ц. Александров,1988, *Изчисляване на икономически изгодната дебелина на допълнителната топлоизолация* [The economically profitable thickness of the additional thermoinsulation calculation]. *J. Constr.*, 7. Sofia (Bul).

Димитров, А.В., 1990, *Определяне на коефициента на естествено осветление с ЦЕИМ* [Computation of the daylighting coefficient]. *J. Build. Act.*, 1. GUSV, Sofia (Bul).

Димитров, А.В., и колектив, 2003, *Подобряване енергийните характеристики на сграда "Студентско общежитие №2" на ВТУ "Тодор Каблешков"* [Energy characteristics of the "Student housing №2" of TU "Todor Kableshkov" improvement, Scientific report of contract №2075/30.07.2003. DES, Sofia (Bul).

Димитров, А.В., В. Недев, 2004, *Стъпкова минимизация на размерите на прозорците при саниране на жилищни и обществени сгради* [Step minimizing the size of the windows in the renovation of residential and public buildings, *XIV International Scientific Conference—Transport 2004*, TU "Todor Kableshkov," Nov. 11–12, 2004. Sofia (Bul).

Димитров, А.В., 2006, *Метод на крайните елементи, приложен за оценка на преноса на топлина и влага през плътна стена с дифузия* [Method of finite elements, applied to estimation of heat and moisture transfer through solid walls], *International Scientific Conference VSU 2006*. Sofia (Bul).

Димитров, А.В., 2008a, *Енергийна ефективност на сградите, техните системи и инсталации, монография—част първа* [*Energy Efficiency of the Building, Their Systems and Installations—First Past*, Monograph. TU "Todor Kableshkov," Sofia (Bul).

Димитров, А.В., 2008b, *Еволюция в енергийните функции на фасадите* [Energy functions of the facades evolution], *International Scientific Conference VSU 2008*, May 18–19, 2008, VSU, Sofia and *J. Power Eng.*, 6, 2008, Sofia (Bul).

Димитров, А.В., Д. Назърски, 2009, *Методика за оптимизиране дебелината на топлоизолацията с използване индекса на енергийна ефективност на сградната обвивка* [Methodology for thickness of insulation optimization by using the index of energy efficiency of the building envelope], *International Scientific Conference UCAG 2009*, Sofia and *J. Constr.*, 3, 2010 (Bul).

Димитров, А.В., Н. Бояджиев, 2010, *Сградната конструкция и предизвикателствата на Наредба №7* [The building structural system and the challenges of code N7], *International Scientific Conference VSU 2010*, VSU, Sofia (Bul).

Димитров, А.В., 2011a, *Теоретично и приложно изледване на интегрираните функции на сградната обвивка спрямо околната среда от енергийна гледна точка (преход от микро към макроскопични оценки)* [Theoretical and applied research of the integrated functions of the building envelope to the environment from energy point of view (a transition from micro to macroscopic evaluation)], Dissertation for awarding the degree "Doctor of Science" presented in Department "Building Structures" of TU "Todor Kableshkov," Sofia (Bul).

Димитров, А.В., 2011c, *Екологическите предизвикателства пред проектирането на Енергийната система на Зелените университетски кампуси в метода "BG_LEED"* [The ecological challenges for the design of energy systems of the Greens university campuses in the method "BG_LEED," *XIth International Scientific Conference VSU 2011*, Sofia (Bul).

Димитров, А.В., Д. Назърски, 2011d, *Приложение на българския метод за проектиране и строителство на екологично съобразени сгради (BG_LEED) за изграждане на Зелени университетски кампуси* [Application of Bulgarian method for design and construction of environmentally sustainable buildings (BG_LEED) to build a green university campuses, *XIth International Scientific Conference VSU 2011*, Sofia (Bul).

Димитров, А.В., Д. Назърски, Н. Николов, 2011f, Portfolio of green university campuses, green universities and schools, *First International Scientific Conference EPU 2011*, EPU, Pernik.

Димитров, А.В., 2012, *Математически модел за оценка на екологическата устойчивост на сградите и енергийните им системи* [A mathematical model for assessments of the environmental sustainability of buildings and their energy systems], *International Scientific Conference "Design and Construction of Buildings and Equipments*, September 13–15, 2012, VFU "Chernorizez Chraber," Varna (Bul).

Димитров, А.В., 2013a, *Сградната енергийна система в условията на екологична устойчивост* [Building energy system in conditions of environmental sustainability], *Sixth International Scientific Conference "Architecture, Constructions—Modernity,"* May 30–June 1, 2013, Vol. II, pp. 345–351, VFU "Chernorizez Chraber," Varna (Bul).

Димитров, А.В., 2013b, *Сградите с близко до нулево потребление на енергия – първа стъпка към екологическа устойчивост* [The near zero energy buildings—First step towards environmental sustainability, *Sixth International Scientific Conference "Architecture, Constructions—Modernity,"* May 30–June 1, 2013, Vol. II, pp. 352–357, VFU "Chernorizez Chraber," Varna (Bul).

Ландау, Л.Д., Е.М. Лифшиц, 1976, *Статистическая физика*, част 1 [*Statistical Physics—First Part*], 3rd edn, Moscow (Russ).

Ларинков, Н.Н., 1985, *Теплотехника* [Heat-Technics], Moscow (Russ).

Назърски, Д., 2004, *Строителни изолации* [*Building Insulations*], Monograph, UCAG, Sofia (Bul).

Назърски, Д., В. Найденова, А. Димитров, 2011a, *Изисквания на Българския метод за проектиране и строителство на екологично съобразени сгради (BG_LEED) към организацията на ландшафта и използваните строителни материали при изграждането на Зелени университетски кампуси* [The requirements of the Bulgarian method for design and construction of environmentally sustainable buildings (BG_LEED) the organization of the landscape and the building materials used in the construction of university], *XIth Internatoinal Scientific Conference VSU 2011*, VSU, Sofia (Bul).

Назърски, Д., В. Найденова, Д. Марков, А. Димитров, 2011b, Evaluation of the index of environmental sustainability for the organization of the landscape, used building materials and indoor environment in the construction of Green university campuses by means the method BG_LEED, *First International Scientific Conference EPU 2011*, EPU, Pernik (Bul).

Марков, А., 1988, *Използване на комплексните функции за оценка на нестационерния топлинен транспорт чрез твърдите стени* [The complex functions application for evaluation no stationary heat transfer through the solid walls], *J. Architect.*, 4, Sofia (Bul).

Тихомиров, К.В., 1981, *Теплотехниика, теплогазоснабжение и вентиляция* [*Heat-Technics, Heat-Gas Supply and Ventilation*], Moscow (Russ).

Addington D.N., D.L. Schodek, 2005, *Smart Materials and New Technology for Architecture and Design Professions*. Architecture Press, Oxford, U.K.

Adkins, P., 1908, *Thermodynamics*. Architecture Press, Oxford, U.K.

Akasmija, A., 2013, *Sustainable Facades: Design Methods for High-Performance Building Envelopes*. John Wiley & Sons, Hoboken, NJ.

American Society for Metals (ASM), 1961, *Metals Handbook, Volume 1, Properties and Selection of Metals, 8th edn., ASM*, Metals Park, OH.

Andrews, F.C., 1971, *Thermodynamics: Principles and Applications*. John Wiley & Sons, New York.

Anisimov, S.I. et al., 1974, Electron emission from metal surfaces exposed to ultrashort laser pulses. *Soviet Phys. JETP*, 67, 375–377.

Aston, J.G., J.J. Fritz, 1959, *Thermodynamics and Statistical Thermodynamics*. John Wiley & Sons, New York.

Atkins, C.J., 1983, *Equilibrium Thermodynamics*, 3rd edn. Cambridge University Press, Cambridge, U.K.

Atkins, P., 2007, *Four Laws, That Drive the Universe*. Oxford University Press, Oxford, U.K.

Attaard, P., 2002, *Thermodynamics and Statistical Physics*. Academic Press, New York.

Back, J.V. et al., 1985, *Inverse Heat Conduction*. A Wiley-Interscience Publ., New York.

Baierlein, R., 1999, *Thermal Physics*. Cambridge University Press, Cambridge, U.K.

Bailyn, M., 2002, *A Survey of Thermodynamics*. AIP Press, New York.

Ball, P., 1999, *Made to Measure: New Materials for 21st Century*. Princeton University Press, Princeton, NJ, pp. 103–105.

Bauman, R., 2002, *Modern Thermodynamics with Statistical Mechanics*. Clarendon Press, Oxford, U.K.

Bejan, A., 1993, *Heat Transfer*. John Wiley & Sons, New York.

Benjamin, P., 1998, *A History of Electricity (The Intellectual Rise in Electricity) from Antiquity to the Days of Benjamin Franklin*. John Wiley & Sons, New York.

Bennet, C.O., J.E. Myers, 1984, *Momentum, Heat and Mass Transfer*, 3rd edn. McGraw-Hill, New York.

Bevensee, R.M., 1993, *Maximum Entropy Solutions to Scientific Problems*. PTR Prentice Hall, NJ.

Bird, J., 2007, *Electrical and Electronic Principles and Technology*, 3rd edn. Newness, Portsmouth, U.K.

Bird, R.B., W.E. Stewart, E.N. Lightfoot, 1960, *Transport Phenomena*. John Wiley & Sons, New York.

Boelter, L.M.K. et al., 1965, *Heat Transfer Notes*. McGraw-Hill Company, New York.
Bohr, N., 1913, On the constitution of atoms and molecules, part 1. *Philos. Mag.*, 26, 1–29.
Bohr, N., 1958, *Atomic Physics and Human Knowledge*. John Wiley & Sons, New York.
Bosworth, R.C.L., 1952, *Heat Transfer Phenomena*. John Wiley & Sons, New York.
Brisken, W.R., G.E. Reque, 1956, Thermal circuit and analog computer methods, thermal response. *ASHRAE Trans.*, 62, 391.
Brorson, S.D. et al., 1990, Femtosecond room temperature measurement of the electron-phonon coupling constant in metallic semiconductors. *Phys. Rev. Lett.*, 64, 2172–2175.
Buchberg, H., 1955, Electrical analog prediction of thermal behavior of an inhabitable enclosure. *ASHREA Trans.*, 61, 339.
Buftington, D.E., 1975, Heat gain by conduction through exterior wall and roofs-transmission matrix method. *ASHRAE Trans.*, 81(2), 89.
Burch, D.M., J.E. Seem, G.N. Walton, B.A. Licitra, 1992, Dynamic evaluation of thermal bridges in a typical office building. *ASHRAE Trans.*, 98, 294.
Butler, R., 1984, The computation of heat flows through multi-layer slabs. *Building Environ.*, 19(3), 197–206.
Challis, L.J., 2003, *Electron-Phonon Interactions into Closed Space Structures*. Oxford University Press, Oxford, U.K.
Carrington, G., 1944, *Basic Thermodynamics*. Oxford Publication, Oxford, U.K.
Carrod, C., 1995, *Statistical Mechanics and Thermodynamics*. Oxford University Press, Oxford, U.K.
Carslaw, H.S., J.C. Jaeger, 1959, *Conduction of Heat in Solids*. Oxford University Press, London, U.K.
Carter, A.H., 2001, *Classical and Statistical Thermodynamics*. Prentice Hall, Upper Saddle River, NJ.
Cengel, Y.A., 1998, *Heat Transfer*. WCB/McGraw-Hill, New York.
Chapman, A.J., 1981, *Heat Transfer*, 5th edn. McGraw-Hill, New York.
Cheng, Y.C., 2006, *Macroscopic and Statistical Thermodynamics*. World Scientific Publishing, Singapore, p. 121.
Clarke, B., 2004, *Energy Forms (Allegory and Science in the Era of Classical Thermodynamics)*. University of Michigan Press, Ann Arbor, MI.
Clausius, R., 1976, On application of the theorem of equivalent of transfer to the internal works of a mass of thermodynamics. In *The Second Law of Thermodynamics* (ed. Kestin, J.). Dover Hutchinson & Ross Inc. Publishing, Stroudsburg, PA.
Clear R., V. Inkarojrit, E. Lee, 2006, *Subject Responses to Electrochromic Windows*. LBNL 57125.
Clough, R.W., 1960, Finite element method in plane stress analysis. *Proceedings of the 2nd Conference of Electronic Computations*, ASCE, Pittsburgh, PA, pp. 345–378.
Craig, N.C., 1992, *Entropy Analysis*. VCH, New York.
Dampier, W.C., 1905, *The Theory of Experimental Electricity*. Cambridge University Press, Cambridge, U.K.
DeBroglie, L., 1953, *The Revolution in Physics*. Noonday Press, New York.
Debye, P., 1928, *Polar Molecules*. The Chemical Catalog Company, New York.
Desai, P.D. et al., 1976, Thermophysical properties of carbon steels. CINDAS, Special Report, Purdue University, West Lafayette, IN.
Dimitrov, A.V., 2006, An approach and criteria for heat insulation efficiency estimation. LBL Seminar—Report. EETD, Lawrence Berkeley National Laboratory, Berkeley, CA, March 2006.

Dimitrov, A.V., 2008, A new statistical methodology for Delta_Q method collected data manipulation. *HVAC Res. J.*, 14(5), 707–718, September 2008.

Dimitrov, A.V., 2009, Evolution in the energy functions of the building envelope. EPU Seminar—Report, May 22–27, 2009. Center for Environmental Energy at EPU, Bulgaria.

Dimitrov, A.V., 2013, *Practical Handbook of FEM for Civil Engineers*. EPU, Pernik, Bulgaria.

Dimitrov, A.V., 2014a, The inside of the modern heat and energy technologies. EPU, Pernik, Bulgaria.

Dimitrov, A.V., 2014b, Engineering Installations in Buildings. EPU, Note in Lectures.

Dimitrov, A.V., 2014c, The role of the spherical standard of the building envelope in preliminary investment design. In *International Conference on Civil Engineering Design and Construction (Science and Practice)*, Varna, Bulgaria, September 11–13, 2014.

Dimitrov, A.V., P.K. Kolev, 1990, A new method for solar screen sizing. *N.S. Conference*, MS-Burno, Czech Republic.

Dodge, R.F., 1944, *Chemical Engineering Thermodynamics*. McGraw-Hill Book Company, New York.

Duffin, W.J., 1980, *Electricity and Magnetism*, 3rd edn. McGraw-Hill, New York.

Duffve, J.A., W.A. Beckman, 1980, *Solar Engineering of Thermal Processes*. John Wiley & Sons, New York.

Duncan, A.B., G.P. Peterson, 1994, Review of microscale heat transfer. *ASME J. Appl. Mech. Rev.*, 47, 397–428.

Eckert, E.R.G., R.M. Darke, 1972, *Analysis of Heat and Mass Transfer*. McGraw-Hill, New York.

Edminister, J., 1965, *Electric Circuits*, 2nd edn. McGraw-Hill, New York.

Einstein, A., 1907, Plancksche Theorie der Strahlung und die Theorie der Spezifischen Wärme. *Annalen der Physik (ser. 4)*, 22, 180–190.

Einstein, A., 1919, Zum quantensatz von sammerfeld und Epstein. *Verhandlungen der Deutschen Physikalischen Gesellschft*, 19, 82–92.

Einstein, A., 1934, *Essays in Science*. Philosophical Library, New York.

Elgendy, K., 2010, Comparing Estidama's Perl rating system to LEED and BREEAM. *Carbon*, Middle East Sustainable Cities. *The 4th Biennial Subtropical Cities Conference, Braving A New World—Design Interventions for Changing Climates*, October 17–19, 2013, Fort Lauderdale, FL, http://www.academia.edu/5026017/Attia.

Fagan, M.J., 1984, *Finite Element Analysis*. John Wiley & Sons, New York.

Farouki, O.T., 1981, Thermal properties of soil. Monograph 81-1, U.S. Army Corps Cold Regions Research and Engineering Laboratory, Hanover, NH, December 1981.

Fast, J.P., 1962, *Entropy*. McGraw-Hill Book Comp., New York.

Fay, J.A., 1965, *Molecular Thermodynamics*. Addison-Wesley, Reading, MA.

Fermi, E., 1936, *Thermodynamics*. Dover Publ. Inc., New York.

Fermi, E., 1938, *Thermodynamics*. Black & Sons Ltd, New York.

Fine, P.C., 1939, The normal modes of vibration of body—Centered cubic lattice. *Phys. Rev.*, 56, 355–359.

Finn, C.B.P., 1993, *Thermal Physics*, 2nd edn. Chapman & Hill, London, U.K.

Fong, P., 1963, *Fundamentals of Thermodynamics*. Oxford University Press, Oxford, U.K.

Fuchs, H.U., 1996, *The Dynamics of Heat*. Springer, Berlin, Germany.

Galerkin, B.G., 1915, Series occurring in some problems of elastic stability of rods and plates. *Eng. Bull.*, 19, 897–908.

Gebhart, B., 1993, *Heat Conduction and Mass Diffusion*. McGraw-Hill Inc., New York.

Greiner, W., L. Neise, H. Stocker, 1995, *Thermodynamics and Statistical Mechanics.* Springer, New York.
Grenault, T., 1995, *Statistical Physics.* Chapman and Hall, London, U.K.
Grevin, A., G. Keller, G. Warnecke, 2003, *Entropy.* Princeton University Press, Princeton, NJ.
Griffith, D.J., 1999, *Introduction to Electrodynamics*, 3rd edn. Prentice Hall, Upper Saddle River, NJ.
Guggenheim, E.A., 1957, *Thermodynamics.* North-Holland Publ. Company, Amsterdam, the Netherlands.
Guha, E., 2000, *Basic Thermodynamics.* Alpha Science, Pegboard, U.K.
Guyer, R.A., J.A. Krumhansi, 1966, Solution of the linearized Boltzmann equation. *Phys. Rev.*, 148, 766–778.
Gyftopoulos, E.P., G.P. Beretta, 1991, *Thermodynamics: Fundamentals and Applications.* MacMillan Publishing Company.
Haar D., 1966, *Elements of Thermodynamics.* Addison-Wesley Publ. Company, Reading, MA.
Halliday, R.R., J. Walker, 2005, *Fundamentals of Physics*, 7th edn. Wiley & Sons, Hoboken, NJ.
Hammond, P., 1981, *Electromagnetism for Engineers.* Pergamon, New York.
Hang, K. 1987, *Statistical Mechanics.* John Wiley, New York.
Harkness, E.L., M.L. Mehta, 1978, *Solar Radiation Control in Buildings.* Applied Science Publishers Ltd., London, U.K.
Harrison, W.A., 1989, *Electronic Structure and the Properties of Solids: The Physics of Chemical Bond*, Dover Publication, New York.
Hatspulos, G.N., J.H. Keenan, 1968, *General Principles of Thermodynamics.* John Wiley & Sons, New York.
Helmholtz, H., 1883, Ueber die thatsächliche Grundlagen der Geometrie, *Wissenschaftliche Abhandlungen*, pp. 610–617, Leipzig.
Hill, T.L., 1986, *An Introduction to Statistical Thermodynamics.* Dover, New York.
Hirschfelder, J.O., C.F. Curtiss, R.B. Bird, 1954, *Molecular Theory of Gases and Liquids.* John Wiley & Sons, New York.
Holman, J.P., 1997, *Heat Transfer*, 5th edn. McGraw-Hill, New York.
Holman, J.P., 1969, *Thermodynamics.* McGraw-Hill Book Company, New York.
Hopkinson, B., 1913, *The Laws of Thermodynamics.* Cambridge University Press, Cambridge, U.K.
Hopness, G.V.R., 2009, Sustaining our future by rebuilding our past. *ASHRAE J.*, 8, 16–21, August.
Hook, J.R., H.E. Hall, 1991, *Solid State Physics*, 2nd edn. John Wiley & Sons, New York.
IEA, 1999, Daylighting simulation: Methods, algorithms and recourse. A reports of IEA SHC Task 21/ECBCS Annex 29, December, LBL Report 44296, LBL.
Incropera, F.P., 1974, *Introduction to Molecular Structure and Thermodynamics.* John Wiley & Sons, New York.
Incropera, F.P., D.P. DeWitt, 1985, *Introduction to Heat Transfer.* John Wiley & Sons, New York.
Incropera, F.P., D.P. DeWitt, 1996, 2002, *Fundamentals of Heat and Mass Transfer*, 3rd edn. John Wiley & Sons, New York.
Inoue, K., K. Ottaka, 2004, *Photonic Crystals.* Springer, Berlin, Germany.
Jackson, J.D., 1999, *Classical Electrodynamics*, 3rd edn. Wiley, New York.
Jaluria, Y., K.E. Torrance, 2003, *Computational Heat Transfer*, Taylor & Francis, New York.

Jancovici, B., 1973, *Statistical Physics and Thermodynamics*. John Wiley & Sons, New York.

Jocky, J., 1965, *The Thermodynamics of Fluids*. Heinemann Ed. Book Ltd., London, U.K.

Jones, J.C., 2001, *The Principles of Thermal Science*. CRC Press, Boca Raton, FL.

Joseph, D.D., L. Preziosi, 1989, Heat waves. *Rev. Modern Phys.*, 61, 41–73.

Joseph, D.D., L. Preziosi, 1990, Addendum to the paper on heat waves. *Rev. Modern Phys.*, 62, 375–391.

Kaganov, M.I. et al., 1957, Relaxation between electrons and crystal lattices. *Soviet Phys. JETP*, 4, 173–178.

Kakas, S., 1985, *Heat Conduction*. Hemisphere Publishing, New York.

Kakas, S., Y. Yener, 1993, *Heat Conduction*. Taylor & Francis, Washington, DC.

Kauzmann, W., 1966, *Theory of Gases*. W.A. Benjamin Inc., New York.

Kelly, D.C., 1973, *Thermodynamics and Statistical Physics*. Academic Press, New York.

Kelvin, W.T., 1872, *Reprint of Papers on Electrostatics and Magnetism*. Macmillan, London, U.K.

Kersten, M.S., 1949, Thermal properties of soils. Engineering Experiment Station Bulletin 28, University of Minnesota, Minneapolis, MN, June 1949.

Kestin, J., J.R. Dorfman, 1971, *A Course in Statistical Thermodynamics*. Academic Press, New York.

Kittel, C., 1996, *Introduction to Solid State Physics*, 7th edn. John Wiley & Sons, New York.

Kittel, C., H. Kroemer, 1980, *Thermal Physics*. W. H. Freeman, San Francisco, CA.

Klassen, F., 2003, The malleability of matter. *On Site Rev.*, 10, 48.

Klassen, F., 2004, Material innovations: Transparent, lightweight, malleable and responsive. *Transportable Environment, Proceedings of the Third International. Conference on Portable Architecture and Design*, Ryerson University, Toronto, Ontario, Canada, April 28–30, 2004, pp. 54–65.

Klassen, F., 2006, *Material Innovations: Transparent, Lightweight and Malleable Transportable Environment*. Spoon Press, London, U.K., pp. 122–135.

Kolesnikov, I.M. et al., 2001, *Thermodynamics of Spontaneous and Non-Spontaneous Processes*. Nova Science Publishers Inc., New York.

Kondeppud, D., 2008, *Introduction to Modern Thermodynamics*. John Wiley & Sons, New York.

Kraus, J.D., D.A. Fleisch, 1999, *Electromagnetic*. McGraw-Hill, New York.

Kreith, F., W.Z. Black, 1980, *Basic Heat Transfer*. Harper and Raw, New York.

Kreith, F., M.S. Bohn, 2001, *Principles of Heat Transfer*, 6th edn. Harper Row Publishers, New York.

Kutateladze, S.S., 1963, *Fundamentals of Heat Transfer*. Academic Press, New York.

Lawden, D.F., 1986, *Principles of Thermodynamics and Statistical Mechanics*. John Wiley & Sons, New York.

Lay, J.E., 1990, *Statistical Mechanisms and Thermodynamics*. Harper & Row, New York, Chap. 5.

Lee, J.C., 2002, *Thermal Physics: Entropy and Free Energy*. World Scientific, Hackensack, NJ, Chap. 5.

Lee, J.F., F.W. Sears, D.L. Turcotte, 1963, *Statistical Thermodynamics*. Addison-Wesley, Reading, MA.

Lee E.S., M. Yazdanian, S.E. Sellcovitz, 2004, *The Energy Saving Potential of Electrochromatic Windows in the US Commercial Building Sector*, LBNL 54966.

Lienhard, J.H., 1981, *A Heat Transfer Textbook*. Prentice Hill, Englewood Cliffs, NJ.

Lstlburek, J.W., 2009, The evolution of the walls. *ASHRAE J.*, Volume 6, June.

Lucas, K., 1991, *Applied Statistical Thermodynamics*. Springer-Verlag, Berlin, Germany.
Luikov, A.V., 1968, *Analytical Heat Diffusion Theory*. Academic Press, New York.
Macaulay, W.H., B. Hopkinson, 1913, *The Laws of Thermodynamics*. Cambridge Engineering Tracts. Cambridge University Press, Cambridge, U.K.
Mackey, C.O., 1946, Periodic heat flow-composite wall or roofs. *ASHRAE Trans.*, 52, 283.
Mackey, C.O., L.T. Wright, 1944, Periodic heat flow-homogenous wall or roofs. *ASHRAE Trans.*, 50, 293.
Mackey, M.C., 1993, *Times Arrow: The Origins of Thermodynamics Behavior*. Springer-Verlag, New York.
Majumdar, A., 1993, Microscale heat conduction in dielectric thin films. *ASME J. Heat Transfer*, 115, 7–16.
Martin, C.M., 1986, *Elements of Thermodynamics*. Prentice Hall, New York.
Martin, R.M., 2004, *Electronic Structure: Basic Theory and Practical Methods*. Cambridge University Press, Cambridge, U.K., Chap. 2.
Maxwell, J.C., 1878, *A Treatise on Electricity and Magnetism, Vol. 1*. Clarendon Press, Oxford, U.K.
McQuarrie, D.A., 1973, *Statistical Thermodynamics*. Harper & Row Publishing, New York.
Mitalas, G.P., 1968, Calculation of transient heat flow through walls and roofs. *ASHRAE Trans.*, 74(2), 182.
Mitalas, G.P., 1978, Comments on the Z-transfer function method for calculating heat transfer in buildings. *ASHRAE Trans.*, 84(1), 667–674.
Mitalas, G.P., D.G. Stephonson, 1967, Room thermal response factors. *ASHRAE Trans.*, 73(2), III.2.1.
Morely, A., E. Hughes, 1994, *Principles of Electricity*, 5th edn. Longman.
Morse, P.M., 1969, *Thermal Physics*, 2nd edn. W.A. Benjamin Inc., New York.
Münster, A., 1969, *Statistical Thermodynamics*. Springer-Verlag, Berlin, Germany.
Naidu, M.S., V. Kamataru, 1982, *High Voltage Engineering*. Tata McGraw-Hill, New Delhi, India.
Nilsson, J., S. Riedel, 2007, *Electric Circuits*. Prentice Hall, Upper Saddle River, NJ.
Nottage, H.B., G.V. Parmelce, 1954, Circuit analysis applied to load estimation. *ASHRAE Trans.*, 61(2), 125.
Obert, E.F., R.L. Young, 1962, *Thermodynamics and Heat Transport*. McGraw-Hill Book Comp., New York.
O'Connell, J.P., J.M. Haile, 2005, *Thermodynamics*. Cambridge University Press, Cambridge, U.K.
Özisik, M.N., L.F. Schutrum, 1960, Solar heat gain factors for windows with drapes. *ASHRAE Trans.*, 66, 288.
Özisik, M.N., D.Y. Tzou, 1994, On the wave theory in heat conduction. *ASME J. Heat Transfer*, 116, 526–535.
Paschkis, V., 1942, Periodic heat flow in building walls determining by electronic analog methods. *ASHRAV Trans.*, 48, 75.
Patankar, S.V., 1980, *Numerical Heat Transfer and Fluid Flow*. McGraw Hill Book Company.
Patterson, W.C., 1999, *Transforming Electricity: The Coming Generation of Change*. Earthscan, London, U.K.
Pepper, D.W., J.C. Heinrich, 1992, *The Finite Element Method-Basic Concepts and Applications*. Taylor & Francis, Washington, DC.
Pitts, D.R., E.L. Sisson, 1998, *Heat Transfer*, 92nd edn. McGraw-Hill, New York.

Plank M., 1945, *Treatise Thermodynamics*, 755–763, Dover Publication Inc., U.K.
Poulikakos, D., 1994, *Conduction Heat Transfer*. Prentice Hall, Englewood Cliffs, NJ.
Qiu, T.Q., C.L. Tien, 1993, Heat transfer mechanisms during short-pulse laser heating of metals. *ASME J. Heat Transfer*, 115, 835–841.
Reif, F., 1965, *Fundamentals of Statistical and Thermal Physics*. McGraw-Hill, New York.
Reiss, H., 1996, *Methods of Thermodynamics*. Dover Publishing Inc., New York.
Reynolds, W.C., 1965, *Thermodynamics*. McGraw-Hill Book Company, New York.
Richet, P., 2001, *The Physical Basis of Thermodynamics*. Kluwer Academic, New York.
Roald, K.W., 1986, *Electromagnetic Fields*, 2nd edn. Wiley, Hoboken, NJ.
Rohsenow, W.M., H. Choi, 1961, *Heat, Mass, Momentum Transfer*. Prentice-Hall Inc., Englewood Cliffs, NJ.
Rohsenow, W.M., J.P. Hartnett, H. Choi, 1998, *Handbook of Heat Transfer*, 3rd edn. McGraw-Hill, New York.
Roll, K., 1980, *Introduction to Thermodynamics*. Charles E. Merrill Publ. Company, Columbus, OH.
Russell, L.D, G.A. Adebeliyi, 1993, *Classical Thermodynamics*. Saunders College Publishing, Philadelphia, PA.
Saad, M.A., 1966, *Thermodynamics for Engineers*. Cambridge University Press, Cambridge, U.K.
Schelkunoff, S.A., 1963, *Electromagnetic Fields*. Blaisdell Publishing Company, New York.
Schneider, P.J., 1955, *Conduction Heat Transfer*. Addison Wesley, Reading, MA.
Schrodinger, E., 1967, *Statistical Thermodynamics*. Cambridge University Press, Cambridge, U.K., p. 95.
Schwarz, S.E., 2001, *Electromagnetic for Engineers*. Saunders College Publ., Philadelphia, PA.
Scurlock, R.G., 1966, *Low Temperature Behavior of Solids: An Introduction*. Dover, New York.
Sears, F.W., 1950, *An Introduction to Thermodynamics*. Addison-Wesley Publ. Company, Reading, MA.
Sears, F.W., G.L. Salinger, 1986, *Thermodynamics Kinetic Theory and Statistical Thermodynamics*. Addison-Wesley Publ. Company, Reading, MA.
Seem, J.E., S.A. Klein, W.A. Beckman, J.W. Mitchell, 1989, Transfer functions for efficient calculation of multi-dimensional heat transfer. *J. Heat Transfer*, 111, 5–12.
Shavit, A., G. Guffinger, 1995, *Thermodynamics*. Prentice Hall, New York.
Silver, R.S., 1971, *An Introduction to Thermodynamics*. Cambridge University Press, Cambridge, U.K.
Sissom, L.E., D.R. Pitts, 1972, *Elements of Transport Phenomena*. McGraw-Hill, New York.
Smith, N.O., 1982, *Elementary Statistical Thermodynamics*. Plenum Press, New York.
Smith, R.A., 1952, *The Physical Principles of Thermodynamics*. Chapman & Hall, London, U.K.
Solo, S.L., 1967, *Analytical Thermodynamics*. Prentice-Hall Inc., Englewood Cliffs, NJ.
Sonntag, R.E., G.J. Van Wylen, 1966, *Fundamentals of Statistical Thermodynamics*. John Wiley & Sons, New York.
Sowell, E.F., D.C. Chiles, 1985, Characterization of zone dynamic response for CLF/CLTD calculations. *ASHRAE Trans.*, 91(2A), 179–200.
Spanner, D.C., 1964, *Introduction to Thermodynamics*. Academic Press, New York.

Spitler, J.D., D.E. Fisher, 1999, Development of periodic response factors for use with the radiant time series method. *ASHRAE Trans.*, 105(2), 491–509.

Srednici, M., 2007, *Quantum Field Theory*. Cambridge University Press, Cambridge, U.K.

Stampone, J., 2010, Green building rating systems around the world, October 8, 2010. http://astudentoftherealestategame.com/.

Stephenson, D.G., G.P. Mitalas, 1971, Calculation of heat transfer functions for multi-layer slabs. *ASHRAE Trans.*, 77(2), 117–126.

Stocker, G.N., 1988, *Thermodynamics and Statistical Mechanics*. Springer-Verlag, Berlin, Germany.

Stöcker, G.N., 1996, *Thermodynamics and Statistical Mechanics*. Springer-Verlag, Berlin, Germany.

Stocker, W.F., 1993, Design of Thermal Systems, 3rd edn. McGraw-Hill, New York.

Stowe, K., 1984, *Introduction to Statistical Mechanics and Thermodynamics*. John Wiley & Sons, New York.

Strosicio, M.A., M. Dutta, 2005, *Phonons in Nanostructures*. Cambridge University Press, Cambridge, U.K.

Sucec, J., 2002, *Heat Transfer*. W.M.C. Brown Publishers, Dubuque, IA.

Sychev, V.V., J.S. Shier, 1978, *Complex Thermodynamics Systems*. Consultations Bureau, New York.

Szokolay, S.V., 1981, *Environmental Science Handbook*. The Construction Press, Lancaster, U.K.

Taine, J., J.P. Petit, 1993, *Heat Transfer*. Prentice Hall, Englewood Cliffs, NJ.

Tassion, D.P., 1993, *Applied Chemical Engineering Thermodynamics*. Springer-Verlag, New York.

Thomas, L.C., 1992, *Heat Transfer*. Prentice Hall, Englewood Cliffs, NJ.

Tien, C.L., G. Chen, 1994, Challenge in microscale conductive and radioactive heat transfer. *ASME J. Heat Transfer*, 116, 799–807.

Tien, C.L., J.H. Lienhard, 1985, *Statistical Thermodynamics*. Hemisphere Publishing Corp., New York.

Tipler, P., 2004, *Physics for Scientists and Engineers: Mechanics, Oscillations and Waves, Thermodynamics*, 5th edn. W.H. Freeman, New York.

Tipler, P., R. Llewellyn, 2002, *Modern Physics*, 4th edn. W.H. Freeman, New York.

Tisza, L., 1977, *Generalized Thermodynamics*. MIT Press, Cambridge, U.K.

Todd, J.P., H.B. Ellis, 1982, *Applied Heat Transfer*. Harper and Row, New York.

Tonchev, N., A. Dimitrov, 2009, An Aaerogel "intelligent" membrane facade technology development a planned experiment. *International Conference on Mechanics and Technology of the Composed Materials*, September 22–24, 2009, Varna, Bulgaria.

Touloukian, Y.S., C.Y. Ho (Eds.), 1972, *Thermophysical Properties of Matter*. Plenum Press, New York.

Touloukian, Y.S., C.Y. Ho (Eds.), 1976, *Thermophysical Properties of Selected Aerospace Materials*. Purdue University, West Lafayette, IN.

Treyba, R.E., 1980, *Mass Transfer Operations*, 3rd edn. McGraw-Hill, New York.

Tribus, M., 1961, *Thermostatic and Thermodynamics*. Van Nostrand, Princeton, NJ.

Trubiano, F., 2012, *Design and Construction of High-Performance Homes: Building Envelopes, Renewable Energies and Integrated Practice*. Routledge, New York.

Tzou, D.Y., 1992, Thermal shock phenomena under high-rate response in solids, *Annual Review of Heat Transfer*, Ed. C.L. Tien. Hemisphere Publishing Inc., Washington, DC.

Tzou, D.Y., 1995, A unified field approach for heat conduction propagation. *ASME J. Heat Mass Transfer*, 117, 8–16.

Tzou, D.Y., 1997, *Macro-to Microscale Heat Transfer.* Taylor & Francis, Washington, DC.

Wangsness, R.K., 1986, *Electromagnetic Fields,* 2nd edn., John Wiley & Sons, New York.

Wark, K., 1983, *Thermodynamics*, 4th edn. McGraw-Hill Book Company, New York, Chap. 9.

Welty, J.R., 1974, *Engineering Heat Transfer.* John Willy & Sons, New York.

Wilkinson, D.S., 2000, *Mass Transport in Solids and Fluids.* Cambridge University Press, Cambridge, U.K.

Wolf, H., 1983, *Heat Transfer.* Harper and Row, New York.

Van Carey, P., 1999, *Statistical Thermodynamics and Microscale Thermophysics.* Cambridge University Press, Cambridge, U.K.

Vild, D.J., 1964, Solar heat gains factors and shading coefficients. *ASHREA J.*, 6(10), 47.

Von Ness, H.C., 1964, *Classical Thermodynamics of Non-electrolyte Solutions.* The McMillan, Company, New York.

Ziegler, H., 1983, *An Introduction to Thermomechanics.* North-Holland Publishing Company, Amsterdam, the Netherlands.

Zemansky, M.W., 1998, *A Modern Course in Statistical Physics.* Cambridge University Press, Cambridge, U.K.

Zemansky, M.W., R. Dittman, 1981, *Heat and Thermodynamics.* McGraw-Hill, New York, Chapter 19.

Zemella, G., A. Faraguna, 2014, *Evolutionary Optimisation of Facade Design: A New Approach for the Design of Building Envelopes.* Springer, London, U.K.

Index

A

Acoustic phonon, 23, 51
Adiabatic barriers, 214
Artificial lighting, 214–216
Atomic model of Bohr, 36
Atom vibration, 49

B

Banned area, 41
  barrier coefficient, 39, 48, 50
  definition, 25–26
  electron trajectory, 38
  energy barrier, 26, 34–35, 89
  narrow, 33
  width of, 32, 34, 40, 42, 67
BES, *see* Building energy system (BES)
Bifunctional systems, 80, 91–92, 94, 259
Bohr' model, 24
Boltzmann–Plank form of entropy, 66
Bose–Einstein statistics, 23
BREEAM, *see* Building research establishment environmental assessment method (BREEAM) Building energy system (BES)
  components, 2, 251
  design classification, 2
  economy of energy, 1
  energy consumption, 1
  envelope system and energy functions, 3
  fossil fuel resources, 1
  integral conceptual design, 2
  internal physical environment, 2
  operation, 251
Building research establishment environmental assessment method (BREEAM)
  building assessments, 249–250
  identical ecological factors, 249–250
  vs. LEED, 255–257
  weighting coefficients, 248–249
Building–surrounding energy interactions, 11
Building system for automatic control and monitoring (BSAC and M), 2, 9

C

CDL, *see* Coefficient of daylight (CDL)
Class of energy efficiency (CEE), 290
  building preliminary design, 202–203
  construction, 227–229
  correlation table, 289
  table of correspondence, 233
  thermal bridges, 223–224
  thermal resistance, 195–197
  total thermal conductivity coefficient, 201
Coefficient of daylight (CDL), 261
  cosine law of Lambert, 219
  depth of periphery area, 218
  distribution, 220–222
  geometry of illumination area, 219–220
  inequality, 218
  reduction of glazed area, 221
  stepwise minimization of overall window dimensions, 220–221
  stereometry of interior natural illuminance, 219
Crystal lattice structure, 17, 32

D

Deadband, 216
Debye waves, 21–22, 27, 31
"Delta-*Q*" method, 243, 261
  Darcy exponential functional form, 234
  data collection and manipulation
    air rate measurement, 238
    depressurization–pressurization, 239–240
    distribution Stata plot, 239–240
    handler off and handler on, 239

"POLY_INTERPOLATE" of ACC DERIVE 6, 241
pressure difference measurement, 239
primary data, 239
quasi-periodicity, 239
statistical model, 240
two fans–type blower doors, 238
data normalization
advantage of data presentation, 241
"DERIVE 6" plot, 243–244
internal scales, 241
Q-p prediction, 243–244
Stata graph, 241–242
statistical model, 242–243
elements, 235
equipment arrangement, 235–236
infiltration–exfiltration, 236
Differential relations
accumulation term, 116
Cartesian coordinate system, 114
dependent and independent variables, 113
derived differential equation, 113
divergent term, 116
by equalities, 114
FEM, 118
general equation of transfer, 117
gradient form, 116–117
inflowing and outflowing quantities balance, 115
matter conservation at local level, 113
method of finite differences, 118
1D unsteady transfer, 117
Poisson's condition, 117
recuperation term, 116
source density, 115
3D steady transfer, 118
time–space continuum, 114
transfer density, 115
Dirichlet boundary condition, 153, 179–180

E

Ecological standard
BREEAM and LEED rating systems
advantages and disadvantages, 257
application, 257
BG_LEED, 256
building assessments, 249–250
comparison of levels, 256
correspondence table, 253
ecological model, 250–252
energy and atmosphere, 251
general-functional criteria, 256–257
indoor environmental quality, 250–251
new versions, 249–250
number of buildings certified, 1998–2010, 248
other systems rating, 255
weighting coefficients, 248–249, 251–252
general index, 246–247
index of building ecological sustainability, 245, 247
mathematical expression, 246
weighting factors, 246–247
Einstein waves, 19–20, 22, 31, 68
Electrothermodynamic system (ETS), 59
bifunctional, 86–87
differential equations, 113, 118–120
electrical and thermal work done, 87–88
energy exchange, 194
entropy, 99
free energy function, 99, 106 (*see also* Weighted residuals methodology (WRM))
free energy potential, 81, 84
free energy structure (*see* Free energy, structure)
isothermal surface, 55
monofunctional, 79
state parameters, 84–85, 97, 106, 118
state variable, 183
trifunctional, 79, 86
Electrothermomechanical system (ETS), *see* Electrothermodynamic system (ETS); Macroscopic state parameters
Energy band orbits, 26
Energy barrier, 7, 32, 261
banned area, 26, 34–35, 89
bilateral, 192

envelope operation, 192
unilateral, 192
Energy efficiency, 2
CEE (*see* Class of energy efficiency (CEE))
energy labeling, household appliances, 223
energy standard of the construction (EE_{Const})
building facades, 227, 229
buildings flat pattern, 229–230
characteristics, 227–228
construction effect, 232
index and class of energy efficiency, 233
index of energy efficiency of constructions (IEE_{Constr}), 228–229
influence of structural system, 232
method, 229
modified building structure, 231
total length of linear bridges, 228
GDP, 223
heat transfer, solid inhomogeneous multilayer walls
components with different heat conductivity, 224–225
heat flux, 226
principle of superposition, 226
technological errors, 224–225
tribute of thermal bridges, 226–227
IEE (*see* Index of energy efficiency (IEE))
indicative targets, household energy savings, 224
PEC, 223
Energy-exchange models, 50, 53
energy impacts, 75, 80
within envelope, 258
ETS, 194
generalized forces, 81–82
gross potential, 80
heat and moisture transfer, 80
hybrid illumination systems, 80
isolated systems, 75
law of conservation (*see* Law of conservation, energy-exchange models)
law of energy conservation, 80
monofunctional potential fields
differential form of work done, 78
displacement trajectory, 78
generalized coordinates, 76–78
generalized force, 77
physical forces, 76
types of energy impact, 79
open systems, 75
related studies, 76
system state parameters, 80
thermal and mechanical interactions, 80
three-functional medium, 79
trifunctional ETS, 79
work done, building envelope
control volume, 81
entropy transfer (*see* Entropy transfer)
free energy potential, 81
at macrolevel, 76
trifunctional power field, 81
Energy flux, 30
adiabatic and filter envelopes, 192
capacitive and volume resistance, 193
control techniques, 7
direction and intensity, 186
energy barrier, 192, 194
energy filter, 193
energy-filtering function, 8
facade spectral determination, 193
inflowing and outflowing, 32, 192–193
1D, 36
solar, 86, 258
surroundings and envelope components, 48
unwanted external impact, 191
Energy standard of the construction (EE_{Const})
building facades, 227, 229
buildings flat pattern, 229–230
characteristics, 227–228
construction effect, 232
index and class of energy efficiency, 233
index of energy efficiency of constructions (IEE_{Constr}), 228–229
influence of structural system, 232
method, 229

modified building structure, 231
total length of linear bridges, 228
Energy Strategy 2020, 223–224
Energy system, 2, 7, 203
Entertaining system, 2
Entropy transfer, 7
and electric charges
bifunctional ETS, 87–88
electrical and thermal work done, 86–87
electric charge variation, 86
entropy at second limit state, 87
internal temperature variation, 88
local Lagrange factor, 88
with/without mass transfer, 88–89
Envelope system, 2, 193
architectural design, 3
building envelope components, problems, 101
building–surrounding energy interactions, 11
energy conversion, 6
energy-filtering function, 8–9
energy flux control techniques, 7
energy-related functions, 5–6
energy transfer through envelope component, 8
envelope materials, 3
envelope–surroundings interaction, 6
envelope type, 4
environmental energy flows, 8
facade function, 9–10
free energy function, 97
glazed components, 6
hybrid lighting system, 4–5
illumination system, 9
intelligent energy membrane, 9–10
price per 1 m^2 of surrounding walls, 6
PV panels, 3–4
solid building components, 6
spectral determination, facade structures, 9
spontaneous/controlled energy transformations, 3
spontaneous energy inflow, 7
wind turbines, 4–5
Environmental sustainability, mathematical model
certification level, 253–254
design level, 253
ecological factors, 244–245
ecological harmonicity, 244
ecological standard (*see* Ecological standard)
factorial indexes, 254–255
factor indices, 247
internal comfort and healthy environment, 245
local priorities, 245
profit function, 245
superposition of ecological effects, 244
ETS, *see* Electrothermodynamic system (ETS)

F

Fermi energy level, 95
Finite element method (FEM), 158, 260
differential relations, 118
distribution density modeling, 129
matrix equation (*see* Galerkin method, FEM)
shape functions, 130
First law of Kirchhoff, 211
First-order boundary condition, 179–180
Fracpoly, 240, 242
Free energy, 101
distribution
continuous scalar function, 97
control volume, 97
coordinate systems, 98
free energy potential field characteristics, 97–98
normal vector, 97
state parameters, 99–100
function (*see* Free energy function)
orbitals, 62–63
structure
bifunctional system, 94
differential equation of the state, 91
electric and mechanical impacts, 92
electric and mechanical interactions, 94
electric and thermal interactions, 94

entropy and system basic parameters, 95–97
ETS state, 92–93
function of ETS free energy, 93
generalized characteristic of system state, 95
Gibbs potential/Gibbs formula, 95
operational form, 92
second-order differential equations, 92
state parameters, 89–90
system volume, electric charge, and entropy, 93
thermal and electric impacts, 91
total differential of free energy functional, 90–91
Free energy function, 53
envelope control volume, 83
first-order boundary condition, 179
gradient, 106, 175, 192
WRM (*see* Weighted residuals methodology (WRM))
Free energy potential, 43, 81, 84

G

Galerkin method, FEM
Cartesian coordinate system, 145
cylindrical coordinate system, 145
equality, 144
Green's theorem, 143
matrix form of FE integral equation of transfer, 143
Newmann's boundary condition, 143
specific flux, 144
weighting function, 142
Geier and Kumhasi model, 28–29
Generalized matrix of element conductivity, 265

H

Hybrid lighting systems, 4–5, 215–216
Hypothetical physical model, phonon generation
internal ionization and polarization
absorption and emission of photons, 34
absorption, reflection, and conduction, 33–34
banned area width, 32
conductors, 33
crystal lattice structures, 32
electric eccentricity, 35
electromagnetic moment pulsing, 35
energy band width, 32
insulators, 32–33
oscillation frequency and amplitude, 35–36
photons and valence band electrons interaction, 35
semiconductors, 33
valence band electrons, 33
lagging temperature gradient
atomic structures ionization and polarization, 41–42
atom vibration, 49
barrier coefficient of banned area, 48
cascade mechanism of phonon generation, 49
distribution function, 43
energy of inherited electrons, 47
energy of photon flux, 47
energy of released electrons, 47
energy transfer within stereometric angle, 45–46
equation of balance, 42
habitation of control volume, 42
high-energy orbitals, 51
intensity of photon emission, 45
Lagrange multiplier, 51
macroscopic distribution of state density, 44
model of, 258
number of free electrons, 48
phonon–photon interaction, 49
propagation of photons, 45–46
solar photons, 41–42
temperature gradient lag, 50
thermal and acoustic phonons, 51
thermal charges, 50
thermal electron flux, 48
time–space/frequency continuum, 46–47
total amount of energy, 42

total energy balance, 44
total energy function, 44
total microstate energy, 42–43
physical pattern of energy transfer
atomic model of Bohr, 36
cascade isotropic transfer, 40
dependence of electric potential, 39
dynamics of electron motion, 36–37
electron trajectory within banned area, 38
high-frequency vibration of electrons, 39
infrared spectra, 39
integral effect of photon flux, 40
isotropic scatter, 37
operation mechanism, 38
oscillations of atomic structures, 41
phonons behavior, 40
photonic gas, 36
photon unidirectional motion transformation, 41
qualitative transformation, 36
scattering photon emission, 37
solar photons, 39
spontaneous photon emission transformation, 41

I

Impurity scatter, 24
Index of energy efficiency (IEE), 290
building preliminary design, 202–203
construction, 227–229, 231–232
correlation table, 289
structural system, 232
table of correspondence, 233
thermal bridges, 223
thermal resistance, 195–200
total thermal conductivity coefficient, 201
Integral forms, 184, 258
balance of transfer fluxes, 119
border surfaces of global area, 119
Cartesian coordinate system, 120
cylindrical coordinate system, 120
Gauss theorem, 120
integral equations of transfer, 119
law of conservation of matter, 119–120
law of energy conservation, 83
parabolic character, 122
Poisson and Laplace equations, 122
spherical coordinate system, 120
steady transfer, one-dimensional finite element (*see* Steady transfer, 1D finite element)
sum of accumulating flux, 121
Intelligent membrane, 2, 9–10, 36, 102, 193, 260–261

K

Kaganov model, 27–28

L

Lagrange multiplier, 84, 95
Kaganov model, 28
lagging temperature gradient, 49–51
macroscopic state parameters
electric potential field, 61–63
entropy, 69–71
temperature field and gradient, 57–58
Law of conservation
energy-exchange models
differential and integral forms, 83
ETS state parameters, 84–85
free energy potentials, 84
mechanical displacement of macrobodies, 82
nonisolated systems, 82
phonon degeneration, 84
potential difference, 83
total energy (TE), 82
trifunctional ETS, 86
work of the field forces, 83
of matter, 119–120, 181
Leadership in energy and environmental design (LEED)
advantages and disadvantages, 257
BG_LEED, 256
vs. BREEAM, 255–257
building assessments, 249–250
correspondence table, 253

ecological model, 250–252
energy and atmosphere, 251
general-functional criteria, 256–257
indoor environmental quality, 250–251
new versions, 249–250
number of buildings certified, 1998–2010, 248
weighting coefficients, 248–249, 251–252
Leak assessment, air-conditioning systems
"Delta-*Q*" method (*see* "Delta-*Q*" method)
fan pressurization, 233
leak locating methods
blower door subtraction tests, 233
characteristics, 234
disadvantages, 234
duct pressurization tests, 233
internal scales, 234
modified balance equation, 236–238
LEED, *see* Leadership in energy and environmental design (LEED)
Load vector
convection to surrounding matter, 266
direct flux, 266
circular cross section, 267
matrix equation (*see* Matrix equation, design and solution)
square cross section, 267
modified matrix equation, 154
recuperation sources, 265
steady transfer
1D finite element, 157–158
3D simple finite element, 172–173
2D finite element, 163–165, 168–169
Logistics system, 2

M

Macroscopic characteristics of transfer
amount of electricity, 104
amount of energy, 103–104
amount of fluid, 104
generalized quantity of matter, 105
general law of transfer, 106–107
quantitative characteristics
transfer densities, 103, 105
transfer rates, 102, 105
transfer vectors, 102
transmission phenomena
potential function, 109–110
stream function, 110–111
streamlines, 108–109
stream pattern, 109–111
stream pipes, 109
stream surfaces, 108
vector product, 108
visualization of vector field, 108
Macroscopic state parameters
electric potential field
definition equality, 66
distribution function values, 62
electrical and mechanical works, 64
energy of electron gas, 61, 64
Fermi-tail, 62–63
free electrons cloud formation, 61
gradient vector, 66
Hamiltonians, 65
Lagrange multiplier, 61–63
number of electrons in free area, 63
scalar function, 65
scalar quantities ratio, 65
state of degeneracy, 61
types of transfer, 63
energy-exchange processes, 53
entropy, 95
barrier coefficient, 68
Boltzmann–Plank form, 66
degeneration of phonons, 66
degree of photon degeneration, 72
dominant energy orbital, 70
eigen energies, 68
eigenvalue of energy, 68
first-generation phonons, 68
first mechanism of degeneration, 71–72
Lagrange multiplier, 69–71
objective laws, 67
photon gas degeneration, 69
radiation intensity, 69
residual entropy, 68
second-generation photons, 68
second mechanism of degeneration, 72
second mechanism of entropy, 69–70

thermal work, 73
third mechanism of entropy, 70–71
total potential, 67
free energy functions, 53
general methodological approaches, 73–74
pressure field
absolute pressure, overpressure, and vacuum pressure, 59
defined by equality, 58
gradient, 59–60
Hamiltonians, 60
static and dynamic pressure, 59
total internal energy, 60
types of, 58
temperature field and gradient
Cartesian, cylindrical, and spherical coordinate systems, 55–56
coordinate systems, 56–57
distribution in 2D areas, 54
equilibrium of control volume, 58
on flat vertical external wall, 54–55
gradient magnitude, 57
internal temperature of photons, 57
isothermal surfaces, 54–55, 57
Lagrange multiplier, 57–58
phonon distribution functions, 58
scalar function, 53
special types of, 54
visual mapping, 54
volume derivative, 57
Majumdar mechanism, 29–30
Matrix equation, 150
design and solution
circular cross section, 267–268
energy efficiency class, 285–292
finite element method, 268–270
3D finite element with node coordinates, 276–278
transfer in 2D area, 281–285
transfer in 1D global area, 279–281
2D finite element, 270–276
Galerkin method (*see* Galerkin method, FEM)
modified matrix equation
absolute coordinates, 153–154
approximating functions, 151–152
boundary conditions, 152–153
Cartesian coordinates, 154
Dirichlet's type, 153
element conductivity, 152
[G^e]-conductivity matrix, 150
load vector, 154
matrix of surface properties, 151
Neumann's type, 153
surface conditions, 151
1D simple finite element, 155, 158–159
3D simple finite element (*see* Steady transfer, 3D simple finite element)
2D simple finite element, 165–166, 169
Matrix of conductivity, 144, 150, 156, 168, 263–264
Matrix of surface properties, 264–265
Micro–macroscopic assessment
macroscopic state parameters (*see* Macroscopic state parameters)
microscopic canonical ensemble
collective macrostate, 52–53
entrance–exit operation, 52
minoritarian macrostates, 53
Schrodinger equation, 51
statistical thermodynamics, 52
time–space continuum, 51
transitory disposition, 51
Minimal-admissible light-transmitting envelope apertures
CDL calculation (*see* Coefficient of daylight (CDL))
energy and visual comfort
adiabatic barriers, 214
aperture dimensions, 218
artificial lighting, 215–216
bound and ambiguous effect, 218
dark sides of, 214
hybrid lighting systems, 215–216
illuminance and brightness standard, 214
illumination discomfort, 214
interior comfort, 213
interior natural illuminance, 218
nullband, 216
power regulation, 216–217
thermal installation, 216
total glazing coefficient, 218

transparency of glazed structural elements, 215
window lighting system, 215
Mixed boundary conditions, 180–181
Model of lagging temperature gradient, 48, 258
Modified matrix equation
absolute coordinates, 153–154
approximating functions, 151–152
boundary conditions, 152–153
Cartesian coordinates, 154
Dirichlet's type, 153
element conductivity, 152
[G^e]-conductivity matrix, 150
load vector, 154
matrix of surface properties, 151
Neumann's type, 153
surface conditions, 151
Monofunctional potential fields
differential form of work done, 78
displacement trajectory, 78
generalized coordinates, 76–78
generalized force, 77
physical forces, 76
types of energy impact, 79

N

Neumann boundary condition, 153, 180
Nullband, 216

O

Ordinary differential equations, 146
Oscillation damping, 20–21

P

Photonic gas, 36
Physical models, microscopic levels
acoustic and optical phonons, 23
actualization of engineering methods, 32
angular frequency and wave number, 21
atomic structures of solids, 25
banned area, 25–26
Bohr' model, 24
Bose–Einstein statistics, 23
crystal lattice structure, 17
crystal nonmetals, 27
Debye *vs.* Einstein transfer frequencies, 22–23
Debye waves, 21–22, 27
Einstein waves, 19–20
electron angular speed and energy, 25
electron orbit radius, 24
energy band orbits, 26
energy transfer, 16
free electrons migration, 16
Geier and Kumhasi model, 28–29
hypothetical model of lagging gradient, 17
hypothetical physical model (*see* Hypothetical physical model, phonon generation)
isotropic scatter of electrons, 24
Kaganov model, 27–28
kinetic thermal theory, 17
lattice internal ionization, 26
lattice vibration waves, 16
Majumdar mechanism, 29–30
micro-transfer, 30
model of phonon radiative transfer, 17
model of scattering phonons, 16
oscillation damping, 20–21
phonon transfer, 23
quasiparticles propagation, 31
specific heat capacity, 17–18
structural thermodynamics and thermal conduction task, 31
theory of heat transfer, 18
thermal characteristics of solids, 26–27
two-step model, 16
valence band, 24
vibration of atoms, 20
virtual physical mechanisms, 30
zone theory of Borissov, 24
Primary energy sources, 191

R

Recuperation of matter, 115–117, 152, 168, 172, 265
Residual entropy, 68

S

Safeguard system, 2
Sanitary system, 2
Schrodinger equation, 51
Second law of Kirchhoff, 211
Second-order boundary conditions, 177, 179–180
Solar shading devices (shield), 207–208, 261
computerized design, 207
coordinates of cross points, 206–207
equation of shading devices surface, 204
heat exchange, window and surroundings
algebraic equations, 211
algebraic system, 211
electric resistance values with restrictions, 210
electrothermal analogue model, 211
equivalent electrical analog system, 209–210
mathematical model, 212–213
radiant heat exchange, 208–209
secondary radiant flux, 209
thermal and electric quantities, 211
inversion of designed object, 203–204
light transmitting aperture, 204–206
shadow masks, 203
solar flux, 205
stereometry selection, 204
Solid structural elements, 224
heat and moisture transfer, 80
initial and boundary conditions
balance equations of matter conservation, 182
convective heat exchange, 182
convective transfer, wall internal surface, 183–185
first-order boundary condition, 179–180
radiative heat exchange, 182
second-order boundary condition, 179–180
third-order boundary condition, 179–181
wall external surface (*see* Wall external surface, boundary conditions)
spectral determination, facade structures, 9
thermal resistance calculation
absolute adiabatic insulation, 195
axisymmetric objects, 197
classes of energy efficiency (CEE_{Env}), 195–197
contra-argument and restrictions, excessive insulation, 194–195
cost-effective coefficient, 199
economically profitable thickness, 195
energy-related characteristics, 202
esthetic and ecological issues, 194
ETS energy exchange, 194
index of energy efficiency (IEE_{Env}), 195–196, 199
insulating layer thickness, 198
maximal admissible thickness, thermal insulation, 196–197
normative values, heat transfer coefficient, 198
spherical standard, 200
temperature gradient, 194
thickness of pipe insulation, 197–198
total thermal conductivity coefficient, 201
variants of building preliminary design, 202–203
Solid structural materials
alloys, 13–14
coefficient of thermal conductivity, 13–14
energy characteristics, 13
insulating systems, 13–14
macroscopic levels, 14–15
microscopic levels, 14
operational temperature, 15
physical models (*see* Physical models, microscopic levels)
pure metals, 13–14
transfer, absorb, filter, and control external energy, 15

Solid wall element; *see also* Solid structural elements
wall dehumidification, mass (moisture) transfer, 175
adiabatic processes, 178
negative pressure gradient, 176
vapor–air mixture, 175–176
wall cooling under moisture extraction, 177–178
wall heating under moisture equilibrium within, 177
wall humidification, 175, 177
adiabatic conditions, 178–179
wall heating under air cooling and moisture release, 178
Spectral-sensitive characteristics, 193
Stata graph, 240–242
Stata model, 243
Steady transfer, 1D finite element
Cartesian coordinate system, 147
cylindrical coordinate system, 147
boundary conditions, 155
Cartesian coordinates, 159
direction of transfer fluxes, 159
load vector, 157–158
matrix of conductivity, 156
matrix of surface properties, 156
integral form of balance of energy transfer
advantages of WRM, 150
balance equation, 148
facial FE, 148–149
one-dimensional global area, 147–148
modified matrix equation (*see* Modified matrix equation)
1D transfer, 146
ordinary differential equations, 146
spherical coordinate systems, 147
Steady transfer, 2D finite element
in Cartesian coordinates
adjugate matrices, 161
boundary conditions, 162
conductivity matrix, 161, 165
linear 2D polynomial functions, 161
load vector, 163–165
matrix of boundary conditions, 163
surface integrals, 162
three-noded triangle, 161
in cylindrical coordinates
approximation functions, 166
boundary conditions, 166
conductivity matrix, 167–168
load vector, 168–169
matrix equation, 169
matrix of surface properties, 167
surface and volume integrals, 166
discretization, 160
operational regimes improvement, 161
patterns/photos design, 160
structure improvement/ optimization, 161
Steady transfer, 3D simple finite element
conductivity matrix, 171
interpolated values, 170
linear approximating functions, 170
load vector, 172–173
matrix of surface properties, 171
3D matrix, 172
Stream function, 110–111
Streamlines, 108–110
Stream pattern, 109, 111
Stream pipes, 109
Stream surfaces, 108, 111
Structural system, 2, 232–233
Superposition of ecological effects, 244

T

Technological system, 2
Thermal charges, 40, 42, 50, 72
Thermal phonon, 23, 32–33, 51, 70
Third-order boundary condition, 179–181
Transfer vectors, 58, 102, 108
Transport evaluation, 113
Trifunctional systems, 259

V

Valence band electrons, 50, 61–62
density of, 39
high kinetic activity, 39
inflowing photon flux, 48–49
and photons interaction, 35
solar photons, 49

Vector-gradient characteristic function, 138–139
Virtual physical mechanisms, 30
Visual mapping, temperature field, 54

W

Wall dehumidification, mass (moisture) transfer, 175
  adiabatic processes, 178
  negative pressure gradient, 176
  vapor–air mixture, 175–176
  wall cooling under moisture extraction, 177–178
  wall heating under moisture equilibrium within, 177
Wall external surface, boundary conditions
  of building cover, 188
  change due to building aerodynamics, 186
  engineering methodologies, 188–189
  gradient direction, 185–186
  nonisothermal distribution, 188
  pressure gradient and energy flux direction and intensity changes, 186–187
  stack building effect, 188
  types of wall elements, 186
Wall humidification, 175, 177
  adiabatic conditions, 178–179
  wall heating under air cooling and moisture release, 178
Wall internal surface, convective transfer
  divergence transfer equation, 183–184
  estimation of matter transfer, 185
  ETS state variables, 183
  Gauss theorem
    Cartesian coordinate system, 184
    cylindrical coordinate system, 184
    gradient form, 184
    spherical coordinate system, 185
    vector form, 184
  initial mass concentration, 185
Weighted residuals methodology (WRM)
  FEM (*see* Galerkin method, FEM)
  hypotheses
    errors, ETS parameter estimation, 122–123
    integral relations, 122
    total error in global domain, 123–125
  linear and nonlinear differential equations, 122
  stages of application
    absolute coordinate system, 133
    approximating functions, 134–135, 137–138
    Cartesian, cylindrical, and spherical coordinate systems, 139
    discrete analogue, 125, 132–134
    discretization of compound physical area, 126
    internal spatial distribution, 134
    interpolation function, 128–129, 135–138
    limited number of subdomains, 125
    linear finite element, 131–132
    linear, quadratic, and cubic shape functions, 130
    modeling physical domain, 127–128
    natural coordinate system, 133
    one-dimensional simple finite element, 140
    physical objects, 126
    polynomials, 130–131
    shape function, 136–137
    tetrahedron, 137–138
    3D physical domains, 125
    three-dimensional simple finite element, 141–142
    two-dimensional simple finite element, 140–141
    types of finite elements, 126–128
    vector-gradient characteristic function, 138–139

steady transfer
1D finite element (*see* Steady transfer, 1D finite element)
3D simple finite element (*see* Steady transfer, 3D simple finite element)
2D finite element (*see* Steady transfer, 2D finite element)
Window lighting system, 215
WRM, *see* Weighted residuals methodology (WRM)

Z

Zero energy band, 216
Zero power demand zone (ZPDZ), 216

Date Due

BRODART, CO. Cat. No. 23-233 Printed in U.S.A.